ENGINEERING CONTRACTS
AND SPECIFICATIONS

ENGINEERING

CONTRACTS

AND

SPECIFICATIONS

ROBERT W. ABBETT
Partner
Tippetts Abbett McCarthy Stratton, Engineers

THIRD EDITION

NEW YORK   JOHN WILEY & SONS, INC.
LONDON

THIRD EDITION
Fifth Printing, March, 1961

Library of Congress Catalog Card Number: 54–6939

PREFACE TO THE THIRD EDITION

The present edition follows the same scope and arrangement as the previous editions of this book. The text has been revised and expanded throughout. The legal considerations in the preparation of contracts and specifications and the presentation of legal rights and obligations in construction work which were formerly covered in one chapter have been amplified and subdivided into five new chapters to make the book more convenient for reference and college textbook purposes. The chapters on cost-plus-fixed-fee contracts and contracts for engineering and architectural services have been completely rewritten and enlarged in accordance with current professional practice. The chapters on specification writing have been rearranged, and one new chapter has been added. In other chapters new material has been incorporated to bring the text up to date.

In the revision of the text on the legal aspects of engineering, Mr. Maurice Grushky contributed much legal research, and his assistance is gratefully acknowledged. Appreciation is also expressed to Mr. Edward R. Higgins of The Surety Association of America, who contributed helpful suggestions for the revision of the chapters on construction insurance and surety bonds.

ROBERT W. ABBETT

March, 1954

PREFACE TO THE SECOND EDITION

The first edition of this book was written during the war and was designed to meet the requirements of the large volume of construction work in progress at that time, much of which was executed by the government. In the second edition, increased emphasis is placed upon contract procedure for private work, although the original material on government methods has been retained and brought up to date. The adoption of the first edition as a textbook for college classes was greater than expected, and the second edition has been written with the view to increasing its usefulness for this purpose.

In general, the entire text has been expanded and largely rewritten within the scope of the first edition. The material on competitive-bid contracts has been combined in Chapter 5 in lieu of the original two chapters in order to obtain a more unified treatment of this subject. An entirely new chapter on construction insurance has been added, and the sections on prequalification and arbitration have been rewritten and expanded. The original chapter on specification writing has been expanded into three new chapters.

ROBERT W. ABBETT

June, 1948

PREFACE TO THE FIRST EDITION

The purpose of this book is to present some of the legal and business aspects of the engineering profession in compact form suitable for reference and textbook use. The scope of the book was determined by a course in contracts and specifications, which I taught in the School of Engineering, Columbia University, and the text was mimeographed for use by students in war training classes in George Washington University during 1943 and 1944. However, much of the material was derived from experience in the management of a large volume of construction work both before and during the present war. I hope that the book may be useful to practicing engineers, architects, and contractors as well as to college students.

An effort has been made to develop basic principles and methods throughout. Chapter 2 contains a brief summary of the elements of business law which are most significant in contract relationships. This is intended to give the engineer a general understanding and appreciation of the legal considerations in construction work and should not be considered a substitute for the advice and services of a professional lawyer on legal matters related to specific contracts. Legal principles are stated in general terms without derivation, and it should be understood that there are many variations in the interpretation and application of the law.

Contract procedure is developed along the lines of private practice, but attention is directed to matters in which government procedure differs. Unique features of government contracts are explained in detail where they occur in the logical development of the subject. This is considered desirable in view of present and probable future emphasis on government-sponsored construction projects. Furthermore, textbook material relative to government contract procedure has not been available previously in engineering literature.

Examples of various contract documents for both private and public works have been included, primarily to illustrate the text rather than for reference purposes. However, they have been selected as representing the best current practice and should serve

ix

as a guide in the preparation of similar documents for specific projects. In this connection the reader is warned against using these illustrations or any of their clauses as "standards" without due consideration of their meaning and intent, in the light of the particular conditions of their proposed use. Disputes and litigation have frequently resulted from injudicious use of "standard" documents.

With regard to specification writing, it is recognized that skill in this field depends largely on experience and engineering judgment, for which there are no substitutes. The instructions given herein are intended to provide a basis for the systematic application of such knowledge and to indicate a form for its presentation. In addition, common flaws collected from a large number of actual specifications are cited in the form of precautions to prevent their recurrence.

ROBERT W. ABBETT

Washington, D. C.

CONTENTS

xi

—Eminent Domain—Right-of-Way—Mechanic's Liens—The Contractor as a Lienor—The Subcontractor as a Lienor—Engineers and Architects as Lienors—Priority of Liens

CHAPTER 5   PARTNERSHIPS, CORPORATIONS, LABOR LAWS

Partnerships—Partners Liability to Third Parties—Fiduciary Character of a Partnership—Dissolution of a Partnership—Corporations—Public Corporations—Private Corporations—Stockholders —Limited Liability of Stockholders—Management of Corporations—Corporate Powers—Ultra Vires Acts—Dissolution of a Corporation—Corporate Business in a Foreign State—Corporations for the Practice of Engineering and Architecture—Labor Laws in Construction Work—Employer's Liability Acts—Workmen's Compensation Acts—Social Security Act—National Labor Relations Act—Fair Labor Standards Act—Labor-Management Relations Act—Labor Laws Unique to U.S. Government Contracts.

CHAPTER 6   LITIGATION, EQUITY, ARBITRATION

Litigation—The System of Courts—Disputes under Private Contracts—Disputes under U.S. Government Contracts—Pleading and Practice—Examination of Witnesses—Evidence—Best Evidence Rule —Secondary Evidence—Real Evidence—Direct Evidence—Circumstantial Evidence—Hearsay Evidence—Admission—Res Gestae Statements—Parol Evidence Rule—Opinion Evidence—Opinion Evidence of Expert Witnesses—The Engineer as an Expert Witness—Equity—The Maxims of Equity—Equity Jurisdiction in Engineering and Construction Work—Arbitration of Disputes—Arbitration Procedure—The Arbitration Agreement—The Submission—The Selection of Arbitrators—The Hearing—The Award—The Confirmation —The Engineer as an Arbitrator—The Advantages of Arbitration

CHAPTER 7   TYPES OF CONSTRUCTION CONTRACTS

Competitive-Bid Contracts—Lump-Sum Contracts—Unit-Price Contracts—Combined Lump-Sum and Unit-Price Contracts—The Contract Documents—Negotiated Contracts—Cost of the Work—Cost-Plus-Percentage-of-Cost Contracts—Cost-Plus-Fixed-Fee Contracts —The Cost-Plus-Fixed-Fee Contract with a Profit-Sharing Clause —The Cost-Plus-Fixed-Fee Contract with a Bonus Clause—Contracts Based on Cost Plus a Sliding Scale of Fees—The Cost-Plus Contract with a Guaranteed Ceiling Price—Management Contracts—Architect-Engineer-Management Contracts—Combined Engineering and Construction Contracts—Joint Venture Contracts—Incentive Type Contracts for Work outside the United States—Selection of Type of Contract—Subcontracts—Renegotiation of Contracts and Subcontracts—Equipment Rental Contracts

Test Pits, and Test Piles—Sequence of the Work—Progress of the
Work—Control of Materials—Rates of Wages at the Site—Lines and
Grades—Space for Construction Purposes and Storage of Materials
—Water, Power, Light, and Other Utility Services—Facilities for
the Engineer—Warranties by Contract Bond—The Owner's Right
to Use Completed Portions of the Work—Special Precautions during
Construction

I

BUSINESS AND PROFESSIONAL RELATIONS IN ENGINEERING

This book deals with the business, professional, and legal relations between the principal parties involved in the accomplishment of construction work. The parties concerned are the *owner,* for whom the work is to be done; the *engineer,* who, representing the owner, furnishes the professional and technical skill required in the planning, administration, and construction of the project; and the *contractor,* who furnishes the materials, labor, and equipment necessary for the physical accomplishment of the work. The character of the relations between these parties is influenced by many factors, depending upon the methods and procedures adopted from the inception to the completion of the work. A clear understanding of the possibilities inherent in these relations is a necessary part of the professional skill of the engineer because a misstep at any point may void all the precautions and provisions of the design as well as create controversies, excessive costs, and possibly litigation.

The Construction Industry. During periods of normal or average business activity, the construction industry is one of the most important, second only to agriculture in the number of workers engaged and the volume of capital expended. It produces a large portion of the capital wealth of the country in the form of buildings, highways, railroads, dams, utilities, and similar works. Since the building of these facilities is essentially a manufacturing process, construction should be considered as a manufacturing industry, with all the problems of economics, management, labor, and production which are implied.

Construction differs from other types of manufacturing in one notable respect, however. In the production of most commodities, models and manufacturing methods are more or less standardized to facilitate mass production in a centrally located manufacturing plant, and the finished product is transported to its place of use. The control of the quality and cost of the product is greatly sim-

1

plified by this standardization of methods. In the construction industry, on the other hand, the product is manufactured at its place of use, and an individual plant is necessary for each construction project. Furthermore, each product is individual in character, custom-built according to the desires of the owner. This is fundamental to the economics of construction and places important responsibilities on the engineer in connection with the management of construction operations and the control of the quality of the work.

Public and Private Construction. Construction work may be performed for a government agency, municipal, county, state, or national, or for a private individual, company, or corporation. All government construction is classed as *public works*. Common types of public works are highways, bridges, sewers, water supplies, land reclamation projects, and public buildings. Private construction includes such work as railroads, buildings, and utilities.

Building construction, with the exception of some types of industrial buildings, is classed as architectural construction, and the various remaining types of public and private works are known as engineering construction. For simplicity, it is understood in this book that any reference to the owner may mean a private individual, company, corporation, or a government agency. In a similar manner, reference to the engineer may mean either a professional engineer or architect as the case may warrant.

Financing Construction Projects. The creation of public and private works is not a matter of whim or accident. Basic economic laws are involved. Construction work requires the expenditure of capital. Before a structure can be built a need for it must be evident, and the need must be sufficiently urgent to justify its cost. Furthermore, funds must be available either in the form of capital or credit. These will be furnished by the sponsor of the project, a government agency in the case of public works, and an individual, company, or corporation for private works.

The necessity for a public works project may be self-evident; it may arise by public demand; or it may be determined by expert engineering study in which its merits must be brought to light and made evident to the public. Since the necessity for private work usually can be established only on the basis of a reasonable return on the investment required to produce it, an estimate of cost and an engineering analysis and forecast of the probable income to be derived are required. A comparison of the cost of the project with the estimated benefits to be obtained by its construction will usually indicate whether the venture is sound from the viewpoint of economics.

Public works are financed by appropriations from annual operating budgets, by special appropriations from taxation receipts, or by special bond issues in which provision is made for the retirement of the bonds by future taxation or by revenue from the use of the structure. Private works are financed by direct expenditure of capital, by loans, or by corporate bond issues in which provision is made for retirement of the bonds by revenue from the completed work. The financing of construction is a matter of promotion as distinguished from the determination of its necessity and type, which are problems in engineering economics.

Construction by General Contract. Practically all important public and private works are constructed by firms who specialize in this type of work and who are known as contractors or constructors. The owner and the contractor enter into an agreement whereby the contractor furnishes the required construction services and receives a specified compensation. The contract may be informal or even oral, but, at the opposite extreme, it may be a lengthy formal document. The type of ownership will have an important bearing on the contract form and procedure.

For public works, the method of awarding a contract for construction work is specified by law. The law requires that public advertisements be circulated advising all who may be interested that competitive bids will be received for the work and that the contract will be awarded to the lowest qualified bidder. The legal control of contract letting is designed to prevent fraud and collusion and to insure free and open competition between those interested in performing the work. For private works the procedure may follow the desires or judgment of the owner or the engineer without restriction, although some form of bidding is generally adopted.

Many of the operations necessary for the construction of a comprehensive engineering project require a high degree of specialization for which the general contractor may not be equipped or qualified. He will then enter into a subcontract with a specialty subcontractor for the accomplishment of such work. In general, specialty subcontract work reflects an economy to the owner because the subcontractor, having mastered the details of his specialty, is usually able to perform such work at a lower cost than can a general contractor whose interests are more diversified. In his relations with the owner the general contractor assumes all obligations of the subcontractor and is in turn held responsible by the owner for the satisfactory completion of the subcontractor's work. As compensation for this responsibility, the general contractor usually includes in his bid price the cost of the subcontract, plus a sufficient amount to cover any of his own expenses in connection with that portion of the work, and a reasonable profit for assuming the responsibility of its administration. The owner and the engineer should maintain close control over subcontract relations to prevent the practice of "construction brokerage," in which a pseudo-contractor may let out all the work to subcontractors, thus obtaining a profit without any effort on his own part.

Construction by Force Account. As an alternative to the employment of a construction contractor, the owner may elect to do the work with his own forces. This is known as the *force account* or *day labor* method. Under the force account system, the owner maintains direct supervision over the work, furnishes all materials and equipment, and employs workmen under his own payroll or force account. Though most construction work is performed under contract, many important projects have been accomplished by force account. The Panama Canal and some of the Tennessee Valley Authority projects are outstanding examples of public works constructed by this method.

The functions of the engineer on force account work are about the same as for work by contract although some of the formalities may be omitted. The owner must have available a complete construction force, including a competent superintendent. Frequently, however, the engineer may combine his duties with those of the superintendent and serve as the owner's construction manager.

Construction by Separate Contracts. Under the general contract method of construction, it is understood that the contractor assumes responsibility for all problems of management, administration, and coordination of construction operations, including those of subcontractors, whereas under the force account method all these matters are handled by the owner. A third method for accomplishing construction is by separate contracts. Under this method, the owner lets contracts for various portions of the work direct to specialty contractors, thus assuming the managerial functions of the general contractor. When this phase of the work is handled by the engineer, he, in addition to his engineering duties, in effect becomes the owner's agent for the purpose of administrating and coordinating the work. By an alternative scheme a general contractor may be retained in a managerial capacity with the definite understanding that all work will be handled by separate contracts which may be on a subcontract basis. This method has many of the advantages of the force account method without the corresponding charges for invested capital. It is not to be confused with "construction brokerage."

Selection of Construction Method. When the work to be done is small in scope and simple in character, or when the owner's organization includes a trained and skilled construction force, construction by force account may be advantageous. In this case the owner would be saved the expense of formal contract procedure, and there would be a reduction in construction cost by the elimination of the contractor's profit. Additional saving would come from the reduction in engineering and inspection costs. Inasmuch as the work is done under the direct supervision of the owner, the plans, specifications, and inspection need not be as elaborate or as detailed as those required for contract work.

On the other hand, these advantages may be more apparent than real when compared with contract work under favorable conditions. A qualified contractor is skilled in the administration of construction operations, including the purchasing of materials and the management of labor. He maintains an organization of trained supervisors, mechanics, and workmen and has available the most advanced construction equipment. These specialized production features result in efficiency, and the corresponding reduction in costs usually cannot be matched on force account work, even when the contractor's profit and other apparent advantages of the force account system are taken into consideration. Excepting minor work which can be handled readily by a maintenance crew, it is the general opinion of the construction industry that in most instances the advantages of work by contract outweigh those for work by force account.

Construction by separate contracts may be advantageous to the owner when the specialty work required is restricted to a relatively few types of construction, when the owner has available a com-

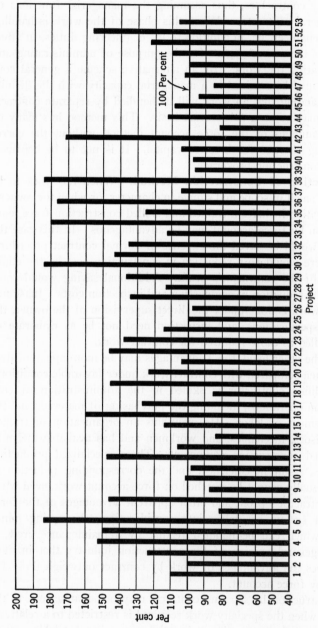

Fig. 1.   Comparison of contract and force account costs based on contract cost being 100 per cent.

petent construction manager, or when his engineer is experienced in handling contracts. About the only economy possible under this system as compared with general contract work is the saving of the general contractor's profit for managing and coordinating the work of subcontractors. This is not a very important item when compared with the total cost of the work, and unless the owner's manager is thoroughly experienced and competent in contract administration, the possible saving will be more than consumed by the excess costs of inefficient management.

In general an experienced contractor charged with the responsibility for an entire project will be able to perform the work at considerably less cost to the owner than is probable by other methods. In an effort to determine the relative efficiency and economy of the contract system versus the force account method, the Bureau of Public Roads in 1933 required that each state undertake to construct one or more sections of highway with forces employed directly by the state, comparing it with similar work executed under contract. Forty-six states constructed fifty-three sections of highways, totaling 244 miles, that were considered representative. These were bid upon by contractors in the usual manner. The work was later withdrawn from contract and performed by force account, adhering as closely as possible to the same standards as required for contract work.

On 75 per cent of the projects the cost by force account exceeded the bid prices by 31 per cent. The remaining 25 per cent of the projects were constructed for less than the bid prices. The total cost of constructing the fifty-three projects by force account was approximately $4,000,000. This was an increase of $593,000, or 18 per cent on the average, over the total of the bid prices submitted under the contract system.

Development of Construction Projects. When the need for a construction project is indicated, preliminary engineering studies are required to determine the feasibility and scope of the project which can be justified by the public need or by the economic benefits which would be derived. The results of these studies are prepared in the form of a preliminary engineering report which serves as the basis for the authorization and financing of the project in the case of private works. For public works, the preliminary report provides the information needed by legislative bodies to judge the merits of the project and to pass the acts necessary to authorize the work and to appropriate public funds to cover its cost. For self-liquidating projects, private or public, such as toll bridges and power development projects, the preliminary report also determines the probable income which may be expected when the completed project is placed in service and thus serves as the basis for the issuance and sale of revenue bonds to finance the construction work.

The second stage in the development of a construction project consists of the detailed design and the preparation of construction drawings and specifications. At this time the planning and physical characteristics of the work are determined in sufficient detail to define all aspects of the construction work necessary to complete the project. Decisions are made as to the method for accomplishing the work, that is, whether by contract or by force account. If by contract, the type is selected and arrangements are made for the award of the contract to the selected contractor.

In the third stage of project development, plans and time schedules for the construction operations are prepared. Materials and equipment are ordered and delivered to the site as required. Superintendents, foremen, mechanics, laborers, and administrative workers are mobilized, and the construction of the project proceeds under the supervision of the engineer.

Of the three stages in the development of a construction project, the preliminary report and the design and preparation of construction drawings and specifications are performed by professional engineers, whereas the third stage, the actual construction work, is performed by a construction organization, usually by contract, under the technical supervision and direction of a professional engineer. The engineering services for all three stages preferably should be furnished by the same engineer or engineering organization, but it is not uncommon for different engineers to be employed in the various stages. Generally, the operation and maintenance of the completed project are matters to be handled by the owner. Occasionally, however, the engineer or the construction contractor may be retained for this purpose.

Preliminary Investigations and Report. The purpose of the preliminary investigations, sometimes called preliminary survey or survey report, is to determine if a project is feasible from an engineering viewpoint and if it can be justified economically by the benefits which would result. In the course of the preliminary investigations the most obviously impracticable schemes are eliminated from further consideration and the most promising methods are disclosed, thereby avoiding the waste of work and funds in the development of negative studies. Maximum use is made of already available information such as maps, reports, soil data, aerial photographs, and the experience of others.

Frequently, preliminary engineering studies result in the determination of the most feasible scheme, the most economical types of construction, and accurate estimates of cost. In other instances, alternative schemes may be so closely competitive that more detailed studies are necessary before final decisions can be reached. Occasionally, it may be necessary to carry such studies to the point required for final design or even to take bids on the alternates in order to decide.

The economic benefits which will result from a completed construction project are estimated from past experience in similar situations and from predictions as to the future use of the improvements. To justify a toll bridge or highway project traffic surveys are required to determine existing traffic volumes and trends of traffic growth. From these and a proposed schedule of toll fees, estimates are made of the probable revenue which may be expected throughout the life of the facility. Surveys of past flood damage will indicate the benefits to be derived from a flood control project. Market surveys serve as the basis for estimating the power demand and probable revenue to be obtained from a power development project. For multi-purpose projects, such as a combined flood control, irrigation, hydroelectric power, and navigation project, the determination of economic benefits becomes very complex and controversial, particularly when public funds and government policy are involved. Otherwise, the principles of economic analysis will usually lead to a reasonably accurate estimate.

The ratio of economic benefits to costs will indicate the economic justification for a construction project. For this purpose, costs are reduced to an annual cost or a capitalized cost basis, which includes first cost, estimated operation and maintenance costs, depreciation and amortization costs. Likewise, economic benefits are analyzed on an annual basis. In general a benefit-cost ratio greater than 1 is necessary to show justification for a public works project. For improvements to be financed by revenue bonds, a benefit-cost ratio from 2 to 2.5 will usually be required to guarantee adequate coverage of the bond debt including the reserves necessary to protect the investment through periods of low revenue.

Certain types of improvements are frequently authorized because of public necessity without regard to the economic benefits which would result. Street pavements, public highways, sewer systems, garbage incinerators, and public parks are examples of this classification. Such projects are usually financed by direct taxation when approved by public referendum or properly authorized administrative acts.

Design Drawings. Normally the design of a structure is completed in detail before construction work is begun. This is essential when the work is to be done under a competitive-bid contract. When the force account or negotiated contract method is to be used, the drawings may be of a general or even preliminary nature, sufficient to define the scope of the work and to indicate its magnitude, and the detail drawings can be developed later as the work progresses.

For the preparation of final construction drawings, accurate detailed surveys, maps, foundation borings, and soil tests will be required. In the design the most economical and most suitable types of construction are determined and the functional requirements, layout, and dimensions of the most feasible scheme are shown on the drawings or described in the specifications. Types of construction are selected on the basis of suitability, structural stability, cost, appearance, available materials, and construction limitations. The plans and specifications which evolve from these studies define the requirements of the project in detail and show how it is to be constructed.

The drawings consist of plan views, elevations or side views, and cross sections. These are designated general drawings as distinguished from detail drawings which are drawn to a larger scale and show the details of various parts of the structure, how they are assembled, and the types of construction to be used. Usually all these drawings are prepared by the engineer, and, together with the specifications, they are the basis of the contractor's estimate and of the construction contract itself.

Shop Drawings. After the contract is awarded and construction begins, additional drawings become necessary. Shop or working drawings are prepared in connection with the use of prefabricated materials. Whereas the design or contract drawings define the component parts of a structure and how they are assembled to form the whole, shop drawings show how each piece is cut, punched, and connected in the prefabrication process. Structural steel work and various types of carpentry and wood trim assemblies are typical of prefabricated construction requiring shop drawings. Erection drawings, showing how the prefabricated parts are assembled in the field, are necessary for some types of construction, notably structural steel work. All these drawings are normally prepared by the contractor subject to the approval of the engineer. In addition, it may be necessary to show certain features of the work in full-scale drawings or by models for the purpose of making patterns or to demonstrate some intricate part of the structure such as the form of a masonry moulding or cornice. Full-scale drawings or models may be prepared by either the engineer or the contractor, depending on whether the purpose is a matter of design or construction.

Specifications. The specifications are written instructions which accompany and supplement the plans. Whereas the drawings show the physical characteristics of the structure, the specifications cover the quality of the materials, workmanship, and other technical requirements. Together with the plans, they form the guide and standard of performance which will be required in the construction of the project. While the plans are technical in nature, the specifications and the contract of which they comprise a part have many legal implications in addition to their technical content. Moreover, in the event of a controversy during construction, these documents may be subjected to legal interpretations where the manner in which they are written may result in decisions contrary to their technical intentions. Therefore, it is necessary that the engineer be fully informed in the legal aspects of specification writing.

Supervision and Inspection of Construction. Upon the award of the contract, the engineer assumes direction of the work as the authorized representative of the owner. It is the function of the engineer to approve all construction methods and to inspect all operations in behalf of the owner. Close inspection is necessary at all times to make certain that the work is free from defective materials and workmanship and that it meets the requirements of the plans and specifications. Usually the engineer is represented on the job by a resident engineer whose staff maintains direct supervision over the details of the work, including the testing and inspection of materials and workmanship. At specified intervals, usually monthly, an estimate is made of work completed, and this forms the basis for progress payments to the contractor in accordance with the terms of the contract. When the work is completed, it is the duty of the engineer to perform a final inspection of the entire project preliminary to its acceptance by the owner and the final payment to the contractor.

One of the most important functions of the engineer during construction is that of interpreting the requirements of the plans and specifications on matters that are questionable or controversial. As an arbiter of disputes within his jurisdiction, his decisions are final and should be rendered with fairness and impartiality to the owner and contractor alike.

It is important to note that although the engineer may have prepared the plans and specifications in the first place, after the contract is signed he is just as rigidly bound by their requirements as the contractor. It is his responsibility to enforce their provisions throughout the work. Any variation from the plans and specifications requires a change in the contract and a corresponding change in the contractor's compensation. These are contractual matters which are usually outside the scope of the engineer's authority. Unless he has been specifically authorized to act as the owner's agent in contractual matters, it is necessary for the owner himself to order any deviation from the plans and specifications. It is the engineer's duty, however, to advise the owner in such matters and to recommend any changes he deems desirable in the best interests of the work.

Distribution of Construction Costs. The cost of a construction project varies with the cost of materials, wage rates, taxes, and other factors of similar nature. Frequently, in preparing plans, specifications, and contracts, as well as in estimating, it is desirable to know the relative costs of these elements for the commonly used types of construction. The data in Table I represent a breakdown of material, labor, and overhead costs on a percentage basis, as collected from the 1930 construction census by the Bureau of Census. The distribution percentages were derived from information on construction work aggregating $5,550,000 in 1929, and it is believed that they also approximately represent present-day conditions.

TABLE I *

DISTRIBUTION OF CONSTRUCTION COSTS

In Percentages of Total Costs

Item	Classes of Contractors				Building Construction					Highways	Heavy Construction Classifications															
	All classes	Operative builders	General contractors	Sub-contractors	All building	Bldg.—not specialized	Bldg.—commercial	Bldg.—mfr. & ind.	Bldg.—residential	Highways	Bridges	Grading	Street paving	Sewer & water lines	Dams & reservoirs	Waterworks	Dredging	Levees	Railroads	Foundations	Power plants	Air transport	Refuse disposal	Oil & gas lines	Subways	Telephone & misc.
1. Materials: Gen. Contr.	33.0	13.8	28.2	45.0	24.5	24.1	18.3	36.8	26.2	38.0	39.9	8.3	38.6	35.4	28.8	40.6	22.0	12.5	24.5	31.1	34.8	25.3	25.0	30.6	28.8	29.0
Subcontr.	10.8	29.8	13.8	1.5	18.9	18.8	22.2	15.1	19.0	3.8	3.1	0.5	3.7	1.8	3.4	2.9	2.9	3.9	3.8	4.2	15.3	5.9	5.8	6.9	3.7	1.2
Total materials	43.8	43.6	42.0	46.5	43.3	43.1	41.2	51.9	45.2	41.8	43.0	8.8	42.3	37.2	32.2	43.5	24.9	16.4	28.3	35.3	50.1	31.2	30.8	37.5	32.5	30.2
2. Labor: Gen. Contr.	23.6	8.3	21.6	28.9	19.3	19.0	17.5	13.0	16.8	23.0	26.4	36.4	21.7	31.6	30.3	26.2	27.6	28.1	32.9	31.1	23.9	35.4	31.2	28.0	32.3	32.7
Subcontr	7.1	19.2	8.9	1.0	12.2	12.1	14.7	9.8	12.2	2.7	2.9	2.9	3.0	1.9	1.7	1.5	2.2	4.4	4.3	4.7	5.3	3.6	3.7	2.8	2.4	0.3
Total labor	30.7	27.5	30.5	29.9	31.5	31.1	32.2	22.8	29.0	25.7	29.3	39.3	24.7	33.5	32.0	27.7	29.8	32.5	37.2	35.8	29.2	39.0	34.9	30.8	34.7	33.0
3. Salaries: Gen. Contr.	5.6	2.8	5.0	7.2	4.5	4.6	4.2	4.7	3.9	5.2	5.4	7.0	5.6	6.1	6.3	6.9	6.3	10.3	5.8	8.1	6.4	10.1	6.2	4.3	5.7	7.3
Subcontr.	1.8	4.8	2.2	0.2	3.0	3.0	3.6	2.4	3.0	0.9	0.6	0.7	0.7	0.2	1.0	0.9	0.8	0.7	0.7	0.8	0.9	0.9	0.9	0.4	0.6	0.3
Total salaries	7.4	7.6	7.2	7.4	7.5	7.6	7.8	7.1	6.9	6.1	6.0	7.7	6.3	6.3	7.3	7.8	7.1	11.0	6.5	8.9	7.3	11.0	7.1	4.7	6.3	7.6
4. Misc. overhead: Equip. rental: Gen. Contr	0.4		0.5	0.3	0.1	0.1	0.2	0.2		1.4		2.7	1.4	1.1	0.6	0.6	1.2	3.1	1.2	0.7	0.2	2.0		0.1	0.5	2.7
Subcontr.	0.1	0.2	0.1	0.1	0.1	0.1	0.1	0.2	0.1			0.9														
Plant charges, rent, interest, bonds, insurance, etc.: Gen. Contr.	8.0	5.2	7.7	8.8	4.0	3.9	4.3	4.2	3.6	15.1	11.1	26.6	13.0	12.8	14.8	12.6	27.8	25.2	14.7	11.0	5.5	12.6	14.9	16.0	11.4	13.8
Subcontr.	2.1	5.6	2.6	0.2	3.6	3.6	4.3	2.7	4.1	0.7	2.1	1.9	2.2	0.4	0.1	0.3	1.4	2.4	1.7	1.9	1.3	1.1	1.1	1.3	0.7	0.3
Total overhead	10.6	11.0	10.9	9.4	7.8	7.7	8.9	7.3	7.8	17.2	13.2	32.1	16.6	14.3	15.5	13.5	30.4	30.7	17.6	13.6	7.0	15.7	16.0	17.4	12.6	16.8
Total Items 1 to 4	92.5	89.7	90.6	93.2	90.5	90.6	90.1	89.1	88.9	90.8	92.5	88.4	90.0	91.3	87.0	92.5	92.2	90.6	89.6	93.6	93.6	96.9	88.8	90.4	86.1	87.6
5. Balance of charges, incl. undistributed expense and profits: Gen. Contr.	5.7	2.1	5.6	6.4	4.3	4.1	3.5	6.5	5.8	7.2	6.9	10.4	9.2	7.8	11.6	6.4	7.0	7.6	8.6	5.5	5.5	2.0	10.2	9.0	13.4	11.4
Subcontr.	1.8	8.2	3.8	0.4	5.2	5.3	6.4	4.4	5.3	2.0	0.6	1.2	0.8	0.9	1.4	1.1	0.8	1.8	1.8	0.9	0.9	1.1	1.0	0.6	0.5	1.0
Total balance	7.5	10.3	9.4	6.8	9.5	9.4	9.9	10.9	11.1	9.2	7.5	11.6	10.0	8.7	13.0	7.5	7.8	9.4	10.4	6.4	6.4	3.1	11.2	9.6	13.9	12.4
Total all items	100	100	100	100	100	100	100	100	100	100	100	100	100	100	100	100	100	100	100	100	100	100	100	100	100	100
Percentage of work sublet	23.7	67.8	31.4	3.4	43.0	42.8	52.0	43.9	43.2	10.0	9.2	8.6	9.6	5.2	7.8	6.7	8.1	12.5	12.3	13.5	23.7	12.6	12.5	12.0	7.9	3.1

* From *Engineering News-Record*, March 21, 1935.

13

For all classes of construction, 43.8 per cent of the total contract cost is for material, 30.7 per cent for labor wages, 7.4 per cent for salaries of executive and supervisory forces, 10.6 per cent for general overhead and plant charges, and the remaining 7.5 per cent is classed as undistributed expense and profit. Table I shows this breakdown as well as corresponding breakdowns for three classes of contractors, operative builders, general contractors, and subcontractors. The table also reveals that, on the average, 23.7 per cent of construction work of all classes is let to subcontractors.

The same breakdown of material, labor, and overhead costs is carried through various classes of heavy construction. Naturally, grading work shows the lowest ratio of material costs while bridges, waterworks, and power plants show the highest. The ratio of overhead costs, including plant charges, is 10.6 per cent for all construction but jumps to more than 30 per cent on grading, dredging, and levee work. The balance of costs, other than materials, labor, salaries, and overhead, which includes profits, is highest on subway construction, almost twice that of the general average of 7.5 per cent.

Table II shows a separate breakdown of building costs divided into the various trades classifications usually found in subcontract work.

The Law in Engineering Relations. In any transaction the law serves as a guide for action. When there is a misunderstanding as to the intended meaning of the law it is necessary that an interpretation be made in the light of the conditions involved in the transaction. This usually means a suit in a court of law or submission of the dispute to a board of arbitration.

The laws of the United States are derived from three sources, namely, common law, statute law, and the Constitution. Common law originated in England and was adopted as a basis for jurisprudence by the United States early in the history of this country. It is derived from traditional usage and custom as representing what is right and what is wrong in human relationships. The principles of common law have been established by gradual adoption and verified by decisions in the courts of justice, without legislative action. The recorded court decisions in previous controversies form the basis of the decision when a similar controversy arises, and thus precedent is a very important factor in defining the principles of common law. Common law is recognized by all

TABLE II *

DISTRIBUTION OF SUBCONTRACTORS' EXPENSES ON BUILDING WORK

In Percentages of Total Value of Subcontracting Work

Class of Subcontractor	Work Sublet	Mate-rials	Wages	Salaries	Equip. Rental	Bonds, Insur-ance, Plant Charges, Rent, etc.	Undis-tributed Balance Incl. Profits
Carpentry and wood floors	7.5	40.5	30.0	7.8	0.6	7.2	6.4
Concrete	3.9	34.8	36.3	8.0	0.7	9.0	7.3
Electrical	0.8	44.0	31.2	9.1	0.2	8.6	6.1
Elevators	2.2	65.7	16.8	2.9	0.1	3.3	9.0
Plumbing and heating	3.2	52.0	24.4	7.1	....	7.4	5.9
Masonry	3.2	40.1	38.6	5.6	0.5	6.3	5.7
Painting and decorating	1.0	20.6	52.0	9.2	0.1	9.0	8.1
Glass and glazing	2.2	52.0	18.4	8.5	0.1	11.8	7.0
Lath and plaster	5.8	26.0	51.0	6.2	0.3	4.8	5.9
Roofing and sheet metal	1.6	44.0	25.6	9.9	0.1	12.4	6.4
Steel erection	8.5	50.7	19.3	5.0	0.2	9.5	6.8
Stonework	4.9	36.8	32.8	8.2	0.4	9.9	7.0
Marble and tile	1.2	44.5	30.4	7.2	0.1	9.2	7.4
Wrecking	1.7	1.0	53.3	13.3	3.0	20.0	7.7
Excavating	6.5	6.2	35.8	8.2	6.2	27.8	9.3
Ornamental iron	3.8	48.3	18.6	7.6	....	10.4	11.3
Soundproofing	1.6	52.5	18.4	7.3	....	9.1	10.1
Flooring	1.1	47.3	26.4	7.4	....	12.1	5.7

* From *Engineering News-Record*, March 21, 1935.

courts, but it differs in many details from state to state. Thus the line of decisions on a certain point in Massachusetts might result in a decision on a given case in that jurisdiction exactly opposite to the decision that would be reached on the same case in Illinois.

Statutory law consists of acts or statutes established by legislative action. Statutes may be passed to clarify the common law, to remedy a defect or an injustice in common law, to repeal some tenet of common law which is no longer applicable, or to impose a penalty for a transgression. Statute laws must not conflict with laws of higher precedence. For instance, a statute passed by a state legislature is not enforceable if it conflicts with a federal law.

The constitutions of the United States and the individual states are another form of written law. Statute law must not conflict with constitutional provisions. Every state statute must conform to the constitution of the state as well as to the Constitution and laws of the United States.

Laws also may be classified as to the fields of activity or subject matter covered. Those of most general application in engineering and construction relations are under the classification of business law, particularly the laws of contracts, torts, agency, sales, insurance, and the like. Although it is not expected that the engineer will master the details and technicalities of the practice of law, it is essential that he have an appreciative understanding of some of the basic principles of business law, particularly those directly related to engineering contracts.

Professional Ethics in Engineering Relations. Although the law governs the formal and more important rights and privileges of individuals in their relations with each other, there are many aspects of human behavior which are not subject to legal control. Moreover, it is recognized that the law only approximates equity and justice, being unduly harsh in some instances and failing to apply in others. In a civilized society the principles of morals and ethics provide recourse for the inadequacies of the law and, in fact, form the fundamental basis for social and business relations. Thus, in the engineering and architectural professions codes of ethics have been established by custom and usage and are recognized as a general guide for professional conduct. The ethical principles which have been adopted by the members of the engineering and architectural professions are summarized in the "Canons of Ethics for Engineers" prepared by the Engineer's Council for Professional Development.

CANONS OF ETHICS FOR ENGINEERS

Foreword

Honesty, justice, and courtesy form a moral philosophy which, associated with mutual interest among men, constitutes the foundation of ethics. The engineer should recognize such a standard, not in passive observance, but as a set of dynamic principles guilding his conduct and way of life. It is his duty to practice his profession according to these Canons of Ethics.

As the keystone of professional conduct is integrity, the engineer will discharge his duties with fidelity to the public, his employers, and clients, and with fairness and impartiality to all. It is his duty to interest himself in public welfare, and to be ready to apply his special knowledge for the benefit ot mankind. He should uphold the honor and dignity of his profession and also avoid association with any enterprise of questionable character. In his dealings with fellow engineers he should be fair and tolerant.

Professional Life

SEC. 1. The engineer will co-operate in extending the effectiveness of the engineering profession by interchanging information and experience with other engineers and students and by contributing to the work of engineering societies, schools, and the scientific and engineering press.

SEC. 2. He will not advertise his work or merit in a self-laudatory manner, and he will avoid all conduct or practice likely to discredit or do injury to the dignity and honor of his profession.

Relations with the Public

SEC. 3. The engineer will endeavor to extend public knowledge of engineering and will discourage the spreading of untrue, unfair, and exaggerated statements regarding engineering.

SEC. 4. He will have due regard for the safety of life and health of the public and employees who may be affected by the work for which he is responsible.

SEC. 5. He will express an opinion only when it is founded on adequate knowledge and honest conviction while he is serving as a witness before a court, commission, or other tribunal.

SEC. 6. He will not issue *ex parte* statements, criticisms, or arguments on matters connected with public policy which are inspired or paid for by private interests, unless he indicates on whose behalf he is making the statement.

SEC. 7. He will refrain from expressing publicly an opinion on an engineering subject unless he is informed as to the facts relating thereto.

Relations with Clients and Employers

SEC. 8. The engineer will act in professional matters for each client or employer as a faithful agent or trustee.

SEC. 9. He will act with fairness and justice between his client or employer and the contractor when dealing with contracts.

SEC. 10. He will make his status clear to his client or employer before undertaking an engagement if he may be called upon to decide on the use of inventions, apparatus, or any other thing in which he may have a financial interest.

SEC. 11. He will guard against conditions that are dangerous or threatening to life, limb, or property on work for which he is responsible, or, if he is not responsible, will promptly call such conditions to the attention of those who are responsible.

SEC. 12. He will present clearly the consequences to be expected from deviations proposed if his engineering judgment is overruled by nontechnical authority in cases where he is responsible for the technical adequacy of engineering work.

SEC. 13. He will engage, or advise his client or employer to engage, and he will cooperate with, other experts and specialists whenever the client's or employer's interests are best served by such service.

SEC. 14. He will disclose no information concerning the business affairs or technical processes of clients or employers without their consent.

SEC. 15. He will not accept compensation, financial or otherwise, from more than one interested party for the same service, or for services pertaining to the same work, without the consent of all interested parties.

SEC. 16. He will not accept commissions or allowances, directly or indirectly, from contractors or other parties dealing with his client or employer in connection with work for which he is responsible.

SEC. 17. He will not be financially interested in the bids as or of a contractor on competitive work for which he is employed as an engineer unless he has the consent of his client or employer.

SEC. 18. He will promptly disclose to his client or employer any interest in a business which may compete with or affect the business of his client or employer. He will not allow an interest in any business to affect his decision regarding engineering work for which he is employed, or which he may be called upon to perform.

Relations with Engineers

SEC. 19. The engineer will endeavor to protect the engineering profession collectively and individually from misrepresentation and misunderstanding.

SEC. 20. He will take care that credit for engineering work is given to those to whom credit is properly due.

SEC. 21. He will uphold the principle of appropriate and adequate compensation for those engaged in engineering work, including those in subordinate capacities, as being in the public interest and maintaining the standards of the profession.

SEC. 22. He will endeavor to provide opportunity for the professional development and advancement of engineers in his employ.

SEC. 23. He will not directly or indirectly injure the professional reputation, prospects, or practice of another engineer. However, if he considers that an engineer is guilty of unethical, illegal, or unfair practice, he will present the information to the proper authority for action.

SEC. 24. He will exercise due restraint in criticizing another engineer's work in public, recognizing the fact that the engineering societies and the engineering press provide the proper forum for technical discussions and criticism.

SEC. 25. He will not try to supplant another engineer in a particular employment after becoming aware that definite steps have been taken toward the other's employment.

SEC. 26. He will not compete with another engineer on the basis of charges for work by underbidding, through reducing his normal fees after having been informed of the charges named by the other.

SEC. 27. He will not use the advantages of a salaried position to compete unfairly with another engineer.

SEC. 28. He will not become associated in responsibility for work with engineers who do not conform to ethical practices.

Questions

1. What are some of the economic benefits which might reasonably justify the construction of the following projects?

   a. A toll bridge.
   b. An hydroelectric development.
   c. The widening and deepening of a navigation channel.
   d. An express highway.
   e. A sewage disposal plant.
   f. A multi-storied office building.

2. Explain some of the conditions which might make it advisable for the Tennessee Valley Authority to adopt the force account method for the construction of a large dam during an economic depression in lieu of accomplishing the work by contract.

3. Railroad companies normally make maintenance repairs to tracks and bridges with their own forces whereas on new construction projects the work is usually accomplished by contract. Discuss the reasons leading to this procedure.

4. Why was the force account method considered advisable for the construction of the Panama Canal as compared with the employment of contractors?

5. Explain why it is usually to the best interests of the owner to adopt the general contract method for construction projects in lieu of any other method.

6. Define common law, statute law.

7. Why is it considered unethical to bid on a price basis for engineering service contracts?

2

THE LAW OF CONTRACTS

A contract may be defined as an agreement enforceable at law. Only the parties to the agreement are bound by its terms, and no right or liability can accrue directly to a person who is not a party. However, if the contract is made expressly for the benefit of a third person, as for instance in a life insurance contract, such benefit is enforceable at law in a suit initiated by the third person. Indirect or indefinite benefits to a third person are not included in this category and are not necessarily enforceable. Third persons may also acquire rights under a contract, subsequent to its execution, by assignment, succession in office, receivership, and inheritance. Contract laws vary from state to state, and the law governing a specific contract will be determined by the place where the contract was made or by the place of residence of the parties unless expressly provided otherwise.

Construction contracts usually stipulate that all controversies and disputes will be decided by the engineer and that his decision shall be final. It should be noted that this provision holds only as long as it is acceptable to both parties on a voluntary basis. No provision in a contract can deprive anyone of his right to submit a dispute to a court for settlement.

Essential Elements of a Contract. In order to be enforceable at law a contract must contain the following essential elements.

a. There must be a real agreement or "meeting of the minds."

b. The subject matter must be lawful.

c. There must be a valid consideration.

d. The parties must be legally competent.

e. The contract must comply with the provisions of the law with regard to form.

The absence of any one of these elements is sufficient to void a contract.

Agreement. If a contract is to exist there must be a mutual understanding and assent as to the terms of the agreement. Inasmuch as this is a mental process, the wording of the contract should express the intended meaning explicitly and clearly. The law requires that there must be a real meeting of the minds. Normally, agreement is indicated by the signatures of the parties. Under certain conditions, however, the contract may be nullified on the basis of *unreality of consent,* when there is sufficient evidence to the effect that there was no real agreement. Unreality of consent may be established if it can be shown that the agreement was reached on the basis of a mistake as to fact. A mistake as to law is not recognized since everyone is assumed to know the law. The following types of mistakes may void a contract.

a. Mistake as to the party.
b. Mistake as to subject matter.
c. Mistake as to the nature of the transaction.

Unreality of consent also may be established if the subject matter did not exist when the "agreement" was made, as when a ship had been sunk before its sale had been agreed upon, or if it can be shown that one party entered into the contract on the basis of misrepresentation, fraud, duress, or undue influence by another.

Lawful Subject Matter. A contract may be declared illegal and unenforceable if its subject matter is contrary to the rules of common law, if it violates a state or federal statute, or if it is in opposition to established public policy. Contracts involving crime, fraud, restraint of trade (Sherman Anti-Trust Act), collusion, and pure speculation or gambling are typical examples of unlawful subject matter. Subject matter contrary to public policy is intangible and difficult to define except in an obvious case of conflict with the interests and welfare of society and the obstruction of justice. Contracts in restraint of competition or collusion in bidding in order to create a monopoly are examples of contracts contrary to public policy. The fact that a person enters into an illegal contract unknowingly does not offer a basis for relief in case of loss or damage suffered thereby. Contracting parties are assumed to have informed themselves of all legal implications in advance of signing the contract.

The Consideration. In the legal sense, consideration is defined as the act or forbearance of one party in return for an act, forbearance, or promise of the other. If the consideration is real and valuable, the requirements of the law of contracts are satisfied, it being understood that the function of the law is to assure the consideration on which the contract was based but not to pass on its fairness or adequacy. That is to say, the courts will not attempt to make a bargain for the parties but will merely try to determine from the evidence what the bargain actually was. The only exception to this rule is in the matter of liquidated debts. One cannot satisfy an established debt by the payment of some lesser sum, even though an agreement was made to that effect, unless there is some *additional* consideration, such as an advance in the date of payment. When a contract requires a promise to give something other than money, one dollar in addition is customarily required in order to establish a real consideration. Unilateral promises, gratuitous promises, and impossible promises usually are not enforceable by contract law because of the absence of a real consideration. Likewise, a consideration made impossible by an act of God may be sufficient reason for the cancellation of the contract. Consideration may be stated explicitly in a contract, as, for instance, a definite promise to pay a sum of money or to do some specific thing, or it may be implied, in which case the existence or the value of the consideration may be difficult to prove.

Competent Parties. The laws relative to competency of contracting parties differ in various states. In general, anyone acting in good faith may enter into a binding contract except minors and insane and drunken persons. A contract made with a minor is not enforceable against the minor but may be enforceable against the other party. After a minor has reached twenty-one years of age, he may affirm a contract made previously, thus making it binding. If a minor obtains possession of goods through an illegal contract he cannot deny the legality of the contract and still keep the goods. The laws of some states consider a contract with a drunken or insane person to be legal if made in good faith and if the person knew what he was doing. Other states void such contracts without reservation. Formerly, married women were considered incompetent to contract, but this restriction has been removed universally except in some states where a married woman may not enter into a contract with her husband, marriage contracts excepted.

The matter of competency to contract is of particular importance to engineers, however, in connection with corporations. Since a corporation is a person created by law, it has only such powers as the law confers upon it, and the types of contract which a given corporation may enter into may be strictly limited. For example, a railroad may legally contract for the construction of a bridge for its tracks but may have no right to contract for the construction of a pleasure resort to attract passenger traffic. Again, the manner in which a corporation may contract may also be strictly prescribed by statute or by its by-laws, and failure to observe these forms may void a contract. Assuming that a contract with a private corporation has been found to be void because of "incompetency," the other party may still be able to recover something at law for his services if the corporation has actually made use of them; the case, however, may be costly and long drawn out. Moreover, in case of a void contract with a public corporation, for example, a city, school board, or a state, there is no possibility of recovery even in a lawsuit. It therefore behooves the engineer or contractor to avail himself of competent legal advice before entering into contractual relationship with any corporation on matters of more than casual importance.

Competency to contract is also of importance to engineers in connection with government agencies. Contracts with municipal, county, state, and federal government agencies must be made within the authority of the contracting officers representing those agencies. Contracts made outside such delegated authority are not binding on the agencies concerned even though the contracting officer was acting in good faith when he exceeded his authority. Persons entering into contracts with government agencies do so at their own risk inasmuch as no recovery is possible for work or materials furnished for a government contract in which the contracting officer acted outside his legal authority.

Offer and Acceptance. A contract may be *express* or *implied*. Express contracts involve a definite offer by one party of the terms upon which he is willing to enter into contract with another. The form of the offer is immaterial from the legal viewpoint; it may be by letter or telegram, or it may be oral. The only requirement, in case of a controversy, is that there must be evidence that the offer was made, and therefore from the standpoint of good business practice the offer should be written and signed, since this is the best form of evidence. The acceptance must be equally definite, unqualified, and unconditional. A conditional acceptance terminates the offer and thus precludes a later acceptance. The conditional acceptance may, however, be considered as a new offer, which the original offeror has the option of accepting. A contract does not exist until the offer is accepted, and the offer may specify the method or other details of the acceptance. For instance, the time and place of acceptance may be made conditions of the offer. The agreement comes into existence at the instant the acceptance is communicated to the offeror or to his agent. If the offeror has used the mail to communicate his offer, the postal service becomes his agent; thus an acceptance creates an agreement the instant it is dropped in the mailbox (communicated to the agent), and it is immaterial whether the letter of acceptance ever actually reaches the offeror. At any time prior to the acceptance an offer may be withdrawn or revoked; offers may also cease to exist because of other reasons, such as the passing of a "reasonable" period of time or the death of the offeror.

An advertisement for bids by a corporation or a government agency is not in itself an offer, but a bid in response to such an advertisement is an offer which, however, creates no legal rights until accepted. Even though a municipal charter expressly requires that a contract be awarded to the lowest bidder, a contract is not formed until the lowest bid is in fact accepted. Even if the municipality can make a contract with no other than the lowest bidder, it is not compelled to make a contract with him.

A series of circumstances may be sufficient to establish an implied contract without a specific offer and acceptance, if the facts involved indicate that the parties showed a mutual intention to contract. Questions relative to implied contracts usually arise only in cases of controversies and disputes and in most cases are settled in a court of law.

Provisions of the Law with Regard to Form; Statute of Frauds. This refers particularly to the requirement that certain types of agreements must be in writing, "signed by the party to be charged," before they can be enforced at law. The old English statute known as the Statute of Frauds sets forth the types of contracts that must be written, and this statute has been adopted with minor modifications by all states of the Union. Principal provisions of interest to engineers, architects, and contractors are that the following types of contracts among others must always be evidenced in writing.

a. A contract for the sale of land or any interest in land, or a lease of more than 1 year.

b. A promise by an executor or administrator to pay a claim against an estate under his charge out of his own money.

c. A promise to answer for a debt, default, or miscarriage of another person.

d. Any agreement which by its terms is not to be performed within the space of 1 year from the making thereof.

e. A contract for the sale of goods, wares, and merchandise in excess of a certain dollar value (varying from $25 to $500 in various states) unless the buyer accepts part or pays part or gives part payment.

It will be noted that many important types of contracts, such as for building a house or a bridge, need not be in writing as far as the Statute of Frauds is concerned. In other words, such contracts would be legal and binding even if oral. However, good business practice dictates that all but the most trivial contracts be written and signed by the contracting parties for the purpose of record in case of disputes. This is particularly true for construction contracts where the services to be furnished are usually complex and there are many promises and conditions in the agreement. The written document should delineate the scope of the services to be performed, the consideration to be given for their performance, the time in which they are to be performed, and the signatures of the parties. Other conditions of primary importance may be included, or they may be set forth in separate documents and, by stipulation, made a part of the agreement.

The signatures to the contract should be the exact names or legal titles of the parties and should be used with the addresses of their places of business. In the case of partnerships the name of the firm and that of each partner should be stated in the agreement. The signature of the firm name by one of the partners will bind each of the partners, but additional emphasis is given if the contract is signed by each of the partners.

When a corporation is a contracting party, the name of the corporation should be followed by the signature of the officer authorized to obligate the corporation through a contract. This signature should be attested by the proper officer, usually the secretary of the corporation, that the contracting officer is duly authorized to execute the contract. In contracting with a corporation it is important to ascertain that the corporation has the right to enter into the contract and that it is doing so by proper legal action.

To the signature may be affixed the seals of the signing parties. The use of the seal presumes a consideration, and legal requirements in this connection vary. In general, unless the seal is required by law, it is better to stipulate the consideration clearly in the contract in which case it becomes unimportant whether or not the seal is used. The seal may be stamped or impressed on the paper, or it may be of the adhesive type.

Assignment of Contracts. A party to a contract may transfer his rights or obligations to a third person unless assignment is expressly prohibited by the terms of the agreement. Such assignment of a contract may not relieve the assignor of his responsibilities under the contract, however, except by agreement between the two original parties. In general, all contracts are assignable with the exception of contracts for the personal skill or services of one of the parties. Engineering and architectural services may fall within this category if it is clear that the personal skill of the contracting engineer or architect was intended when the contract was made. Otherwise, such contracts are assignable. Contracts for construction work almost always are assignable. Usually, however, the terms of the contract will require the owner's approval before any assignment is permitted. Payments due under a contract may be assigned to a creditor in security of a loan or debt unless prohibited by the terms of the agreement.

It is customary to provide for the death or incompetence of either of the parties to a contract by making his heirs, executors, successors, and assigns parties to the agreement. An executor is held to be the representative of a party after his death, and the benefits and liabilities of the party can be enforced against his estate through the executor unless the contract stipulates to the contrary.

Changes in Contracts. A contract may be modified or changed by subsequent agreement of the parties. Moreover, most contracts for engineering services and construction work contain clauses which provide for the owner to make any change or alteration in the work to be performed, as he may see fit, subject to a corresponding change in the contract amount. Inasmuch as such changes are frequently made as the work progresses and cannot be foreseen when the contract is signed, it is necessary that the amounts to be added to or deducted from the contract price be determined by negotiation when the changes are ordered. Controversies and lawsuits frequently result when all details of changes and extra work are not settled in advance.

It should be noted that the agreement to change a contract must conform to the same basic requirements as an original contract with regard to competency of the parties, consideration, legality of subject matter, and mutual agreement. Likewise, a change order may be explicit or implied, and it is in the implied type that disputes are more likely to arise unless the scope and cost of the change are definitely established before the work is performed. For this reason, change orders should be in writing with the scope of the extra work clearly defined and the mutually agreed change in the contract amount definitely stated. The change order should also cover any change in the contract time of completion because of the extra work. This is particularly important when the contract includes a liquidated damage clause based on the specified time of completion of the work.

An alternative method for arriving at the cost of changes provides for the contractor to be paid the actual out-of-pocket expense in connection with the change plus a stipulated percentage of the cost to cover overhead and profit. This procedure frequently leads to disputes, however, and it is not recommended in general.

Time of Completion. If the time in which the work is to be completed is not specified in the contract, the courts will hold that it shall be performed within a "reasonable time"; this is also true when the contract states that the work shall be done "as soon as possible," "at once," or "without delay." A "reasonable time" is indefinite and difficult to determine; therefore, it is important that the contract stipulate a definite date before which the work is to be completed or a definite number of calendar days after receipt of the notice to proceed. The latter provision should include Sundays and holidays and thereby eliminate that uncertainty as a possible source of controversy.

Liquidated Damages. When time is declared to be "of the essence of the contract," it is implied that the time of completion is a part of the consideration in the agreement, and failure of one party to complete his obligations within the time specified may make him liable to the other for damages. The amount of the damages may be determined by agreement or by a suit in a court of law. Such damages are difficult to determine exactly, and in construction contracts the usual practice is to stipulate a definite amount per day of delay as *liquidated damages* in lieu of a determination of the actual damages suffered. This amount, agreed to by both parties, is enforceable at law, provided that it is a reasonable measure of the actual damages suffered. The injured party, therefore, must be prepared to justify the amount; otherwise it may be changed or disallowed by the court in case of a lawsuit even though it was agreed to in the original contract. The burden of proof is on the defaulting party, however. Unjustifiable liquidated damages imply a penalty, and the courts will not assess a penalty. That is to say, time is a sufficient consideration for the actual damages suffered or a reasonable measure thereof, but there is no consideration for any amount in excess of the actual damages unless the contract contains a corresponding provision for the payment of a bonus for performance better than that anticipated in the contract.

The contractor may be excused from liquidated damages, or he is entitled to an extension of the time of completion if the owner adds extra work to the contract. Likewise, if the contractor is delayed by acts of the owner, he is entitled to an extension of time. Delays caused by other contractors on the work ordinarily are not grounds for extension of time inasmuch as most contracts require the work to be coordinated with that of other contractors in progress concurrently. Bad weather is usually assumed to be a normal hazard of construction work to be anticipated in advance and is not considered sufficient reason for waiving liquidated damages unless it can be shown that such weather was unprecedented and was therefore an extraordinary hazard.

Penalty and Bonus Clauses. Provisions for the payment of a bonus of a specified amount per day for completion earlier than the date stated in the contract and the assessment of a corresponding amount per day as a penalty for delay are sometimes included in a contract as an incentive to the contractor when the completion of the work is urgent. The courts have repeatedly held that a penalty is not enforceable without the corresponding bonus provision. Although penalty and bonus clauses are frequently encountered, their use is not recommended inasmuch as they tend to encourage speculation and usually work to the disadvantage of both parties. Such clauses are prohibited in U.S. Government contracts.

Termination of Contracts by Performance. Ordinarily a contract will be terminated when both parties have performed their obligations under the agreement. A contract may be terminated by *specific performance* in which the terms of the agreement are carried out to the letter. Work under a construction contract completed exactly in accordance with the plans and specifications and paid for at the contract price constitutes specific performance by both parties and terminates the agreement.

A contract also may be terminated by *substantial performance* when, due to conditions beyond the control of either party, specific performance is impossible but one party has performed his obligations to substantial completion and is entitled to receive payment in full or with minor deductions, provided that no damage has been suffered by the other party. Substantial performance will not be recognized when the contractor has not attempted to comply with the specifications, when he has willfully or fraudulently departed from the plans, or when his work has been performed in a careless or negligent manner.

Termination by Agreement. In lieu of performance, the parties may agree to the termination of a contract at any time in its life. The termination agreement may be based on a *waiver* where the parties agree to waive all their rights and obligations under the contract at a specific time. It may be based on a *substituted agreement* where a new contract replaces the original agreement. It may be based on *payment in lieu of performance* where one party pays a negotiated amount of money to the other in lieu of the performance of his required obligations. A contract termination agreement may also be based on *accord and satisfaction* in which one party agrees to accept a substitution in the obligations of the other.

It should be noted that all U.S. Government contracts contain a clause providing for the termination of the contract at any time it appears to be in the government's interests to do so. Many private contracts have similar clauses. This constitutes termination by agreement (in advance of actual termination) in which an incomplete portion of the work under the contract is accepted in lieu of complete performance. The work completed at the time of such termination is paid for on the basis of a negotiated price satisfactory to both parties or on the basis of a prearranged method of settlement set forth in the contract.

Termination by Breach. If one party either fails or refuses to perform his obligations under the contract or when his acts make performance impossible, the contract is terminated by *breach*. The defaulting party thereby becomes liable for damages incurred by his failure to perform his obligations. Moreover, the other party may be relieved of all liabilities and obligations because of the breach of the contract. Settlement of a breached contract may be accomplished by negotiation and agreement (accord and satisfaction), or it may require litigation in a suit for damages. When the performance of a contract is guaranteed by a surety, such as a bonding company, the surety becomes liable for any damages or for the performance of the contract in case of a breach by the bonded party. In the case of a *conditional contract*, that is, a contract the performance of which is conditioned on the happening of some act or event, a breach of the condition by the promising party excuses the other party from further performance under the contract and, in effect, breaches the entire contract. Accordingly, the injured party may recover any damages suffered thereby.

Termination by Impossibility of Performance. A contract may be terminated if conditions unforeseen when the agreement was made preclude performance. For example, subsoil characteristics previously assumed to be satisfactory for a bridge pier foundation may be determined unsuitable after the signing of the contract, making the construction of the pier impossible, and such a contract could not be enforced. On the other hand, bad weather may preclude the construction of asphalt paving, but unless it can be definitely shown that such bad weather was abnormal and unprecedented for the specific location and unforeseeable when the contract was made, the courts would normally find it to be a hazard of the contract rather than an impossible condition. Impossibility of performance is usually difficult to substantiate, and the burden of proof is the responsibility of the defaulting party.

The general rule is that the mere fact that a party cannot perform his promise because it has become impossible for him to do so will not excuse him. However, the attitude of the courts appears to be in a state of change with regard to this rule, and many exceptions have been allowed. There are four well recognized exceptions when impossibility excuses.

a. Where performance becomes impossible by operation of law.

b. Where the contract relates to a specific subject matter and that subject matter ceases to exist before the time for performance arrives.

c. Illness or death of one of the parties to a contract for personal services.

d. If the contemplated means of performance are destroyed or cease to exist.

Force Majeure. The doctrine of force majeure refers to the effect on an active contract of an overwhelming force, such as war, revolution, and change of government, which would prevent performance under the contract. It may also apply in the case of supervening illegality where a change in the law would result in practical impossibility of performance under the contract. The effect of impossibility or illegality of performance in this sense is to discharge the duty of further performance. If after the execution of the contract, facts or conditions arise which the parties had no reason to anticipate and for the occurrence of which the party whose performance has been rendered impossible is not to blame, his duty of performance is discharged unless a contrary intention has been manifested by the parties. Similarly the duty of further payment by the other party is discharged. Temporary impossibility merely suspends the obligations of the parties, however.

The field in which contractual language can cover force majeure is quite limited. Of particular importance, however, are the methods of suspension or termination of the contract and the general conditions under which such action is to be considered justified. The basis for payment of all outstanding costs and fees and the costs of suspension or termination of the contract should be covered in detail. This is extremely important in contracts with foreign governments where the remedy for any breach of contract must be pursued before the courts of that government since it is a rule of international law that one country cannot be sued without its consent in the courts of another country.

Questions

1. What are the essential elements of a valid contract? Are they required by common law or statute law?

2. Can the terms of a contract between two parties be enforced against a third?

3. What is meant by the expression "Ignorance of the law excuses no one"?

4. What types of mistakes may void a contract?

5. Give an example of unreality of consent.

6. Give an example of unlawful subject matter.

7. What is meant by an implied contract? an express contract?

8. What precautions should be observed when entering into a contract with a government agency?

9. What types of contracts are required to be in writing under the Statute of Frauds?

10. What precautions should be observed when entering into a contract with a partnership? a corporation?

11. What is the significance of the use of a seal in connection with contracts?

12. What is meant by "a real consideration"?

13. What happens when one of the parties dies before the completion of a contract?

14. What types of contracts may not be assigned?

15. How may a contract be changed?

16. Explain the meaning of the expression "Time is of the essence of the contract."

17. What is the difference between a penalty and liquidated damages?

18. What are the methods of terminating a contract?

3

TORTS, AGENCY, THE INDEPENDENT CONTRACTOR

TORTS

A tort may be defined as a civil wrong which, in violating the private personal rights of an individual, inflicts injury or damage to his person or his property. It may be an act of one person against another or an omission when there is a duty to act. In such cases the law gives the injured person the right to bring civil suit for damages against the offender as compensation for the violation. It should be noted that a tort is distinguished from a crime or a misdemeanor in that it is an offense against an individual rather than against the state, and the injured party must himself seek redress. However, a tort may also be a crime as in the case of criminal negligence or physical assault.

Whereas the law of contracts states that a person must do what he has agreed to do, the law of torts may prohibit him from doing certain things which violate the rights of others. These natural rights may be defined by either common law or statute law, or they may not be defined specifically, in which case it will be up to a court to determine whether or not a wrong has been done.

Since the purpose of the law is to provide a legal remedy for every legal wrong, there are many forms of tort actions. Some of those which are of special interest to engineers and contractors are discussed in the following.

Slander and Libel. Torts against the reputation are called slander or libel. Slander consists of a false report maliciously circulated orally and tending to injure the reputation of another. Libel consists of the writing or publication of any statement or representation without just cause or excuse and tending to expose another to public hatred, contempt, or ridicule.

Trespass. The tort of trespass is the unlawful disturbance of one's property, real and personal. In the case of land, the disturbance of possession is generally an unlawful physical entry onto the land of another.   Theoretically, any invasion is a trespass, no matter how far above or below the surface, for the ownership of land includes the air and space above the land infinitely upward and the soil beneath it.   However, the law is not concerned with trifles, particularly where the demands of progress justify overlooking them, thus aeroplanes flying over property without landing are not liable for damages.   Further, mere accidental invasion, as where stones are carried by a storm, is not actionable, for there is no human culpability.   Since intent is not a factor in this tort, the following have been held actionable: the inundation of premises by a defective sewer, the accidental throwing of stones and debris on the land of another in blasting and turning water onto land. In some cases the law does permit an entry, as in the case of an impassable highway, when one may lawfully enter upon adjoining land.   Also where private property is taken for public purposes under the process of eminent domain, entry on the land becomes lawful.

Infringement of Patents, Trade-Marks, and Copyrights. A patent is a grant to an inventor by the Government of the exclusive right to make, to use, and to sell the invention defined by the patent for a period of seventeen years.   The grant may be renewed only by an act of Congress.   Infringement of patent is a tort consisting of the unauthorized making, using, or selling for practical use, or for profit, of an invention covered by a valid claim of a patent during the life of a patent.

A trade-mark grants to the holder the right to use a certain label, design, slogan, or device as his exclusive trade-mark.   The grant runs for twenty years and is subject to renewal for a like period. Infringement of trade-mark is a tort consisting of the unauthorized use of colorable imitation of the mark, already appropriated by another, on goods of a similar class.

A copyright is a grant conveying the exclusive privilege of printing, publishing, or vending copies of certain articles, and musical and literary productions, for a period of twenty-eight years.   The grant is renewable for a like period.   The tort of infringement of copyright is to copy, more or less servile, a copyrighted work. Mere likeness is insufficient.   There must be appropriation of substantial portions of the copyright matter.

"Patent Applied For" and "Patent Pending" serve as a notice that an application for patent is pending in the U.S. Patent Office. They have no actual legal force but have a practical effect on warning infringers that they may be held liable when the patent is granted.

Nuisance. The tort of nuisance may be defined as the disturbing of the reasonable comfortable use and enjoyment of one's property by another, or when his conduct or the state of things for which he is responsible subjects ordinary persons in the neighborhood to unreasonable discomfort. A nuisance violates a legal right, or results in a material annoyance, inconvenience, or injury, and intent is not a factor. In determining what constitutes a nuisance, the question is whether the nuisance will or does produce such a condition as in the judgment of reasonable men is naturally productive of actual physical discomfort to persons of ordinary sensibility and ordinary tastes and habits.

The location of a cemetery within 6 feet of a water supply system has been held to be a nuisance. Fences over 10 feet in height, pollution of the air, pollution of running water, each may be judged a nuisance. Likewise, it is a nuisance to divert a flowing stream or to dam it up through any appreciable period, or to change the natural condition of one's land so that water drains from it onto that of another person where none did before the change, or so that it drains in greater quantity than it did before the change. The unlawful obstruction of navigable waters, a highway, a sidewalk, or a right of way, or the maintaining of a structure, such as a dam, which is likely to give way and do damage, is a nuisance. Nuisance may also result from the storage of explosive fuels, or any dangerous things, in a place where their escape or discharge will do serious injury, and their presence is a menace to health or reasonable comfort.

Negligence. The tort of negligence is the failure to observe the necessary care, precaution, and vigilance to protect the interests and rights of another person. It is the omission to do something which a reasonable man, guided by those ordinary considerations which ordinarily regulate human affairs, would do, or the doing of that which a reasonable and prudent man would not do. Thus, when an authorized agency tears up a street or digs a trench, it must properly protect the obstruction by a barrier and lighting to prevent injury. Criminal negligence is negligence of such a character, or occurring under such circumstances, as to be punishable as a crime by statute.

Contributory negligence of the injured person is a defense to an action for damages for negligence. There are exceptions as when the injured voluntarily goes into danger for the purpose of saving human life. When a plaintiff puts himself in danger, though negligently, and the defendant, having full notice of the fact, can nevertheless avoid the accident and does not, then the contributory negligence of the plaintiff does not defeat his action for damage.

The burden of proof as to negligence is on the plaintiff. The plaintiff proves merely the accident, and the defendant must explain it to show that he was not negligent as in the case in which the plaintiff is injured by the falling of a building.

Violation of Riparian Rights. The failure to observe the rights of the riparian owner on either side of a non-navigable stream is classed as a tort. Frequently property is owned to the middle or thread of the stream, and the owner has a right to the reasonable use of water flowing past his land. He has no right to interfere with the lawful use of the water by the other owners. In general, he must permit the water to pass on undiminished in quantity and unimpaired in quality. Under its police power the state may establish regulations of streams for the protection of public health and welfare. "Reasonable use" is a highly flexible term, allowing for the solution of each controversy on its own merits. The riparian right to receive water undiminished in quality does not imply water absolutely free of any impurities. Normal requirements preclude the strict doctrine of "water undiminished in quantity." Local adjustments in the rules are made due to local demands and necessities. Domestic use is given priority over agricultural and manufacturing needs.

The law of riparian rights does not apply to ponds and lakes. As to the rights of riparian owners over the beds of navigable waters between high and low water marks, the decisions are somewhat in conflict, although the general rule is that the riparian owner holds the right of access to the water subject to the right of the state to improve navigation, the right to build a pier to the line of navigability, the right to accretions, and the right to a reasonable use of water as it flows past the land.

An owner may appropriate the surface water on his land, since it is not a stream and the law of riparian rights does not apply.

As regards water which percolates below the surface, an owner may draw off the water for any reasonable purpose, even if it should drain his neighbor's well. He cannot contaminate percolating waters by building privies or cess pools.

Violation of the Right of Lateral and Subjacent Support. The failure to observe a land owner's right to have his land supported by the adjoining land or the soil beneath is also classed as a tort. A landowner may excavate indefinitely downward, but he cannot remove lateral support from another's land in its natural state when not loaded with buildings or materials, nor withdraw subjacent support. By ordinance in some cities, a landowner who excavates to a depth of 10 feet or more below the curb is required to seek permission from adjoining landowners to enter upon their land, and, if such permission is given, he must protect any building or structure the safety of which may be affected by said excavation. In some building codes the adjoining owner must protect his own wall, building, or structure when the excavation is 10 feet or less, and he has the right to enter premises where excavation is being made in order to do so.

The right of subjacent support exists between the lower and upper strata of the earth, when they are owned separately by different persons. The lower stratum of mineral land is often granted to another, the grantor retaining the surface. In such case the owner of the mine must not excavate in such a way as to cause a settling of the surface. In underground work the withdrawal of ground water does not constitute a tort even if some settlement of adjoining soil results, but if quicksand is encountered, and it flows into the trench causing subsidence of near-by soil, a suit for violation of right of support may properly be maintained.

Wrongful Removal of Party Walls. When two buildings on adjoining parcels of land have been constructed with a single wall between them, used in common for both buildings and erected on the boundary line so that part of the wall is on each of the two parcels, the common wall dividing the buildings is called a party wall. Each is the owner of so much of the wall as stands on his lot, each being subject to and having the benefit of mutual easements of support, each section or segment of the wall being bound to support the other.

The easement for the use of party walls is lost if the structures are destroyed by fire or storms unless it was created by express grant as a permanent easement affecting all buildings to be erected on the lots. Generally, neither owner may remove a party wall. It must be left undisturbed as long as either of the buildings it serves remains standing. If the wall is torn down with the purpose of building a new wall entirely on one of the lots, excluding the owner of the adjoining lot entirely for its use, the owner who so demolished it is liable to the other, the destruction being wrongful.

Discharge of a Tort. Like a contract, a tort may be discharged by any of several methods. Inasmuch as the affected parties are the only ones concerned the method of discharge is not prescribed by law. The following are the common methods of settlement.

a. By satisfaction of a judgment in a civil suit for damages.

b. By agreement or release.

c. By the death of either or both parties, which discharges some torts. In general, an action to redress a property tort may be maintained or continued by or against an executor or administrator.

d. By bankruptcy of the tort feasor, which discharges some torts. His liability for willful and malicious injuries to person or property continue.

e. By a Statute of Limitations which discharges the tort automatically if no action is started to recover damages within a specified period. This period of time varies in different states. The object of a Statute of Limitations is to require a person who has been wronged to enforce his rights within a reasonable time or to waive them.

AGENCY

The laws of agency deal with the conditions whereby one person may represent another or act as his agent in certain types of business transactions. An agency relationship exists by agreement, and the requirements with regard to competency are the same as those for other types of contracts. An agency may be established by implication, or it may be created formally by the issuance of a *power of attorney* in which case the agent is known as an *attorney-in-fact*. Anyone who is competent to act in his own behalf may be designated as an agent or an attorney-in-fact inasmuch as the latter term does not have professional implications. The person for whom an agent acts is known as the principal.

Authority and Duties of an Agent. It is the responsibility of the principal to define the scope of the agency in clear unmistakable terms, and the authority of the agent is limited to the scope so defined or to the acts necessary to accomplish the purpose of the agency. Anyone may refuse to transact business with an agent, but if he accepts the agent in lieu of the principal, it is his responsibility to learn the limitations of the agency or to suffer the consequence of not doing so. Secret instructions in conflict with the agent's apparent powers do not bind an innocent party doing business with the agent, however. It is the agent's duty to represent his principal with honesty, loyalty, and diligence, employing competent skill and keeping the principal informed on all matters affecting his interests since any notice given to an agent is the same as if given to the principal. Negligence in the performance of his duties constitutes a tort on the part of the agent, and he is liable to a suit for damages. A similar situation exists with regard to an agent's failure to consider the rights of others in carrying out his duties under the agency. In the case of an *undisclosed principal* in which the agent does not reveal the fact of his agency he assumes full responsibility for all his acts as an agent.

Responsibilities of a Principal. The responsibilities of a principal to his agent should be defined by the terms of the agency agreement. They include the obligation to pay the agent for his services and to reimburse him for all expenses incurred in connection with the agency. The principal is liable for the torts, misdeeds, and incompetent acts committed by an agent while acting for the principal within the scope of the agency. Likewise, the principal is bound by contracts made in his name by his agent within the authority of the agency.

The Engineer as an Agent. The engineer as ordinarily defined in construction contracts is an agent of the owner. This is true regardless of whether he is a consultant retained on a professional basis or an employee on a salary. As an agent he is required to exercise care, skill, and diligence in carrying out his specified duties. Questions then arise as to his responsibilities and the required degree of his proficiency under the agency and how to measure his abilities. In general, it may be said that the engineer is expected to possess reasonable skill in the practice of his profession but that perfection is not demanded. If through negligence or lack of diligence errors are allowed to occur, the engineer is liable for damages to his employer inasmuch as this would be considered a tort. A situation of this sort would exist where the engineer failed to inspect construction operations properly or failed to enforce the provisions of the specifications when so charged by his agreement with the owner. The foregoing presupposes that the engineer actually has the necessary skill to accomplish the purposes of his agency but fails to use it. If, on the other hand, he obtained his employment upon his representation of skill which he did not possess, he is liable to his employer for breach of contract, or possibly for the tort of fraud, inasmuch as the possession of skill was one of the conditions of the agreement. As in the practice of medicine and law, the charge of malpractice of engineering usually is difficult to prove, and it is assumed not to have occurred until so proved.

It is also important to note that the engineer may be subject to criminal charges if, through negligence, he is the cause of serious personal injury or death, such as might occur in the event of the failure of a bridge or dam, when the cause of the failure can be traced to his negligence in not applying ordinary engineering skill and care when the structure was built.

THE PRINCIPLE OF THE INDEPENDENT CONTRACTOR

Under an agency relationship, the principal controls or directs the methods and acts of the agent, or when such control is not exercised directly the agent is empowered to act for the principal as if it were. The principal under these conditions becomes responsible for the acts and torts of the agent within the scope of the agency. In contrast to this arrangement the principal may be represented by another person in the accomplishment of certain acts or work under an agreement whereby the principal specifies the results to be obtained but exercises no control or direction as to the methods by which these results are to be accomplished. Such a representative is known as an *independent contractor*. The acts and torts of an independent contractor do not obligate the principal (with certain exceptions listed below), and the principal is not liable for any damages caused thereby. In general, it is the intention of every construction contract to make the contractor an independent contractor and not an agent. Therefore, it is of prime importance that the terms of the contract and the provisions of the specifications do not remove too much of the control of the work from the contractor. Particular reference is made in this connection to the tendency of many specification writers to give the engineer authority to prescribe construction methods and to control the hiring and discharge of the contractor's workmen. If this is carried too far, it may be possible, in case of difficulties, for the contract relationship to be changed from that of the independent contractor to one of agency, thereby shifting the responsibility for the contractor's torts to the owner in opposition to the original intention of the contract. It is general practice for the owner to reserve the right of general supervision over the work, and this is considered permissible under an independent contractor relationship as long as no attempt is made to exert control over the workmen in the method and manner in which they perform their duties. The engineer or an inspector may inadvertently waive the owner's rights by telling the contractor or his employees *how* to do a certain piece of work.

The exceptions to the general rule whereby the principal is held harmless for the torts of an independent contractor are:

a. When the only means for accomplishing a specified result are injurious to a third party.

b. When the principal employs an independent contractor to accomplish a specific result which is injurious to a third party.

c. When the principal employs a contractor professionally unqualified to perform the work required and a tort results from the contractor's incompetence. For example, the principal may be held liable in the case of damages arising from defective plans prepared by an incompetent engineer employed as an independent contractor.

d. When the principal is under a duty to another and engages an independent contractor to perform that duty and the contractor defaults.

The Subcontractor as an Independent Contractor. When a subcontract is executed on an independent contractor basis, the question arises as to the responsibility for torts committed by the subcontractor. Under these conditions it is generally held that the prime contractor is responsible for any loss or damage suffered by the owner or a third party through the methods utilized in construction. This is true even though the damage may be the result of acts of the subcontractor or his workmen. The prime contractor in turn may hold the subcontractor liable, but it is his first responsibility to protect the owner, recovering for himself such compensation for the damages as is possible under the conditions of the subcontract. In general, a subcontractor acting as an independent contractor and not as an agent of the prime contractor is liable to the prime contractor with respect to torts just as the prime contractor is liable to the owner.

Contingent Liability. Although the laws of agency hold an independent contractor liable for torts committed in connection with the performance of the contract, the owner is not completely protected thereby. In a suit for damages he may be held jointly responsible with the contractor, or he may be sued directly by the injured person on the grounds that the contract relationship is one between the contracting parties and is irrelevant to a third person. This is consistent with the law of contracts which states that a contract between two parties cannot obligate in any way a third person who is not a party to the agreement. Thus, under a construction contract, the owner may hold the contractor responsible for any damages arising from the work, but the owner himself may be held jointly or wholly liable by an injured third person. Likewise a prime contractor may be held liable for damages caused by a subcontractor. This is termed *contingent liability*.

Questions

1. How may a contractor protect himself against claims for damages because of inadvertent torts committed by his employees on construction work?

2. What is the status of the supervising engineer under the law of torts when the construction contractor damages the property of a third party?

3. Define an agent. What qualifications must one have in order to become the agent of another person?

4. Discuss the authority and responsibilities of an agent.

5. Distinguish between an agent and an independent contractor.

6. Under what conditions may a principal be held responsible for the acts of an independent contractor? Under what conditions may the engineer as an agent of the owner be held responsible for his own acts?

7. What is meant by contingent liability?

8. An employee of a grading contractor on a highway was injured through careless driving of a truck that was hauling gravel to the job from a pit. The gravel company's only connection with the work consisted in

furnishing gravel to the grading contractor. The careless driver was the employee of the haulage contractor employed by the gravel company to deliver gravel to the job. The gravel company paid the haulage contractor so much per ton for hauling the gravel and merely designated where the material was to be delivered. All details as to loading, unloading, and movement of the trucks were under the control of the haulage contractor. Was the grading contractor, the gravel contractor, or the haulage contractor liable for the accident?

9. If the work which an independent contractor is hired to do constitutes a nuisance, would the person who hired the independent contractor be liable for the latter's negligence?

10. Does an action for property damage caused by negligence abate upon the death of the plaintiff?

4

SALES, REAL PROPERTY, MECHANIC'S LIENS

SALES

A sale is an agreement whereby the seller transfers the property in goods to the buyer for a consideration called the price. A sale agreement is subject to all the requirements of a contract relative to competency of the parties, mutual assent, legality of subject matter, and consideration. If there is an absence of consideration, the transfer is a gift.

The law of sales deals with personal property, or chattels, as distinguished from real property. In a general way, real property consists of land and all rights or interests therein. As distinguished from this, personal property is of a personal or removable nature and includes all rights not included as real property. The old division of property into movable and immovable things expresses the same idea. Movable things are always personal property. It is obvious that rules of law applying to the sale of land must be different from the rules which govern the ownership of movable property. Land cannot be transferred by delivery. Liens on land are classed as personal property, however, because they are incident to the debts which they secure.

In the absence of any provision to the contrary, the transaction will be classed as a sale when specific goods owned by the seller are accepted and paid for by the buyer. When either delivery of the goods or the acceptance and payment therefor is promised in the future, most courts hold that the transaction is a contract to sell, and title does not pass until the conditions of the contract are fulfilled. For goods to be specific they must be in existence, be owned by the seller, and be so described in the agreement as to be readily identified and distinguished from other goods of the same kind belonging to the seller.

A *contract to sell* differs from a sale in that the title is not transferred by the mere agreement; instead the seller assumes a contractual obligation to make the transfer of title, and the buyer assumes a contractual obligation to accept the title at a future date. A contract to sell becomes a sale when the title passes.

The time at which title to the property passes from the seller to the buyer is of great significance in the law of sales. Responsibility for the loss or damage of goods in transit from the seller to the buyer, for instance, is placed on the party who possessed title when the loss or damage occurred. Title passes when the parties intend it shall pass, and sales agreements should be explicit in this regard. If the parties intend title to pass at the time of the bargain, it will pass at that time although delivery may not have been made or the price paid. In conditional sales, possession may be obtained upon the payment of a down payment, for instance, but title to the property is not transferred until the final payment is made. In most states a cash sale implies the immediate transfer of title while C.O.D. sales constitute contracts for sale in which title passes when payment is made.

Distinction must also be made between a sale, a contract to sell, and a *contract for labor and materials.* In many types of construction contracts, the contractor agrees to furnish a complete structure such as a bridge or a building for a stipulated price. The contractor retains a lien on the structure during its construction and is responsible for it in every way until title is accepted by the owner upon its completion and the contractor receives final payment for his work. Although the owner, in effect, buys a completed structure from the contractor in such a transaction the law of sales does not apply. Reasons for this interpretation vary, but one frequently stated is that the labor required during construction is part of the consideration. An important result of this interpretation is that the Statute of Frauds does not apply in contracts for labor and materials. Distinction between sales contracts and contracts for labor and materials is also important in the case of purchases and subcontracts under construction contracts. Subcontracts should always be subject to the approval of the owner, whereas direct purchases by the contractor of materials for construction are usually made independently of the owner, the only condition being that the purchased materials shall meet the requirements of the specifications.

Conditions of Sale. The maxim *caveat emptor* (let the buyer beware) is as old as the common law itself. Today, however, nearly all important trade transactions are based on certain critical conditions, expressed or implied. A sale rarely takes place without some conditions or warranties as to quality, utility, or other characteristics of the commodity sold. Conditions relative to payment and delivery are also common. Where an implied warranty is presumed to exist, the burden of proving that it did not exist is on the seller. Where an express warranty is alleged, the burden of proving its existence is on the buyer. In a contract to sell, as in any other contract, there may be inserted such conditions as the parties agree upon. Sale on approval is one in which it is a condition precedent to the passing of title that the goods prove satisfactory to the buyer.

Responsibilities of the Seller. The conditions of sale which may be enforced against the seller are those related to the delivery of the goods. The property may be a specific article, and, if so, this and no other must be furnished. It may be goods with specific characteristics as when structural steel with definite physical properties is purchased for the construction of a bridge. Unless the specified properties are furnished the conditions of the sale are not fulfilled. Opportunity must be provided by the seller for inspection of the goods. Another common condition of sale is that the article must be suitable for a given purpose. In this case responsibility is placed on the seller unless provision is made for *sale on approval* where the buyer may take temporary possession of the article but may return it to the seller within a specified time if it should prove to be unsuitable.

Responsibilities of the Buyer. The conditions of sale which may be enforced against the buyer are those related to acceptance of the property and payment therefor. Failure to accept and pay for property in accordance with the agreement constitutes a breach of the contract, and the defaulting buyer is liable for damages. Other remedies of the seller include the right to stop the goods in transit if it appears that the buyer may fail to accept or pay in accordance with the agreement, or he may attach the goods in transit. If title has passed, the risk of accidental loss or damage rests upon the buyer.

REAL PROPERTY

Property may be broadly defined as including all those things and rights which are the objects of ownership. Property is classified into real property and personal property, or chattels. Real property is generally considered to be fixed, immovable, and imperishable, whereas personal property is movable and perishable. Real property includes lands, tenements, and hereditaments. Land includes all those things attached to the land permanently, buildings, trees, fences, bridges, mines, and minerals. It also includes standing water and percolations beneath the surface. Running waters are not included, since the owner of the land has only the right to the use of such streams. Tenements include not only land, but rents and other rights and interests issuing out of or concerning land. A hereditament is a right incident to the ownership of real property, or any interest therein, e.g., the right of the owner of riparian land to the natural flow of water in a stream along the land.

Trees, crops, grass, and ore may be real or personal property. The distinction seems to be that before they are severed from the land the natural products, such as trees, which grow without cultivation and labor are parts of the realty. If the products are growing crops which are harvested annually, planted, and cared for, the general rule is that they are personal property, even when attached to the soil. Emblements are the crops which result from the tenant's labor, and which he is entitled to take away after his tenancy has ended. Fixtures are chattels attached to land in a permanent way, with the intention of incorporating the objects annexed with the land permanently. They become part of the land and are no longer personal property but become real property.

Estates. An estate is the interest that one has in land. Estates in real property are divided into estates of inheritance, estates for life, estates for years, estates at will, and at sufferance.

An estate of inheritance is termed a *fee simple*. Estate in fee simple, the highest type of ownership of real property, is the absolute ownership wherein the owner can use, dispose, or grant rights as he chooses. A life estate is measured by the duration of the owner's life or *estate per autre vie* if held for the duration of the life of some other person. He may dispose of his interest or grant a less estate; however, he must not commit waste, that is, allow any permanent and material injury to the property. He may continue to work mines or take gravel from pits that have been previously worked, but if he opens new mines or quarries he is guilty of waste.

Estate for years is an interest in lands which exists by virtue of a contract for the possession of lands for a definite and limited duration. The tenant for years is called the lessee or tenant, and the owner, the lessor or landlord. Leasehold is a possession of real property held by virtue of a lease, which is a contract creating an estate for years. The term of the lease is the time for which it is to run. If the tenant has been in possession under a lease and he retains possession without executing a new lease, he is presumed, in the absence of some agreement, to be a tenant from year to year. Estate at will is one which is terminable at the will of the landlord and also at the will of the tenant and which has no other designated period of duration. Estate at sufferance arises where a tenant wrongfully holds over expiration of his term. Future estates are possessions which are to commence in the future.

Life estates arising by operation of law pertain to curtesy and dower and are of relatively little interest to the engineer. Curtesy is a life estate which the husband has in the real property of his wife. Dower is the estate which the law gives a wife in the real property of her husband.

Equitable estates are those in which the legal title may be in one party, while the equitable title is in another. A trustee is the legal owner of property but the beneficiary is the equitable owner.

In respect of the number and connection of their owners, estates are divided into estates in severalty in which the ownership is in one person and joint estates owned by two or more persons. Joint estates include joint tenancies and tenancies in common. In the case of a joint tenancy, upon the death of one of the joint tenants his interest vests in the survivor or survivors, whereas upon the death of a tenant in common his interest passes to his representatives. In the United States, every estate granted to two or more persons, unless expressly declared to be a joint tenancy, is a tenancy in common. An estate by the entirety is one granted or devised to two persons who are husband and wife in the absence of words creating a different estate. It is an estate in which neither spouse can convey his interest so as to cut off the right of survivorship for his spouse.

Title. Title to real property is the means by which the ownership is acquired and held. Title or ownership of land may be acquired by deed or purchase, inheritance, public grant, adverse possession, foreclosure, and accretion. Acquiring title to real property by deed or purchase involves a valuable consideration. Title by inheritance is either by will or by the law of descent. Upon the death of an owner of real property without lawful heir, the property *escheats* to the state and the state becomes the owner by operation of law. A public grant is the act of creating a title in an individual to lands which had previously belonged to the government. Adverse possession is a claim of title hostile to the other party by open possession of land for a length of time which varies from state to state. Foreclosure is the proceedings by which mortgaged premises are applied to the payment of the mortgage debt. Accretion is an addition to riparian land by water by gradual deposit through the operation of natural causes, e.g., increase in area due to erection of a jetty.

Registry of Deeds. In all states the recording of real property instruments is substantially the same. The recording of an instrument is not compulsory in order to pass title, but, unless an instrument is recorded, it is not binding on a subsequent innocent purchaser who has no actual notice of the existence of the unrecorded instrument. In most large cities a Registry of Deeds or Recorder has been established whose function is the recording and care of real property papers. An index of the deeds is kept which is divided into two parts—grantors and grantees. These indexes are frequently divided by years, and, therefore, in searching it is advisable to have the names of both parties and the approximate date of the transaction.

Title Search. A land sale contract is required by the Statute of Frauds to be in writing. The contract does not convey the land but agrees to convey at some future time. Before a conveyance of real property, a title search of the records is made to ascertain that the title is clear. The chain of title should be complete; that is, the conveyances should form an unbroken chain from the beginning to the party about to convey the property. It may not always be possible to go back to the original grant, but records of conveyance should go back far enough to determine that there is no difficulty beyond the point from which they start.

The title search is also to ascertain that the various conveyances are in proper form and sufficient, so that the claim of title does convey good title. Besides deeds, it is necessary to look for mortgages, taxes, sewer assessments, wills, contracts to convey, judgments of the court, attachments, and whatever else may tend to constitute an encumbrance or impair the title of the estate. After the search an abstract of title is prepared which shows all of the transactions affecting the particular real property.

Title insurance companies may be employed to examine and report on titles. These companies not only search the claim of title from party to party, but also pass upon the question whether each deed is satisfactory and does in fact properly convey what it purports to convey.

Deeds. A deed in real property law is a written contract between competent parties by means of which the grantor conveys real property to the grantee. For a valid deed, there must be property to be conveyed, words of conveyance, description of the property, a writing signed and in some states sealed by the grantor, delivery, and acceptance. The deed must then be acknowledged, witnessed, or registered. There are several types of deeds, namely, full covenant and warranty, bargain and sale, quitclaim, executor's, and referee's.

Under a full covenant and warranty deed, the grantor covenants and warrants: (1) that he owns and possesses the premises in fee simple, and has good right to convey the same; (2) that the grantee shall quietly enjoy the premises; (3) that the premises are free from encumbrances; (4) that the grantor will execute or procure any further and necessary assurance of title to the premises; and (5) that he will forever warrant title to same. Right to convey is an assurance that the grantor has possession of the property granted and has a right to convey it. Quiet enjoyment is to the effect that the grantee, his heirs, and assigns shall not be legally disturbed in their quiet possession of the premises. The covenant against encumbrances assures that there are no outstanding rights in third parties to the land conveyed. Further assurance is an agreement by the grantor to perform any act that may be necessary to perfect the grantee's title. Warranty of title is an assurance that the grantee shall not be evicted from the premises by reason of a superior title in anyone else.

A bargain and sale deed is usually received for property purchased from a bank or an insurance company. It differs from the full covenant and warranty deed in that the grantor includes a covenant as to his own acts but does not give any further guarantee of title. The grantee must correct any defects in title. A quitclaim deed grants only such interests, if any, which the grantor may have. It contains no guarantees or warranties of any sort. An executor's deed is in effect a bargain and sale deed since the grantor states that he is the executor and that he has done nothing to encumber the title. A referee's deed is one executed by the referee at a mortgage foreclosure or by a sheriff at a sale under an execution.

Description of Property. The deed must contain a description of the property sufficient to identify it with reasonable certainty. There is no prescribed form. Real property may be described in a deed by the following ways: a government survey, reference to a plot or map, monuments, courses and distances or metes and bounds, city lots or premises, prior conveyance, or by a statement of the quantity of land being conveyed. In determining the boundaries, intent governs and boundaries fixed and determined by monuments will prevail over those indicated by distances and courses, in the absence of fraud in setting the monuments. When there are conflicting descriptions in the same deed, the most favorable to the grantee prevails, in the absence of intent to the contrary.

Surveyor's Function. It is the surveyor's duty to find the position of the original boundaries of the property and not attempt to correct the original survey even though he may be sure that an error exists in it. This is equitable, since the property was originally purchased with reference to the actual or visible bounds which vest the owner with rights to the property bounded by these lines. Where artificial features are given as boundaries, these take precedence over the recorded measurements of angles and distances, but these marks must be mentioned in the deed in order to have the force or authority of monuments. When the area does not agree with the boundaries as described in the deed, the boundaries control.

Where property is bounded by a highway, the abutters usually own to the center line, but where the boundary is an accepted street, each abutter yields his portion of the street for public use. If, however, the street is abandoned, the land reverts to the original owners.

Waters are often natural boundaries, and it is important to know what water line to use. Generally, the boundary of tidal waters is the line of ordinary high tide. The shore between high and low tides belongs to the states. When the description is "by," "upon," "to," or "along" the shore, the authorities are conflicting as to whether this means high or low water. Where property is bounded by a non-navigable stream, it extends to the thread of the stream.

If the property is described as running to the bank of a river, it is interpreted to mean to the low-water mark, unless otherwise stated.

Mortgages. A mortgage is any conveyance of land intended by the parties at the time of making it to be a security for the payment of money or the doing of some prescribed act. The borrower or mortgagor gives a mortgage in favor of the lender or mortgagee. Although the land itself is the security for the loan, a mortgagor will be required to execute a note and bond, by the terms of which the mortgagor agrees to pay the indebtedness if the property should be insufficient to satisfy the loan.

The covenants of the mortgage define the obligations of the mortgagor. He agrees to pay the indebtedness, to maintain adequate insurance, to pay all taxes and assessments levied against the property, and gives the mortgagee the right to foreclose upon default of payment.

Foreclosure is a court action in which a judgment is obtained, decreeing that the property be sold and the proceeds applied towards the payment of the mortgage debt. In case the property does not sell for enough to satisfy the debt, a personal or deficiency judgment on the note or bond is taken for the balance.

Mortgages must be recorded in order to give notice to subsequent purchasers of the property. When the mortgage is paid, it is discharged or satisfied. In order to cancel the mortgage on the records, a formal discharge is executed and filed.

Second or subsequent mortgages may be placed on a property. The mortgages take priority according to their date; that is, the second mortgagee cannot be paid until the first is paid in full. Any mortgagee can foreclose his mortgage when the mortgagor is in default, but he cannot affect the interest of a prior mortgage by such proceeding, although he may cut off any subsequent mortgagee.

Equity of redemption is the right which a mortgagor has to redeem his estate after the mortgage has become due. The length of time for redemption is defined by statute.

Trusts. A trust results when the creator or settler gives legal title to the property to a trustee who holds it for the benefit of a third party or beneficiary. The beneficiary does not take any legal estate or interest in the property but may enforce the performance of the trust in equity.

Express trusts are those created intentionally by the direct act of the parties, whereas implied trusts are those created by operation of the law. Anyone having dealings with a trustee under an express trust should know the terms of the trust, as the trustee's powers are limited to those created by the deed of the trust. Only such trusts in real property may be created as provided by statutes. However, a valid trust in personalty may be created for any lawful purpose.

Easements. An easement is the right to make use of the land of another for definite and limited purposes. The following are examples of easements

a. Easement of way. A common driveway between two neighboring houses, partly on the land of each, gives each owner an easement on the other's land.

b. Party wall easement. Each adjoining landowner owns the half of the party wall standing on his land and is entitled to an easement of support in the other half standing on the property of the adjoining landowner.

c. Easement of support. A landowner is entitled to lateral support of his lands in their natural state by the lands of adjoining property owners.

d. Easement for burial purposes. The purchaser of a plot in a cemetery has the right to use it for burial purposes but does not acquire a title in fee simple.

The right of support of one's soil against his neighbors is a "natural easement." An express permission or right-of-way across one's land makes an easement by "actual grant." When a rear lot is sold and is accessible only through the front lot, the law provides a right-of-way over the front lot by means of an easement by "implied grant."

Licenses and Franchises. A license is a permission, express or implied, to do an act affecting the property of the licensor which but for the permission would be wrongful, constituting trespass or nuisance. A license is not an interest or estate in land but a mere privilege without enforceable rights, which distinguishes it from an easement giving definite property rights.

A franchise is the right to operate a utility over property or in connection with property that is partly controlled or owned by a municipality granting the franchise. This right is personal property rather than real property.

Eminent Domain. Under the right of eminent domain the state can take land for public use upon the giving of just and adequate compensation. This right is sometimes given to corporations who perform some public function, as railroad and telegraph companies. Streets, sewers, water pipes, and railroads are considered projects for public use.

Highways and streets are created by dedication, prescription, or condemnation proceedings under a statute authorizing their opening. Dedication of land for a street, highway, park, school, public dock, or other public use requires an offer to dedicate followed by an acceptance by the public. Public rights of the same nature may arise by a prescriptive user on the part of the public for twenty years or more. Streets and highways may be opened by condemnation proceedings under the power of eminent domain.

Right-of-Way. A right-of-way is a tract of land acquired for the operation of a highway or public utility in which each parcel of property is dependent upon the other property for its use or operation, as distinguished from a holding for purposes in connection with the utility but not strictly necessary for its operation.

A right-of-way required for a public utility is usually obtained by purchase. Where an agreement cannot be reached and the utility company has the right of eminent domain, a proceeding in condemnation can be brought against the property to be acquired, and a jury fixes the values of the property. A railroad may acquire a right-of-way for its main tracks by the right of eminent domain. However, industrial tracks to plants, docks, or mines are for private use, and the right of eminent domain cannot be invoked to acquire any necessary lands. Where the utility company has no power of condemnation, it must acquire its right-of-way by purchase or lease.

A right-of-way acquired by a municipality or a public utility by the right of eminent domain is only an easement for the purposes for which it is secured. A discontinuance of such use automatically operates as a reversion of the property to the original owners. Where possible the condemning party obtains a quitclaim deed to the properties involved in order to avoid the question of disuse or reversion. Where a deed in fee simple is taken, the utility secures a clear title and may reconvey for any use.

MECHANIC'S LIENS

The laws of mechanic's liens are designed to protect workmen, material suppliers, and, under certain conditions, engineers, architects, contractors, and subcontractors on construction work. They are statute laws and vary from state to state. In general, it is held that anyone who has contributed to or increased the value of real estate through his personal services or by furnishing materials to be incorporated into the property has a claim against the property for priority of payment of the price and value of the services performed and the materials furnished. Thus a mechanic's lien is somewhat of the nature of a mortgage. It should be noted in this connection, however, that a contract, either express or implied, in which the owner agrees to the performance of the work on which the services and materials are furnished is a condition precedent to the enforcing of a lien. A lien may be discharged by the payment of the obligation, but if this is not done the lienor may force the sale of the property in order to recover payment for his services. Therefore, the owner may be made to suffer loss if the contractor should fail to pay his workmen. As a protection against this possibility construction contractors are usually required to furnish security to the owner in the form of a payment bond. Public property is not subject to liens as such action would be contrary to the public interests. Money due a contractor or subcontractor for construction work on a public project is subject to liens, however. Ordinarily lien rights are not affected by changes in ownership of the property, the death of the owner, or the assignment of the property for the benefit of other creditors.

The Contractor as a Lienor. The legal rights of a contractor to claim a lien vary to such an extent in state statutes that it is impossible to define them in general. Usually the question will hinge on whether the labor required to install the materials is relatively large or small in amount, it being necessary to distinguish between original contractor and material vendor. In order to qualify for a lien the contractor must show that the obligations of the contract have been substantially performed and that the work has been approved by the engineer, when the contract so requires.

The Subcontractor as a Lienor. The rights of a subcontractor to protection by a lien are even more complex than those of a prime contractor. In some states a subcontractor may claim a direct lien on the property, but other states require that the prime contractor shall be entitled to a lien as a condition precedent to the enforcing of a lien in behalf of a subcontractor. Default of the prime contract by either the owner or the prime contractor, requirements as to the recording of the prime contract, and the terms of the prime contract are all matters bearing on the lien of a subcontractor and must be investigated in the light of the statute of the state in which the work is performed.

Engineers and Architects as Lienors. Ordinarily the right to the protection of a lien can be extended only to those who participate in the physical improvement of real estate. This principle has been interpreted by many courts to include the services of engineers and architects who superintend construction work. It is generally considered that design and the preparation of plans and specifications without superintendence of construction do not qualify the engineer for lien protection. This interpretation is based on the condition that such services are separate and distinct from the improvement of the property although they may define the methods for such improvement. By agreement between the engineer and the owner, however, it may be possible to extend lien protection to the engineer for services of this character which are not allowable normally under the lien laws. A few states, however, have adopted statute laws to provide lien protection to engineers and architects for work spent on design and the preparation of plans and specifications.

Priority of Liens. Like other questions in connection with liens the matter of priority of payment is treated individually by the various state laws. Usually, however, workmen and material vendors will have priority over other creditors when an owner or a contractor defaults. Also, day laborers and material men will usually have priority over contractors and subcontractors regardless of the respective dates on which the liens are filed. Workmen and material vendors are given priority among themselves in the order in which their liens are filed.

Questions

1. Distinguish between a sale, a contract to sell, and a contract for labor and materials.

2. What is the significance of the expression "caveat emptor"?

3. In the case of a sale, what are the responsibilities of the seller? of the buyer?

4. Why is the sale of real property governed by different laws than those which control the sale of personal property?

5. What is the difference between a lease and an easement?

6. What precautions should be observed during construction work to protect the owner against mechanic's liens?

7. Distinguish between joint tenancies and tenancies in common.

5

PARTNERSHIPS, CORPORATIONS, LABOR LAWS

Businesses are conducted by individuals, partnerships, or corporations. The type of the business organization determines its methods of control, distribution of profits and liabilities. For individual proprietorship no formal documents are required and the undertaking can be discontinued at will. The individual owner has control of the business. He retains the profits, and his liability is unlimited. Partnerships and corporations on the other hand exist under the regulation of state statutes which vary somewhat from state to state. All three types of business organizations are encountered in engineering and construction operations, and the engineer must have a general knowledge of their characteristics.

PARTNERSHIPS

A partnership is an association of two or more competent parties to engage in a common legal enterprise and to share in the profits or losses. The partnership is formed as the result of an agreement which should be in writing setting forth all contractual conditions.

Partners in a business are classified in accordance with the character of their participation. A general or public partner is an active and known party who participates in the management and control of the business and is liable for all of its debts. A nominal partner is held forth as a partner with his own consent and is liable as a partner. As between himself and the true owners there is no actual partnership, but there exists an implied partnership as to third parties. A silent partner is an actual partner as concerns the other members of the firm but takes no active part. He may be known to the public as a partner. A secret partner is in reality a partner but conceals the fact from the public. He may or may not take an active part in the operation of the business. A special or limited partner may be admitted to the partnership, but he has no voice in the management. He may invest money and receive a certain share of the profits without incurring any liability for debts for the firm in excess of the amount he invested. Such a partnership is formed only under state statutes.

Partners Liability to Third Parties. One of the most important points in the law concerning partnership is that each partner, general, nominal, silent, or secret, is liable to third parties to the full amount of all debts of the firm. The creditor may not only take firm property; he may pursue his remedy against an individual member and may reach his personal property. However, the authority of a partner to bind the firm is limited to transactions within the scope of the partnership business. The partnership is not liable if fair notice is given that the firm will not be liable for any particular acts of a partner. The partners are liable for the fraud and false representations of one partner when they are made in the course of the firm business. It is a well established principle that notice to one partner in the course of business is considered to be notice to all.

Fiduciary Character of a Partnership. Partners in a firm stand in a fiduciary relation to each other. No one of them may make private gains at the expense of the others on matters in any way related to the business. If a partner contributes real estate to the partnership, he retains the legal title and holds it as a trustee. Real estate may not be acquired in the firm name, but legal title may be held by the individuals comprising the partnership.

Profits and losses may be divided as the partners may agree. If no agreement has been made, each partner receives an equal share of the profits and bears an equal share of the losses, irrespective of the amount each partner has invested in the business.

Dissolution of a Partnership. A partnership may be dissolved by any one of the following ways.

a. Duration provision in the partnership agreement.

b. Mutual consent of all partners.

c. Act of one or more of the partners. As where the partnership is for no definite period, any partner can terminate the relation by notice to the other parties.

d. Change in the partnership. Where a partner transfers his interest to a stranger.

e. Death of a partner.

f. Court of equity decree.

g. Bankruptcy.

CORPORATIONS

A corporation is a collection of individuals united by authority of law into one body, under special name, with the right of perpetual succession and of acting in many respects as an individual. Corporations are in the eyes of the law separate from the members who compose them. The property of the corporation is owned by it and not by the members of the corporation. Suits for or against a corporation are brought by or against the corporation and not the individuals who compose it. Corporations differ in their purposes and character, and may be classified as public or private corporations.

Public Corporations. Public corporations include municipal corporations created for local government and public benefit corporations which are organized to construct or to operate a public improvement. They receive certain powers to pass laws or ordinances and to take, hold, and sell property in their corporate name. Those organized for the purpose of local government include villages, towns, cities, school districts, counties, and other territorial subdivisions of the state. Public benefit corporations are authorized to construct and operate public improvements such as ports and toll highways.

Private Corporations. Private corporations are created for private purposes and for the management of affairs in which the members are interested as private parties. They may be of either the stock or non-stock type. Stock corporations are authorized to issue shares of stock and distribute dividends to the holders. These include railroad, transportation, business, or cooperative corporations. Non-stock corporations include religious, fraternal, and trade organizations.

Stockholders. The owners of one or more shares of the capital stock in a private stock corporation are termed stockholders. As a general rule, each stockholder is bound by all acts adopted by a vote of a majority of the stockholders, provided such acts are within the scope of the powers and authority conferred by the charter.

Limited Liability of Stockholders. One of the most important attributes of a corporation is that which exempts the stockholders from liability for the debts of the corporation. However, the stockholder of a bank is, by statute, liable for the doubled value of his stock. The stockholder is liable to the corporation on his stock and to the corporation creditors if any part of the capital stock has been unlawfully paid out to him, either directly or indirectly, leaving creditors unpaid.

Management of Corporations. The management of a corporation is vested in a board of directors presided over by the chairman of the board. Directors may or may not be stockholders. The executive officers of a corporation consist of a president, vice-presidents, secretary, treasurer, and general manager. The officers are responsible for the direction and administration of the corporate organization.

Corporate Powers. A corporation has only such powers as are conferred upon it by its charter or articles of incorporation, or are incidental and necessary to carry into effect the purposes for which it was established. Implied powers include: to have perpetual succession during the life of the corporation; to have a corporate name by which to contract, to sue, and to be sued; to purchase and hold real and personal property; to have a common seal; and to make by-laws. A corporation has no implied power to enter into a contract of partnership or suretyship.

Ultra Vires Acts. Those acts performed by a corporation which are not within its authorized powers are termed ultra vires acts. An ultra vires contract is invalid, and if executory it cannot be enforced at law. Therefore, in contract relations with a corporation, it is important to make certain that the officers are acting within their authorized powers.

Dissolution of a Corporation. The effect of the dissolution of a corporation is that thereafter the corporation no longer exists, but statutes make provisions under which the business of dissolved corporations may be liquidated and settled by a receiver. A corporation may be dissolved by expiration or surrender of its charter, an act of legislature repealing its charter under the power reserved by the state when the charter was granted, or by forfeiture of its franchise or charter for misuse or non-use of its powers.

Corporate Business in a Foreign State. A corporation is a creature of state law and does not exist beyond the state of its incorporation. Therefore, it is considered a foreign corporation in every other state. If a corporation wishes to engage in an enterprise in a foreign state, the consent of that state must first be obtained by complying with its statutory requirements. This procedure is known as qualifying to do business. The requirements of the various states differ somewhat in this regard although they are all based on a general policy of protecting citizens of the state against unreliable corporations and of placing foreign corporations within reach of state courts.

Corporations for the Practice of Engineering and Architecture. Although there still exist many corporations empowered to practice engineering and architecture, in most states they were chartered before the enactment of professional registration and licensing laws. These laws generally hold that professional practice is a matter of personal service by the principals of the firm. In some states, corporations are not considered eligible for registration or licensing, inasmuch as officers of a corporation are not personally responsible for the corporate obligations. Partnerships are acceptable because of the personal responsibility of each of the partners.

LABOR LAWS IN CONSTRUCTION WORK

The basic laws governing the relations between employer and employee are derived from the English common law of master and servant. These are devoted to such matters as terms of service, dismissal, responsibility for injuries to employees, and payment of wages. The laws of most significance in connection with construction contracts, however, are Employer's Liability Acts, Workmen's Compensation Acts, the Social Security Act, the National Labor Relations Act, the Fair Labor Standards Act, the Labor-Management Relations Act, and, for U.S. Government contracts, the Minimum Wage Act, the Non-Rebate of Wages Act, the Overtime Compensation Act, and the Buy American Act.

Employer's Liability Acts. Under the old common law an employer was held responsible for injury to an employee only if it could be shown

a. That the employer was negligent.

b. That the injured employee was not negligent.

c. That the accident was not caused by a fellow worker.

d. That the injured employee had not assumed the risk of his employment.

The development of the mechanized methods of modern industry was accompanied by a serious increase in the frequency of industrial accidents. Because of the complexity of the new processes it became evident that the older concept, whereby the employee was assumed to foresee the dangers of his occupation, was no longer practicable. The most frequent causes of accidents were found to be in the operation of machines by careless or improperly trained employees, hazards from improperly guarded equipment, and fatigue arising from the monotonous processes of mass production which cause a loss of alertness in the operator. To overcome the deficiencies of the old common law, employer's liability laws were enacted to place a number of additional responsibilities on the employer. Under these laws the employer was obliged to furnish the employee with safe working conditions, proper tools, and competent fellow workmen. The employer was held liable for damages for personal injuries to an employee caused by defects in plant equipment or by the acts of an incompetent fellow worker. In general, an employer is not liable for a tort committed by an employee unless the employee has been given sufficient authority to class him as an agent.

Although employer's liability laws resulted in some improvement in protection to employees it failed notably in some respects. Contributory negligence on the part of the injured workman still remained to void a part or all of the employer's liability. Furthermore it was usually necessary for the employee to retain a lawyer to litigate his case, and frequently a large part of the damages awarded would be consumed by legal expenses. Fellow employees were reluctant to give evidence which might be objectionable to the employer. Long delays were encountered in obtaining court judgment, and there was no consistency in the various courts as to the amount of damages and the injury suffered. These deficiencies led to the enactment of workmen's compensation laws which have now been adopted by all the states.

Workmen's Compensation Acts. Although workmen's compensation laws vary somewhat from state to state, the essential elements are common in general. These laws recognize that the employer should bear the costs of industrial accidents and provide for prompt settlement of claims, irrespective of whether or not the accident was caused by the employer's negligence and irrespective of any contributory negligence on the part of the employee. In addition to the payment of benefits for injuries, provision is also made for medical and hospital expenses in connection with the injury. The laws cover industrial workmen primarily, and some occupations, such as farming and domestic service, are excluded. Usually some form of workmen's compensation insurance carried by the employer is mandatory so that prompt payment is guaranteed to the employee in case of injury. In this connection an employee is defined as one who works for another who obtains profits from his services. Careful distinction must be made between an employee and an independent contractor in this regard.

Compensation benefit payments usually start after a period of a week of disability after the accident and will usually be from 50 per cent to 66⅔ per cent of the weekly wages in the case of temporary disability. For a permanent partial disability the maximum compensation will run for a specific number of weeks. For a permanent total disability the compensation may be paid for life, or settlement may be made on the basis of the payment of a lump sum. If a worker dies as a result of the accident the compensation is usually paid to his dependents for a period of about six years or on the basis of a lump sum settlement. Intoxication, willful intent to injure himself or another, and failure to use safeguards provided for his protection may disqualify an injured employee from receiving compensation benefits.

The administration and control of a workmen's compensation act is usually vested in an industrial accident board appointed by the governor of the state. Provision is made in some states, however, for workmen's compensation to be within the jurisdiction of a court.

Social Security Act. The Social Security Act, approved August 14, 1935, and its subsequent amendments, provides old age insurance for the benefit of workers who have reached the age of retirement. Under this act, deductions are made from the worker's pay by his employer, and the employer is also assessed a certain amount, on a percentage basis, of his payroll. The combined amount of the employee's and the employer's contributions is paid to the federal government as a tax to cover Old Age and Survivor's Insurance. The act was designed primarily for the protection of industrial workers for whom old age and retirement compensation was not available previously. Non-industrial employees, such as farm workers, domestics, and government workers, were not included in the original act. However, by amendment in 1950 non-industrial employees were included to a limited extent. All other workers who are employed on a salary or wage basis are covered, including the employees of engineers, architects, and contractors. It is the responsibility of the employer to pay to the government both the employer's and the employee's assessment. The operation of the Social Security Act is complex, and the details of its application for any specific employer are best determined by referring to the Social Security Board of the Federal Security Agency under whose cognizance the act is administered.

National Labor Relations Act. The steadily increasing frequency of labor disputes and consequent labor unrest, strikes, and other costly interruptions to the progress of industry and commerce led to the adoption of the National Labor Relations Act, approved July 5, 1935. This act applies to employers engaged in interstate commerce or the production of goods for such commerce and was designed to protect workmen against unfair labor practices, which the act defined as follows.

a. To interfere with, restrain, or coerce employees in the exercise of the right of self-organization, to form, join, or assist labor organizations; to bargain collectively through representatives of their own choosing and to engage in concerted activities for the purpose of collective bargaining or other mutual aid or protection.

b. To dominate or interfere with the formation or administration of any labor organization or contribute financial or other support to it (for the purpose of exerting influence or control by coercion).

c. By discrimination in regard to hire or tenure of employment or any term or condition of employment to encourage or discourage membership in any labor organization. An employer may, however, make an agreement with a labor organization to require, as a condition of employment, membership therein if the labor organization is the authorized representative of the employees for collective bargaining purposes.

d. To discharge or otherwise discriminate against an employee because he has filed charges or given testimony under the National Labor Relations Act.

e. To refuse to bargain collectively with his employees.

For the purposes of collective bargaining, representatives are selected by a majority of the employees in a unit, and the representatives so selected are recognized as representing exclusively all the employees in the unit. Individuals have the right, however, to present grievances to the employer independently.

This act also created the National Labor Board, composed of three members to be appointed by the President and authorized to carry out the provisions of the act. Whenever it is charged that any person has engaged in any unfair labor practice, the Board has the authority to investigate the complaint and take such action as necessary to enforce the policies of the act. This may include petition to any circuit court of appeals which then takes jurisdiction of the case with the power to enforce the act. When supported by evidence the findings of the Board as to facts are conclusive.

Fair Labor Standards Act. The Fair Labor Standards Act of 1938, as amended by the Fair Labor Standards Amendments of 1949, is designed to eliminate, in industries engaged in interstate commerce, conditions detrimental to the maintenance of the minimum standard of living necessary for health, efficiency, and general well-being of workers. The program is in charge of an administrator in the Department of Labor who functions with the advice of various industry committees.

The law requires that every employer pay to each employee not less than a minimum rate of wages per hour as prescribed in the statute or as determined by an order of the administrator. It prohibits employment of an employee for a work week longer than a number of hours prescribed by the statute or determined by collective bargaining. Wage orders in any industry are made by the administrator after consideration and recommendation by the committee for that industry. Every wage order is subject to review by the United States courts.

The administrator is authorized to make general investigations and to inspect the records of a specific employer in order to determine whether the Act has been violated. Employers are required to preserve their records for that purpose. Transportation or sale of merchandise with knowledge that it has been produced in violation of the Act is punishable by criminal prosecution or civil action, and may be restrained by injunction.

Labor-Management Relations Act. The Labor-Management Relations Act of 1947, commonly known as the Taft-Hartley Act, is designed primarily to provide additional methods for the mediation of labor disputes and to equalize the legal responsibilities of labor organizations and employers.

It defines certain practices of labor organizations as unfair, including the following.

a. Interference with or coercion of employees in the exercise of their rights, which the statute establishes, to join or to refrain from joining labor organizations.

b. Causing an employer to discriminate against an employee for non-membership in a labor union.

c. Refusal to bargain collectively with an employer.

d. Engaging in a so-called "secondary boycott" by inducing employees to refuse to handle goods where the object is to compel anyone to join a labor organization, to compel an employer or other person to cease handling the products of another, to require bargaining with a labor organization which has not been certified as the representative of the employees, or to require the assignment of work to employees in a particular labor group.

e. Requiring excessive entrance fees to be paid by workers becoming members of a labor union.

f. Requiring an employer to pay for services which are not performed or not to be performed.

The statute established a new National Labor Relations Board, organized somewhat differently from the previous one, charged with the responsibility of preventing unfair labor practices by both employers and employees through administrative orders and by proceedings in the United States courts. There are safeguards designed to insure free and fair election of collective bargaining agents by employees. Provisions are made for conciliation, through the Federal Mediation and Conciliation Service, of labor disputes in industries affecting commerce.

Whenever in the opinion of the President of the United States a strike or lockout will imperil the national health or safety, he may appoint a board of inquiry to report to him on the issues involved in the dispute. On receiving the board's report he may direct the Attorney General to petition the United States Court to enjoin the strike or lockout. There is provision for settlement of the dispute by agreement of the parties, and for secret ballot of the employees on approval of the settlement.

The Act establishes the legal responsibility of labor organizations for breach of collective bargaining agreements and for misconduct of their agents. Contributions by corporations or unions to political campaigns, as well as so-called "secondary boycotts" and other unlawful boycotts, are prohibited. The Act also requires the filing of certain reports by labor unions, and affidavits by officers of labor unions that they are not members of nor affiliated with the Communist party.

Labor Laws Unique to U.S. Government Contracts. Statute laws have been adopted by Congress imposing conditions on U.S. Government contracts which are not, in general, required under private contracts. These laws were designed to regulate the wages, hours, and employment conditions of workmen on U.S. Government projects. Those having the most important bearing on construction work are as follows.

a. The Davis-Bacon Act of March 3, 1931, as amended August 30, 1935, requires that mechanics and laborers receive not less than prevailing wages as determined by the Secretary of Labor. Decisions as to prevailing wages are made for specific contracts and are not changed during the life of the contract. The payment of the minimum wage rates so determined is made one of the conditions of the contract.

b. The Copeland Act of June 13, 1934, prevents kick-back of wages. This law requires that no employee may be induced to give up or return any part of the compensation to which he is entitled, by force, intimidation, threat of dismissal, or by any other manner whatsoever.

c. The Walsh-Healey Act of June 30, 1936, and subsequent amendments, implements the Fair Labor Standards Act of 1938. It requires that in any contract for the manufacture or furnishing of materials to any U.S. Government agency in an amount exceeding $10,000.00 there shall be included stipulations that all persons employed by the contractor will be paid not less than the prevailing minimum wages for the industry in the locality, as determined by the Secretary of Labor; that no person working on the contract will be permitted to work in excess of eight hours per day or forty hours per week, except pursuant to a collective bargaining agreement; that no child labor and no contract labor will be employed on the contract; and that the materials or supplies will not be manufactured under unsanitary or dangerous conditions. There may be recovery of damages by the United States for breach of these stipulations. The provisions requiring the inclusion of representations with respect to minimum wages apply only to contracts relating to such industries as have been the subject matter of a determination by the Secretary of Labor.

d. The Buy American Act of March 3, 1933, and as amended on October 29, 1949, requires the use of only such unmanufactured articles, materials, and supplies as have been mined or produced in the United States and only such manufactured articles as have been produced in the United States substantially from supplies mined, produced, or manufactured in the United States. Exceptions to this law may be approved by the Government when the materials or supplies in question are not reasonably available in commercial quantities and of a satisfactory quality in the United States.

Questions

1. Explain the procedure in executing a contract when one of the parties consists of a partnership and the other is a corporation.

2. A contractor needing certain construction materials agreed to provide funds to meet the material manufacturer's payroll on condition that he be repaid in the materials at current prices. Did this agreement constitute formation of a partnership between the contractor and the manufacturer, thereby making the manufacturer liable to the contractor's creditors?

3. A group of contractors under a joint venture are engaged on a construction project located wholly within one state. The same contractors under a separate joint venture furnished the materials for the work. More than half of the required materials and labor was secured from outside the state. Were the two groups subject to National Labor Relations Board orders to refrain from unfair labor practices? Was it unfair labor practice to use an employment application form requiring statement as to the trade, professional, or fraternal organizations to which the applicant belonged? (See *Contractors and Engineers Monthly*, July, 1952, page 73.)

4. Were the employees on a project to construct a U.S. military base in Bermuda covered by the provisions of the Fair Labor Standards Act?

5. A watchman on a federal housing construction project assisted in unloading materials brought to the site in interstate shipments. Was he covered by the minimum wage provisions of the Fair Labor Standards Act?

6

LITIGATION, EQUITY, ARBITRATION

LITIGATION

Disputes frequently occur during the performance of construction contracts because of conflicting interpretations of the plans and specifications. A great deal can be accomplished in reducing the causes of disputes by exercising proper care in the preparation of the contract documents, but it is generally recognized that the perfect set of plans and specifications have not been and probably never will be devised. Therefore, disputes are always to be expected, and consideration must be given to methods for their settlement.

The contract often provides that the engineer will be the interpreter of the plans and specifications and that his decisions will be final. When questions arise as to the quality of materials and workmanship, the engineer's decisions are usually accepted, but in disputes relative to extra work, time, liquidated damages, and the like, the jurisdiction of the engineer is limited, and other means of settlement are necessary.

If neither party yields and the dispute cannot be settled by negotiation or by compromise, recourse may be had in a court of law. Lawsuits are expensive to both parties. Moreover, the formalities of litigation result in long delays in the settlement of the dispute and frequently in the performance of the contract as well. Therefore, the engineer should exert every reasonable effort to settle the dispute so as to avoid litigation. This will almost always be to the best advantage of both the owner and the contractor.

In a lawsuit arising from a contract which permits more than one interpretation, the court will try to determine the original intentions of the contracting parties, and these will govern. When the original intentions cannot be clearly determined, the court will attempt to determine all the facts in the case from the available evidence and will render a decision in the light of the law which governs the contract. In general, if more than one interpretation is possible, the contract will be construed most unfavorably against the one who prepared it.

The System of Courts. The administration of the law is accomplished through a system of courts, each classification of which is empowered to hear only a specific line of cases or disputes; cases involving different amounts or arising between different parties or being of different natures will be determined by entirely different courts. The power to hear, to determine, and to enforce its judgment in a case is called the jurisdiction of the court. It is essential that the particular court before which a question is brought for determination shall have jurisdiction. Otherwise, its decision is of no effect and may be set aside. The selection of the court of original jurisdiction before which litigation may be properly initiated varies from state to state, but in each state the jurisdiction of each court is defined by the state constitution or by the Civil Practice Acts. The following are the principal classes of courts in the United States.

Courts of original jurisdiction have authority to hear and to determine questions when they are first presented for judicial determination. Courts of original jurisdiction may be either civil courts or criminal courts. Civil courts hear cases in which the rights and liabilities of individuals toward each other are in dispute. Criminal courts are those which administer criminal law and hear and determine criminal cases. Crimes are classified as felonies and misdemeanors. Felonies are punishable by death or imprisonment. All other crimes are misdemeanors.

An appellate court generally has no power to hear a case when it first arises. It can only review the decision of a lower court when such a decision is brought before it for determination. The taking of a case from a lower court to a higher one is called an *appeal.*

Federal courts have original jurisdiction where the amount in controversy exceeds $3,000 in cases which involve a federal question or in cases based upon diversity of citizenship. Cases raising a federal question are mainly those which arise under the Constitution, laws, or treaties of the United States. Federal courts also have jurisdiction, irrespective of amount, in actions under the patent, copyright, or trade-mark laws, Employer's Liability Acts, the Interstate Commerce Act, Anti-Trust laws, and other actions specified by statute. Cases based upon diversity of citizenship pertain to actions between citizens of different states, between citizens of a state and foreign states, or citizens or subjects thereof.

The judicial power of the United States is vested in the Supreme Court and several inferior courts which are known as circuit courts and district courts. Appeals from the district and circuit courts are taken to the Circuit Court of Appeals and in some special cases directly to the Supreme Court.

State courts have jurisdiction over questions not specifically placed within the jurisdiction of the federal courts. The systems of courts, although varying somewhat in the different states, are essentially the same. A justice court is generally the lowest court and has limited statutory jurisdiction of certain misdemeanors or cases involving small amounts. These courts do not have jurisdiction over actions involving the title of real property. A county court has jurisdiction limited to the county. It has original and appellate jurisdiction in both civil and criminal cases and equity jurisdiction in actions relating to real property situated in the county. Surrogates' or probate courts have jurisdiction to administer and distribute the estates of deceased persons and infants. The Court of Claims hears and determines private claims against the state where the state consents to be sued. The Court of Appeals or State Supreme Court is the court of last resort in the state, and its jurisdiction is exclusively appellate. It reviews only questions of law, except in capital punishment cases when it also reviews questions of fact.

A referee may be appointed by the court to hear the evidence in a case and to report his findings to the court. A referee is usually appointed in cases involving long accounts as in bankruptcy or intricate technical problems.

Disputes under Private Contracts. In private construction, disputes may arise in connection with any phase of the work, but the common sources of controversy are the amounts to be allowed for changes and extra work, the existence of subsurface conditions materially different from those shown on the plans, the extension of time for changes or extra work and for delays due to conditions beyond the control of the contractor in relation to the assessments for liquidated damages. If the decision of the engineer should not be acceptable to one of the parties and if all efforts to arbitrate or compromise the dispute should fail, the only alternative would be a legal action before the local court of original jurisdiction. This will be the district, county, or circuit court, depending upon the local designation.

In a lawsuit questions as to the law of the contract are always decided by the court. Questions of fact are decided by the jury on the basis of evidence submitted. If the court decision is considered erroneous, appeals may be made to the higher courts in accordance with the legal provisions for such action. Similar procedure is followed in disputes under public works contracts with the exception of those with the U.S. Government.

Disputes under U.S. Government Contracts. The standard U.S. Government construction contract form stipulates that the contracting officer of the agency having cognizance over the work shall decide all disputed questions of fact. If the contractor should disagree with the decision of the contracting officer, he has the privilege to appeal in writing within a specified period, usually thirty days, to the head of the department concerned. The decisions of the contracting officer and the head of the department are final with respect to questions of fact. Questions as to the law of the contract may be decided by the Comptroller General or the Court of Claims. If the decision is against the contractor he may, at any time within six years, file suit in the Court of Claims in Washington. If the amount of the claim is less than $10,000 the suit may be filed in a district court of the United States in lieu of the Court of Claims. At any stage of the proceedings for the settlement of a claim, the contractor is entitled to representation by an attorney if he so desires.

It should be noted that no contracting officer has the right to settle claims for damages caused by acts of the government. Claims for damages in connection with work under contracts with the government always have to be settled by the Court of Claims. Congress initiates special legislation to provide an appropriation for their payment. In general, congressional operating appropriations may not be used for the payment of damages by contracting officers.

Pleading and Practice. Courts conduct their business in accordance with certain prescribed rules of practice. A lawsuit is instituted by an *action* or *suit,* which is the legal and formal demand of one's right made upon another party and insisted upon in a court. The party who brings the action is known as the plaintiff, and the one against whom it is brought, as the defendant. An action is commenced by the service of a notice called a *summons* upon the defendant. In some jurisdictions the summons is issued by the judge or clerk of the court, whereas in other jurisdictions it may be issued by the plaintiff's attorney.

After an action has been started, the parties serve their pleadings by which both plaintiff and defendant present their respective versions of the question in dispute.

The *complaint,* sometimes called the petition or declaration, which is the first pleading in a case, consists of a statement of the cause of action in which the plantiff sets forth his reason for seeking the aid of the court against the defendant. The complaint is commonly served with the summons but may be served later. After the complaint has been served upon the defendant or filed with the court, it is then necessary for the defendant within a certain number of days to file or serve a statement of the reasons why he should not comply with the demands of the plaintiff. If such a statement is not filed, the plaintiff is given judgment against the defendant by default. The pleading which is filed by the defendant may be either an answer, a motion to dismiss, or a demurrer.

The answer, or plea, is a statement of the defendant's defense to the matters set up in the complaint. It may deny the plaintiff's claim, or it may admit it and set up other facts by way of counterclaim. When a counterclaim is alleged, the plaintiff, if he wishes to deny it, makes a reply or replication. After the pleadings are served the case comes to trial.

After a jury has heard the witnesses and considered the evidence, it arrives at a decision called the *verdict*. The judge then gives the judgment which is the official decision of the court upon the respective rights and claims of the parties of the action.

After the judgment has been rendered, the law provides a method of procedure, called an *execution,* for the enforcement of the judgment. This is an order by the court to one of its officers, either a sheriff, constable, or marshal, authorizing him to collect the amount named as damages and, if not paid, to take certain property of the party against whom the judgment is given, sell it, and apply the proceeds upon the judgment. The taking of the property under the authority of the execution is called a *levy*. The property after being levied upon is advertised by the officer and sold at public sale to the highest bidder. Certain articles, such as clothing, household articles of a certain value, and tools of trade owned by the debtor, are exempt from execution and sale.

The unsuccessful party in a lawsuit may within a certain time move for a new trial, either because new evidence has been discovered or because of some error of the judge, at trial. The judge may, at his discretion, order a new trial. The party dissatisfied with the judgment of the trial court may take an appeal to a higher court by complying with certain conditions. When the case has been taken to the highest appellate court to which a case of its kind can be carried and this last court affirms the judgment of the trial court, the case is finally determined.

Examination of Witnesses. When a witness is called to testify the examination is called direct examination. During this procedure, the questions asked are intended to bring out those facts which are in issue. A witness may testify as to fact when he has a recollection of his observation or when he has an impression of it or a belief of it. As a general rule, leading questions are not permitted on direct examination. Leading questions are those which are so framed as to openly suggest to the witness the desired answer or to prevent him from telling the story in his own words by calling for the monosyllabic answer of "yes" or "no."

After a witness has testified in direct examination, the adverse party has a right to examine him with a view to discrediting his testimony. The process of testing the story and the credibility of the witness is known as cross-examination. Subject to certain limitations the credibility of a witness may be attacked on his own cross-examination or by independent testimony offered by other witnesses.

Evidence. Evidence consists of the testimony of witnesses who know something about the facts in the case. Written documents and papers and objects pertaining to the case are also admitted as evidence. In order to procure the attendance of the witnesses at the trial, the court issues an order, called a *subpoena*. Refusal to appear or to testify when called as a witness is *contempt of court*. Evidence must be competent, material, and relevant in order to be admissible in court.

The law does not require that to be proven which is apparent to the court. The process by which the court dispenses with proof of facts of common or general knowledge is called *judicial notice*.

Best Evidence Rule. When the contents of a written instrument are offered as proof, the law requires that the party seeking to prove the contents produce the document itself as the best evidence. Where only the existence of the document is in issue, oral evidence may be introduced to prove such existence. The rule requiring such production when the contents of a writing are to be proven is technically known as the best evidence rule.

The best evidence rule applies only to the contents of writings in issue, such as a written contract on which suit is being brought. Those writings which merely recite the happenings of events, such as certificates and receipts, do not come within the best evidence rule.

Secondary Evidence. When a writing is in issue but unavailable, secondary evidence of its contents may be offered. If the party seeking to offer secondary evidence was originally in control of the paper, he must show his good faith by proving, to the satisfaction of the court, that he has exhausted all reasonable avenues of search. If a party seeks to introduce secondary evidence of a paper, which is in the custody of his adversary, he must serve on the attorney for the adversary or the adversary himself a notice to produce.

Real Evidence. Real evidence consists of physical objects which have value in proving a fact. Such items as material used on construction may be offered in evidence and exhibited to the jury.

Direct Evidence. Direct evidence goes to prove the existence of a fact without any inference or presumption.

Circumstantial Evidence. Circumstantial evidence does not prove the existence of a fact directly but gives rise to a logical inference that such a fact does exist. In making these inferences, the determination of the facts involves logic and human experience. Although neither logic nor human experience is unerring in the arrival at a true conclusion, generally the result of this process is satisfactory.

Hearsay Evidence. Hearsay evidence consists in the testimony of a witness in which the oral or written statement of another is offered as evidence of the existence or non-existence of a fact in issue. Hearsay evidence is deemed unreliable and is therefore inadmissible because the person whose observation is repeated in court is not under oath or subject to the test of cross-examination.

The courts sometimes allow hearsay statements as proof of the facts contained therein when there is difficulty in procuring other satisfactory and direct proof of the facts involved. However, when such hearsay statements are received, certain collateral facts must be adduced to aid in guaranteeing the trustworthiness and veracity of the evidence. These collateral facts are the foundation for each hearsay exception and must be presented before any hearsay statement will be received in evidence.

Admission. An admission is any statement made, any act done, or any course of conduct pursued by which a party to the litigation has indicated a contrary attitude to the one which he is taking at trial. Such prior contrary attitude tends to throw suspicion on his claim at trial and is used as independent evidence against him.

Res Gestae Statements. A declaration which accompanies and explains some act in issue is admissible in evidence, when it appears to have been spontaneously made and photographic of the act. For example, a victim falling from a broken scaffold cries out: "The wood was rotten." Such a spontaneous utterance is termed a res gestae statement and is deemed trustworthy in that the declarant expresses what he observes under the excitement of the event and before time is had for deliberation and reflection.

Parol Evidence Rule. In general, the parol evidence rule excludes any verbal evidence which would tend to change the meaning of a written instrument. The written document is assumed to express the full intent of the parties concerned, and any evidence tending to change such expressed intent would defeat the purpose of the writing. The parol evidence rule applies only to writings which are intended to create legal rights, such as contracts, deeds, and mortgages.

Ambiguities may appear in a writing either patently or latently. A patent ambiguity is one which is apparent from the instrument itself. In some instances parol evidence of the surrounding facts is admissible to show the intent of the writer. A latent ambiguity is one which does not appear on the face of the writing but becomes apparent when an attempt is made to apply the terms of the writing. Parol evidence is admissible to discover the writer's intent, but the proof must not change the writing. If the writing is so indefinite that proof of surrounding circumstances would only be speculative as to the writer's meaning, no parol evidence is admissible to explain the meaning.

When a writing is ambiguous, proof of custom or usage may be introduced to ascertain its meaning, provided that the parties concerned entered into the arrangement with knowledge, express or implied, of the custom or usage. Ambiguities in a written instrument will be construed most unfavorably against the one who prepared the document.

Opinion Evidence. The testimony of a witness as to the conclusions he has made in respect to facts which he has observed is called opinion evidence. This type of evidence is generally inadmissible because of the unknown reasoning powers of the witness. Consequently the ordinary witness is required to state the facts which he observed, leaving to the court or jury the function of deducing the conclusions.

Opinion Evidence of Expert Witnesses. An expert witness is one who, by reason of his general superiority in a given field of knowledge, is qualified either to discover the existence of facts not readily discoverable by the average man or to draw special conclusions from the existence of usual facts which the average man by reason of his inexperience would be unable to draw.

An expert is permitted to give his opinions based either on his own observations or on an assumed state of facts presented to him. The cross-examiner is aware of the basis of the opinion and can cross-examine accordingly. An assumed state of fact is made in the form of a hypothetical question. In this question the witness is asked to assume the existence of facts which are in evidence as the basis of his conclusion. This is necessary so that the cross-examiner will be informed as to the foundation of the opinion of the witness. The hypothetical question should state each assumption of fact from which the conclusion is to be made.

The Engineer as an Expert Witness. Engineers are called upon to give expert evidence in many kinds of lawsuits, such as those concerned with the strength and stability of structures, failure of structures, damage to structures, land boundaries, and traffic accidents. In the settlement of claims under construction contracts opinions are usually required relative to the interpretation of plans and specifications, quality of materials and workmanship, quantities of work performed, standard customs and practice, adjustments in the contract price to cover conditions different from those originally contemplated, and like matters on which controversies frequently arise. The basic procedures in the preparation of expert testimony are quite similar, however, irrespective of the type of case.

Before his appearance in court, the engineer should carefully review all aspects of the controversy. He should examine all correspondence, plans, specifications, tests, and other technical records and make a careful inspection of the site. Based on this study, he should prepare a preliminary report to the attorney. The preliminary report should give his opinion together with the facts, calculations, tests, drawings, or other data on which his opinion is based. He should also point out the probable technical arguments of attack by the opponents and otherwise assist the attorney in organizing the case. At this point, if he should feel that the case is ethically indefensible, he should not hesitate to say so and decline to testify. If on the other hand he is convinced that his client's case has merit and his testimony can benefit the case, he should proceed with the detailed preparation of his testimony.

On the witness stand the engineer will usually confine his testimony to the answers to questions asked by his attorney, but when desirable he may make a general statement on a particular subject. His testimony should be presented in a quiet, dignified manner in order to secure the confidence and good will of the court. He should always keep in mind the fact that his testimony must be understood by persons who are not trained technically, and therefore his statements must be made in simple non-technical language. He may make use of notes, calculations, drawings, and the like, and he may cite as the basis of his opinion the works of recognized authors of textbooks. He should carefully consider every question asked of him and avoid hasty answers, taking time to make certain that he understands fully what is intended. He has the right to request sufficient time to make any calculations or researches he considers necessary to arrive at an answer, even to the extent of asking for a recess of the court for that purpose when necessary. The witness should be particularly careful in answering hypothetical questions because they frequently are designed to be misleading to the expert and are usually pointed as to their significance in the controversy.

In addition to the giving of testimony, the expert witness is expected to advise his attorney during the examination of other witnesses and in the cross-examination of opposing witnesses. In this manner he may bring out points of weakness in the opponent's case or discredit the opposing witnesses. In suggesting questions to be asked of witnesses, he should be certain to furnish the correct answers so that his attorney will always know what to expect in reply. Otherwise, he may inadvertently introduce testimony into the case which the attorney would prefer to have omitted. The suggesting of appropriate questions to the attorney should be particularly helpful in attacking the testimony of opposing experts.

EQUITY

It is a matter of record that the common law does not cover every possible case and that in some instances it is inadequate and in others unduly harsh. The common law may provide a remedy after a wrong has been committed, but there may be no procedure under the common law to prevent such a wrong from being committed even when that intention is known. To provide for these deficiencies in the law, recourse may be had in a court of equity. Thus a court of equity may recognize rights not recognizable in common law and may apply remedies for which no provision is made in common law. The primary objective of a court of equity is to see that justice prevails. In general, the same judge may try cases both in law and in equity, the court being converted from one in law to one in equity by declaration when there is no adequate remedy at common law. Some states make no distinction between courts of law and equity, and establish procedures by which any court may grant remedies at law or equity without notice as to the type of action. In a case in common law, it is the duty of the judge to determine the applicable law, and the facts are determined by a jury, whereas in equity it is also the judge's duty to determine the facts. By consent of the court, however, the questions of fact may be referred to a jury. The party who brings suit in a court of equity is called the *complainant* corresponding to the plaintiff in a court of law. The party being sued is called the defendant as in a court of law.

The types of cases within the jurisdiction of a court of equity may be classified as follows.

a. Equitable titles, involving trusts, mortgages, foreclosures, assignments, and the like.

b. Equitable rights, relative to liens, adjustments, mistakes, fraud, etc.

c. Equitable remedies, such as specific performance of contracts, injunctions, partition of land, receiverships, partnership bills, bills of accounting, and protection of infants.

The Maxims of Equity. In equity, maxims of broad general rules or principles are recognized and enforced. In some states they have been incorporated into the state code and made matters of statute law. They are subject to limitations and exceptions, however, and may limit or restrict each other.

The following are some of the maxims of equity.

He who seeks equity must do equity.
He who comes into equity must come with clean hands.
Equity aids the vigilant, not those who slumber on their rights.
Equity acts specifically.
Equity follows the law.
Equity delights to do justice and not by halves.
Equity will not suffer a wrong to be without a remedy.
Equity regards that as done which ought to be done.
Equity regards substance rather than form.
Equity imputes an intent to fulfill an obligation.
Equity is equality.
Between equal equities the law will prevail.
Between equal equities the first in order of time shall prevail.

Equity Jurisdiction in Engineering and Construction Work. The laws of equity are of great importance in engineering relationships. The assignment of contracts and the assignment or subrogation of funds due under a contract to cancel a debt are common business practices within equity jurisdiction. Certain types of mistakes of fact if not caused by negligence may be rectified by a court of equity. There is no relief in equity for mistakes of law.

Frequently injunctions are encountered in engineering and construction work to prevent torts, to cause a change in construction plans, to prevent the violation of water rights, and the like. An injunction is usually temporary, pending a hearing on its merits, and must be obeyed until a final judgment is reached by a court of equity.

Questions pertaining to penalties and liquidated damages in construction contracts will usually be subject to settlement by a court of equity. The rewriting of lost documents, the cancellation of contracts involved in fraud, the infringement of patent rights, and the dissolution of partnerships are among the many other matters within the jurisdiction of equity on which the engineer must be informed.

ARBITRATION OF DISPUTES

Arbitration is a procedure whereby contracting parties may submit a dispute to an impartial board of experts for decision as an alternative to a suit in a court of law. By this method, it is possible to avoid many of the legal formalities, delays, and expenses which result from litigation. Arbitration may be adopted for the settlement of a controversy at any stage of the performance of a contract by agreement of the contracting parties, or the original contract may contain a provision for the settlement of all disputes by arbitration during the life of the contract. If the contract provides for arbitration the requirement is valid and enforceable under the laws of most states, and the decision of the arbitrators properly entered in a court has about the same strength as that of a judgment in a court of law and is subject to all the provisions of the law relative to a judgment.

In states which do not have arbitration laws the decision is valid only in so far as it is acceptable by both parties. If one party is dissatisfied with the decision, he may still resort to a lawsuit even though he signed the contract providing for arbitration. That is to say, every party to a contract has the right under the common law to a hearing in the appropriate court if he has a grievance, and no provision in the contract can cause the forfeiture of this constitutional right against his will.

Arbitration Procedure. Many states have laws which govern the procedure to be followed in submitting any dispute to arbitration. Where these do not exist, it is advisable to specify a definite procedure, such as the Code of Arbitration Practice and Procedure of the American Arbitration Association. The proceedings may be conducted informally or, if desired, may be under the supervision of a court. In general, arbitration requires the following procedure.

a. The agreement to arbitrate.

b. The submission.

c. The selection of arbitrators.

d. The hearing.

e. The award.

f. The confirmation.

The Arbitration Agreement. The agreement to arbitrate disputes may be included in any contract, and it becomes effective for all disputes when the agreement is signed. In the case of a dispute under a contract in which provision for arbitration has not been included, a special or supplemental agreement may be negotiated if desired, providing for arbitration of all existing and future disputes during the life of the contract, or it may be restricted to a specific controversy only. In either event, the agreement must be in writing, signed by both parties, and should specify the procedure to be adopted, as in the following example.

ARBITRATION PROCEDURE RECOMMENDED BY JOINT CONFERENCE ON STANDARD CONSTRUCTION CONTRACTS

ARBITRATION. (*a*) Demand for Arbitration. Any decision of the Engineer which is subject to arbitration shall be submitted to arbitration upon the demand of either party to the dispute.

The Contractor shall not cause a delay of the work because of the pendency of arbitration proceedings, except with the written permission of the Engineer, and then only until the arbitrators shall have an opportunity to determine whether or not the work shall continue until they decide the matters in dispute.

The demand for arbitration shall be delivered in writing to the Engineer and the adverse party, either personally or by registered mail to the last known address of each, within ten days of the receipt of the Engineer's decision, and in no case after final payment has been accepted except as otherwise expressly stipulated in the Contract Documents. If the Engineer fails to make a decision within a reasonable time, a demand for arbitration may be made as if his decision had been rendered against the demanding party.

(*b*) Arbitrators. No one shall be nominated or act as an arbitrator who is in any way financially interested in this Contract or in the business affairs of the owner, or the Contractor, or the Engineer, or otherwise connected with any of them. Each arbitrator shall be a person in general familiar with the work or the problem involved in the dispute submitted to arbitration.

Unless otherwise provided by controlling statutes, the parties may agree upon one arbitrator; otherwise there shall be three, one named in writing, by each party to this Contract, to the other party, and the third chosen by those two arbitrators, or if they should fail to select a third within fifteen days, then he shall be appointed by the presiding officer, if a disinterested party, of the Bar Association nearest to the location of the work.* Should the party demanding arbitration fail to name an arbitrator within ten days of his demand, his right to arbitration shall lapse. Should the other party fail to name an arbitrator within said ten days, then said * presiding officer shall appoint such arbitrator within ten days, and upon his failure so to do then such arbitrator shall be appointed on the petition of the party demanding arbitration by a judge of the Federal Court in the district where such arbitration is to be held.

The said * presiding officer shall have the power to declare the position of any arbitrator vacant by reason of refusal or inability to act, sickness, death, resignation, absence, or neglect. Any vacancy shall be filled by the party making the original appointment, and unless so filled within five days after the same has been declared, it shall be filled by the said * presiding officer. If testimony has been taken before a vacancy has been filled, the matter must be reheard unless a rehearing is waived in the submission or by the written consent of the parties.

If there be one arbitrator his decision shall be binding; if three, the decision of any two shall be binding in respect to both the matters submitted to and the procedure followed during the arbitration. Such decision shall be a condition precedent to any right of legal action.

(c) Arbitration Procedure. The arbitrators shall deliver a written notice to each of the parties and to the Engineer, either personally or by registered mail to the last known address of each, of the time and place for the beginning of the hearing of the matters submitted to them. Each party may submit to the arbitrators such evidence and argument as he may desire and the arbitrators may consider pertinent. The arbitrators shall, however, be the judges of all matters of law and fact relating to both the subject matters of and the procedure during arbitration and shall not be bound by technical rules of law or procedure. They may hear evidence in whatever form they desire. The parties may be represented before them by such person as each may select, subject to the disciplinary power of the arbitrators if such representative shall interfere with the orderly or speedy conduct of the proceedings.

Each party and the Engineer shall supply the arbitrators with such papers and information as they may demand, or with any witness whose movements are subject to their respective control, and upon refusal or neglect to comply with such demands the arbitrators may render their decision without the evidence which might have been elicited therefrom and the absence of such evidence shall afford no ground for challenge of the award by the party refusing or neglecting to comply with such demand.

The submission to arbitration (the statement of the matters in dispute between the parties to be passed upon by the arbitrators) shall be in writing duly acknowledged before a notary. Unless waived in writing by both parties to the arbitration, the arbitrators, before hearing testimony, shall be sworn by an officer authorized by law to administer an oath, faithfully and fairly

* To provide some other agency for appointing arbitrators strike out reference to presiding officer of the Bar Association and insert desired designation. In the vicinity of New York, the Arbitration Society of America, Inc., and the Chamber of Commerce of the State of New York have Arbitration Committees which often act in this capacity.

to hear and examine the matters in controversy and to make a just award according to the best of their understanding.

The arbitrators, if they deem the case demands it, are authorized to award to the party whose contention is sustained such sums as they shall consider proper for the time, expense, and trouble incident to the arbitration, and if the arbitration was demanded without reasonable cause, damages for delay and other losses. The arbitrators shall fix their own compensation, unless otherwise provided by agreement, and shall assess the costs and charges of the arbitration upon either or both parties.

The award of the arbitrators shall be in writing and acknowledged like a deed to be recorded, and a duplicate shall be delivered personally or by registered mail, forthwith upon its rendition, to each of the parties to the controversy and to the Engineer. Judgment may be rendered upon the award by the Federal Court or the highest State Court having jurisdiction to render same.

The award of the arbitrators shall not be open to objection on account of the form of the proceedings or the award, unless otherwise provided by the controlling statutes. In the event of such statutes providing on any matter covered by this Article otherwise than as hereinbefore specified, the method of procedure throughout and the legal effect of the award shall be wholly in accord with said statutes, it being the intention hereby to lay down a principle of action to be followed, leaving its local application to be adapted to the legal requirements of the jurisdiction having authority over the arbitration.

The Engineer shall not be deemed a party to the dispute. He is given the right to appear before the arbitrators to explain the basis of his decision and give such evidence as they may require.

The Submission. When a dispute arises on work under a contract in which an arbitration agreement is in force, the first step towards a settlement by arbitration is the preparation of the submission which is a statement of the matters in dispute and an agreement between the contracting parties to submit the dispute, either in part or in its entirety, to arbitration as provided by the arbitration agreement. The submission must be in writing and signed by both parties. In the case of a corporation, only the officers authorized to sign contracts may sign the submission. In some states, the submission must be reviewed by the court which settles all questions as to its validity. Invalidity of the submission, if discovered at a later date, may be sufficient cause for the court to vacate the award of the arbitration upon the appeal of one of the disputants. Therefore, it is essential that the submission conform to all requirements of the law with regard to validity.

The submission defines the scope of the arbitration, the time within which the award is to be made, and the rules of procedure to be followed. The latter will be the general rules in common usage, such as those of the American Arbitration Association and, in addition, any special rules agreed upon by the disputants to meet the requirements of the particular dispute.

The Selection of Arbitrators. The board of arbitrators may consist of one member who will act as an umpire; or, more frequently, each party selects one member, and the two so chosen select an independent third member. In selecting arbitrators, it is of primary importance that they be qualified technically in the subject matter of the dispute and also that they be familiar with the arbitration procedure and the laws under which it is to be conducted. The arbitrators should be impartial persons with known integrity. They should not be financially interested in the contract or in the business affairs of either of the parties to the dispute. The American Arbitration Association maintains a national panel from which qualified arbitrators from every profession can be drawn. Selections from this panel should insure impartial arbitration by arbitrators who have no personal interest in the dispute or its outcome.

Disinterested engineers, architects, and contractors are frequently called upon to act as arbitrators in construction disputes because of their technical and practical knowledge of construction work. It is important to note, however, that anyone acceptable to the disputants may serve as an arbitrator, and no particular qualifications are specified by law other than the requirement that the arbitrators shall act fairly and make a just award.

Before the arbitration proceedings begin, the arbitrators are sworn to hear and examine all evidence in the dispute and to make a fair award to the best of their ability. There can be no liability to an arbitrator because of a mistake in judgment, however, although he might be held liable for damages in the case of proved fraud or misconduct.

The Hearing. In conducting the hearing the arbitrators have the power to summon witnesses, records, documents, and other evidence as in a lawsuit. Testimony may be taken under oath subject to examination, cross-examination, and re-examination. The complaining party who initiated the arbitration submits his testimony supported by witnesses, and the defending party is entitled to cross-examine. Likewise, the evidence of the defending party is submitted and subjected to cross-examination. After the submission of all evidence, both parties may be permitted to summarize the arguments in support of their respective contentions. At any time during the proceedings the arbitrators or the parties to the dispute may apply to the court for the determination of any questions of law.

The Award. The award representing the decision of the arbitrators must be in writing, signed by the arbitrators or a majority of them, and a signed copy must be delivered to each of the parties. The award must be confined to the scope of the submission and must deal with all elements of the dispute so submitted. It is not necessary, however, to show how the conclusions as to law and fact were arrived at; a simple announcement of the decisions is sufficient. However, it is customary to state concisely the findings of fact and law pertaining to the dispute. The conclusions should be clear so that each party will know his rights and obligations, and there should be no conditions or reservations in the award which might be subject to further controversy. Needless to say, the award may neither direct nor enable any party to do an illegal act.

The Confirmation. In most states any party to an arbitration may apply to the court for an order confirming the award at any time within one year after the award is made. The court must then issue an order either confirming, correcting, or vacating the award. This court order has the same force and effect as the judgment in a court of law. Upon the filing of a request for confirmation of award, the opposing party must be notified in writing before the hearing. He has the right to request that the award be vacated or corrected if he has sufficient evidence to show that the arbitrators' decision was illegal or in error.

The court may vacate an arbitration award if it can be shown that the arbitrators were guilty of collusion, fraud, or partiality, if they refused to hear material evidence, or if they were otherwise guilty of misconduct which adversely affected the rights of any party. An award may be corrected by the court on the grounds of mistakes and miscalculations in the evidence or where the award is contrary to the law. In lieu of vacating or correcting the award, the court may require a rehearing of the arbitration.

The Engineer as an Arbitrator. The basic concept in arbitration is that experts in the field of the dispute shall be chosen as arbitrators; thus the engineer is fitted by professional training and experience to serve in this capacity when controversies arise from construction contracts. Likewise, architects and contractors make excellent arbitrators when the dispute lies in their particular fields of knowledge. It should be noted, however, that technical knowledge alone is not enough to qualify an engineer as an arbitrator. It is necessary that he inform himself on the law under which the arbitration is to be held and on the rules of procedure to be followed in collecting evidence and in arriving at a decision. Otherwise he may find himself confused by technicalities during the proceedings, or his decision may be vacated by the court because of his failure to comply with some requirement of the law.

The Advantages of Arbitration. By accepting arbitration in lieu of litigation in the event of a dispute, many advantages result to both the owner and the contractor. The proceedings are conducted in private; large court costs are avoided; and delays due to crowded dockets are eliminated. Fair and intelligent decisions are to be expected inasmuch as the arbitrators are selected for their expert knowledge of the subject. Decisions based on legal technicalities and questionable legal precedents are avoided, and the presentation of evidence is not hampered by legal restrictions as may occur in court procedure. Settlements of disputes by arbitration are usually amiable and result in no interference or delays in the prosecution of the work.

Questions

1. Outline the system of courts in the United States.

2. What are the differences in court procedure for lawsuits arising out of private construction contracts as compared with those in connection with U.S. Government contracts?

3. Define the following terms.

a. Summons.	*e.* Levy.
b. Plea.	*f.* Appeal.
c. Judgment.	*g.* Evidence.
d. Execution.	*h.* Cross-examination.

4. Why is hearsay evidence considered unreliable?

5. What is an expert witness, and how does his testimony differ from ordinary opinion evidence?

6. Discuss expert testimony in professional engineering practice.

7. Name three common types of controversies in engineering and construction work which are governed by equity.

8. Distinguish between a court of law and a court of equity.

9. Why is arbitration generally considered more satisfactory than litigation as a method for settling disputes under construction contracts?

10. What general class of disputes under construction contracts may be settled by arbitration? Can a dispute on a subject covered by statute law be settled by arbitration?

11. May an attorney on cross-examination ask leading questions?

7

TYPES OF CONSTRUCTION CONTRACTS

Like all types of contracts, those for engineering construction may be written in any one of many different forms, varying from a simple "offer and acceptance" to lengthy legal documents setting forth all the details and technicalities of the work. However, standardization of practice is desirable in the interests of economy of construction and satisfactory completion of the work. A great deal has been accomplished along this line by professional societies, trade organizations, and various government agencies. As a result of the studies of these organizations the general form and content of construction contracts are fairly well established for the conditions most likely to arise.

The indiscriminate use of standardized contracts is dangerous, and, in the past, has led to difficulties during construction and to litigation. Within reasonable limits, however, the use of applicable standardized clauses in a contract may be desirable. By repeated use they have become well understood in general by engineers, architects, and contractors, and, therefore, they are not likely to be misinterpreted. Moreover, many of the standardized clauses have been tested in courts, and their legal implications have been well established. Another consideration is that original composition in a contract can also lead to trouble if not written or reviewed by an experienced lawyer. Conditions may be inadvertently established which are capable of more than one meaning and may be subjected to an interpretation by a court which could not be anticipated when the contract was drawn.

Construction contracts may be classified in two general types, namely, competitive-bid contracts and those that are negotiated with selected contractors. Competitive-bid contracts are usually drawn on a fixed price basis whereas negotiated contracts usually provide for the owner to pay the actual cost of the work plus compensation to the contractor to cover his overhead expense and profit. Each

90

type is designed to meet the requirements of particular conditions, and, since there are variations within each classification, it is necessary that they be clearly understood by the engineer.

Competitive-Bid Contracts. For public works construction all who are so inclined may submit a bid, and the contract will be awarded to the lowest responsible bidder. This is required by law for U.S. Government work, and most state and local governments have similar restrictions. Responsibility in this sense usually means that the bidder must have ample working capital, in so far as it can be determined at the time; that he can furnish the required sureties and guarantees; and that his record is free from defaulted contracts and unjustified litigation. Some evidence of previous experience is desirable but is not always required under existing laws. Private construction may be handled in a more informal manner in that the list of bidders may be selected by the owner or recommended by the engineer, and the contract may be awarded to other than the lowest bidder if this appears to be in the best interests of the owner.

Competitive-bid contracts are of two common types. The *lump-sum* form provides for the compensation of the contractor on the basis of a total amount to cover all work and services required by the plans and specifications. The *unit-price* contract includes a breakdown estimate of the number of units of each type of construction and a price for each unit. Inasmuch as the estimate of quantities usually will not be exact at the time the contract is executed, provision should be made for payment to the contractor on the basis of actual quantities, measured in place when the work is completed, at the unit prices specified in the contract. For the purpose of comparing bids, an approximate total cost is determined from the estimated quantities and the unit prices in the bid. This approximate total cost is considered as a lump sum in the determination of the low bidder. With the exception of the method of payment to the contractor, the wording of the lump-sum and unit-price contracts may be similar. Important differences are found, however, in the methods of bidding and accomplishing the work under the two types and also in the wording of the specifications.

Lump-Sum Contracts. The use of the lump-sum contract is indicated where the types of construction are largely standardized and where a variety of operations is required, making it impracticable to break down the work into units. An important percentage of all building construction is accomplished under this form of contract. As a prerequisite the plans and specifications should be comprehensive and should show in complete detail the requirements of the work. Changes and extra work orders after the contract is signed are expensive and lead to controversies and disputes. Furthermore, when the plans are indefinite, the contractor is forced to gamble on the uncertainties or to increase his bid to cover the worst conditions to be expected. Consequently high costs to the owner will result if complete information is not furnished. If these hazards are avoided, the owner has the advantage of knowing in advance the exact cost of the work and of having the assurance that the contract will be completed at the earliest practicable date since efficiency and speed in completion will usually tend to increase the contractor's profit.

Unit-Price Contracts. When the work requires large quantities of a relatively few types of construction and the volume of work cannot be exactly determined in advance, the unit-price contract has many advantages. It is elastic in that reasonable variations may be made in the amount of work to be done without formal change orders, as long as the changes are restricted to the bid items covered by the contract. The plans and specifications must be complete in that they must show the nature and details of the work, but its limits may be left more or less indefinite, the magnitude and scope of the work being indicated by the engineer's estimate. Under these conditions the contractor is not forced to gamble on uncertain conditions, such as depths of underground foundation work, for example, as he would under lump-sum contracts.

Combined Lump-Sum and Unit-Price Contracts. Frequently, it may be advisable to combine the significant features of lump-sum and unit-price bidding in one contract. In general, this will be indicated when the work includes items, the details of which may be defined and broken down into units but which are somewhat indefinite as to quantities. For work of this character, unit-price payment is advantageous. If the same project includes specialty items or structures involving a large number of types of construction, the requirements of which are exactly known and for which the contractor can prepare an accurate estimate and bid, payment on a lump-sum basis will be simpler for everyone concerned. Thus the contractor's bid will include lump-sum amounts for some items and estimated quantities with corresponding unit prices for others. The combined type is well suited for building construction, for instance. Exact foundation conditions, generally indeterminate until the excavation is well along, can be covered satisfactorily on a unit-price basis, whereas the superstructure is well adapted to lump-sum bidding since it is more definitely planned.

The Contract Documents. The procedure for setting up a competitive-bid contract is rigidly defined by law for public works projects. The same procedure is recommended for privately sponsored projects although some of the formalities may be omitted. The prescribed procedure for both lump-sum and unit-price contracts requires the preparation of the following documents.

a. Advertisement, or *notice to contractors,* which gives a brief description of the work and invites bids from contractors for the accomplishment of its construction.

b. Instructions to bidders, which furnish detailed information required for the preparation of the bid or proposal.

c. Form of bid or proposal. This document is desirable in order that all bids will be uniform and prepared on a common understanding.

d. Contract, the legal document, binding on the signing parties, which states the work to be done, the compensation for the contractor, the time of completion, and the other legal and business features of the project as a whole.

e. Specifications, which cover the engineering or technical requirements of the work and describe the details of construction.

f. Plans, which show the physical details of the work.

The contract documents, excepting drawings, may be typed, mimeographed, or printed, and it is standard practice to bind them all into one volume which, together with the plans for the work, is available to contractors for bidding and construction purposes.

Negotiated Contracts. In contrast to competitive-bid contracts in which the award is the result of open competition among all who are interested, negotiated contracts are awarded to selected contractors after a study of the qualifications, previous experience, and facilities of one or more candidates. Since competitive bidding is required by law for public works projects, negotiated contracts are normally restricted to private construction where they have many advantages under certain conditions. During periods of national emergency or war, legal restrictions against negotiated government contracts are removed by special legislation in order that their advantages may be available for the construction of defense facilities. This is done at the expense of open competition, which is against the spirit of government contracting practice, but of no permanent effect because, during such periods, the construction industry is required to work at full capacity, and competition is not a major consideration. At the end of the emergency normal procedures are restored.

The fact that a contract is negotiated has little bearing on its form. That is to say, lump-sum and unit-price contracts may be negotiated as readily as any other type designed particularly for the purpose. In general, however, negotiated contracts are based on the premise that the owner will pay the actual cost of the work plus some compensation for the services, facilities, and technical knowledge of the contractor. It is only in the provision for the contractor's compensation that the various types of cost-plus contracts differ materially. This arrangement places the contractor more or less on a professional basis where he represents the owner in matters relating to the work. Under these conditions, the contractor's compensation naturally is of the nature of a professional fee rather than the profit anticipated in a business transaction. Except for the payment provision, the content of all types of cost-plus contracts may be similar.

Cost of the Work. Under the common forms of negotiated contracts, the contractor furnishes the working capital required to finance the construction. Later he is reimbursed in the full amount of the actual cost of the construction plus his compensation for the management, coordination, and accomplishment of the work. Therefore the terms of the contract must anticipate methods for the control of expenditures and for determining the actual cost. This is particularly true of certain items of the contractor's overhead and indirect general expenses in his central office. These are difficult to determine exactly, and an agreement as to some approximate method for their distribution should be made. As an alternative, they may be eliminated altogether as items of cost, and the contractor's compensation may be increased a reasonable amount to cover them. This would be a matter for negotiation and is probably the most satisfactory method for handling the problem.

In order to control expenditures, provision should be made for the engineer to approve all payrolls and purchases prior to the placing of orders. Salaries of key men and executives should also be subject to approval. Wage rates will be those prevailing in the community and so stipulated in the contract. The determination of the actual cost of the work can be made only through an adequate cost accounting system, and the accounting methods of the contractor should be subject to the approval of the owner or the engineer.

Cost-Plus-Percentage-of-Cost Contracts. Undoubtedly the oldest form of the negotiated type of contract is that providing for the contractor's profit on the basis of a fixed percentage of the actual cost of the work. In common with all negotiated contracts this type permits the beginning of construction before the plans are completely developed, resulting in an important saving of time in the completion of urgent projects, and the owner may make any desired changes in the plans and specifications as the work progresses. It has the disadvantage to the owner that the contractor's compensation is increased by an increase in construction cost. Therefore, there is no incentive for the contractor to economize during construction. Indeed an unscrupulous contractor may deliberately inflate the construction cost in order to obtain the corresponding increase in his fee. This may be done by padding payrolls, taking commissions on material purchases, and the like. For these reasons, the cost-plus-percentage contract has been outlawed for U.S. Government contracts even during war and other emergencies. It is not recommended for general use, although extra work and change orders may sometimes be handled satisfactorily on this basis.

Cost-Plus-Fixed-Fee Contracts. This type of contract provides for payment to the contractor of the cost of the work plus a fixed amount as a fee for the accomplishment of the work. The amount of the fee is determined from a consideration of the character and scope of the work and its estimated cost. Thereafter the fee remains fixed although the actual cost of the work may vary considerably from the estimate. In order to negotiate such a contract it is necessary for the scope of the work to be clearly defined and for both parties to agree on the amount of the estimate. This type had wide usage on Government work during the war emergency and has also been used extensively in the construction of large, privately owned buildings for which accurate bidding estimates are difficult to prepare.

From the point of view of the relation between the owner, the engineer, and the contractor, the cost-plus-fixed-fee contract more nearly approaches the ideal construction contract than any yet devised. The contractor is selected on the basis of merit and paid a fixed amount for completing specified work. There is no incentive for him to inflate the construction cost since his fee would not be affected. Indeed his profit might be decreased thereby, depending on the provisions for reimbursement of his overhead costs, because overhead costs vary almost directly with construction costs. Therefore the contractor is free in every respect to act in the best interests of the owner.

On the other hand, there is no specific incentive for the contractor to exert his best efforts towards efficiency, and the cost of the work may be increased thereby. This is to be watched particularly in the latter stages of a large construction project when frequently there is a tendency to string out the final operations and the completion of the contract. When this happens excessive overhead costs are to be expected.

The Cost-Plus-Fixed-Fee Contract with a Profit-Sharing Clause. As an added incentive to the contractor to keep the cost of the work at a minimum, a profit-sharing clause is sometimes written into the cost-plus-fixed-fee contract. Ordinarily the amount of the fixed fee is based on a preliminary estimate of the cost of the work. The profit-sharing provision allows the contractor to receive a share of any saving if the actual cost should be less than the original estimate. The contractor's share is usually set at 25 per cent to 50 per cent of the amount saved and is in addition to the fixed fee.

The Cost-Plus-Fixed-Fee Contract with a Bonus Clause. When the completion of the work is urgently desired by the owner, a bonus provision may be written into the contract as an incentive to the contractor to reduce the time of construction to a minimum. This is usually in the form of a fixed amount to be paid to the contractor, in addition to his fee, for each day on which the owner has full use of the completed work before the originally estimated date of completion. The contract may also provide for the assessing of a fixed amount as liquidated damages against the contractor for each day after the originally estimated date of completion required by the contractor to finish the work.

Contracts Based on Cost Plus a Sliding Scale of Fees. Another variant of the cost-plus-fixed-fee type of contract which has been widely used for construction work provides for the contractor's fee to change proportionately to the actual cost of the work in accordance with a sliding scale of fixed fees. As in the profit-sharing type this may be used as an incentive to the contractor to reduce the cost of the work by allowing the fee to be increased in fixed amounts for various increments of cost less than the original estimate and decreased in fixed amounts for various increments of cost greater than the original estimate. In the latter case it is customary to arrive at a minimum fee which would remain unchanged with further increases in the cost of the work, thus guaranteeing a minimum profit to the contractor.

This type of contract also may be considered a compromise between the cost-plus-fixed-fee and cost-plus-percentage forms by allowing the contractor's compensation to increase with the cost of the work in accordance with the desired sliding scale rates for various increments of cost in lieu of a constant percentage of the cost.

The Cost-Plus Contract with a Guaranteed Ceiling Price. One of the principal arguments against cost-plus types of contracts is that the owner has no way of knowing in advance what the work will cost. This objection may be overcome to a considerable degree by placing a maximum limit on the cost of the work. That is to say the contractor is reimbursed for the actual cost of the work plus his fixed fee, provided that the total amount does not exceed the maximum limit established in the contract. If the total amount should exceed the maximum limit the contractor is held responsible for the excess and receives no compensation over the guaranteed ceiling price. This type of contract removes some of the uncertainties from the ordinary cost-plus contract but requires that the plans and specifications for the work be sufficiently developed to permit the establishment of a reasonable ceiling price.

Management Contracts. Negotiated contracts may anticipate that the contractor's forces, equipment, and personnel will be utilized on the work in the normal manner expected for general contract work, or the contractor may be retained in a managerial capacity only. In the latter case the entire work is supervised by the contractor, including the purchase of materials, hiring of labor, and letting of subcontracts. He may or may not elect to do parts of the work with his own forces. If management services only are intended, it should be so stated in the contract.

Management agreements are not to be confused with so-called brokerage contracts in which the contractor without the owner's prior knowledge enters into a contract with the intention of accomplishing all or most of the work by subcontracts with the view of collecting a profit on the work of each subcontractor. These are regarded with disfavor and should not be permitted. Management contractors, to the contrary, are selected on a professional basis for their honesty, technical skill, and administrative ability. Distinction also should be made between management and agency in this connection. Under an agency stipulation the owner is legally bound by the acts of the contractor, whereas the acts of an independent contractor in connection with the work are his own responsibility. Under an independent contractor relationship the owner does not assume responsibility for the work until it is completed and formally accepted. The contract should be explicit in this connection.

Architect-Engineer-Management Contracts. A form of contract whereby all planning, designing, and management phases of construction work are delegated to one firm of architects and engineers was developed and used by the U.S. Government during World War II. This type had its widest use in the construction of military and industrial installations involving both architectural and engineering construction required in connection with the war effort. Responsibility is placed upon the contract architect-engineer-manager for all designs, plans, and specifications, the award of all construction subcontracts, and the inspection and management of construction work. This form of contract is usually drawn on a cost-plus-fixed-fee basis and differs from the usual owner-engineer-contractor relationship in that the architect-engineer-manager is placed in the status of an independent contractor whereas ordinarily he would be an agent of the owner. Also, all construction work under this type of agreement is accomplished by subcontracts between the architect-engineer-manager and the construction contractor rather than by prime contracts between the owner and the construction contractor. The architect-engineer-management contract has the advantage during a war emergency of delegating all matters pertaining to planning and construction to a single organization properly equipped and staffed to carry through all details of the project at a time when limited personnel and an overwhelming volume of work prevent the expeditious handling of the owner's responsibilities by his own organization. When drawn on a cost-plus basis it does not differ materially in effect from the ordinary cost-plus contract for engineering and architectural services for both design and supervision of construction. Any other basis, however, would place the architect-engineer in the position of a construction contractor.

Combined Engineering and Construction Contracts. It sometimes happens that the owner contemplating a construction project desires to deal with only one party for all services, both engineering and construction, in connection with the work; this is a so-called "turn-key" or "package" job. This procedure is, in fact, standard practice in most European and South American countries. A few United States companies also operate in this manner. The contract may be drawn either on a firm price or cost-plus basis, and all planning, design, plans, specifications, and construction services are included under one contract. Combined engineering and construction contracts are contrary to professional engineering practice in the United States and are not ordinarily recommended. Traditionally the engineer is an agent of the owner. In his professional status he inspects and supervises construction work and acts to safeguard the owner's interests during construction. Also, he acts as an interpreter of the plans and specifications and as an arbitrator of disputes between the owner and the contractor. Under a combined engineering and construction contract the engineer is automatically placed in partnership with the contractor, and an objective professional viewpoint towards the owner's interests is impracticable. Many years of experience in the United States have proved that the owner's interests are best served both in economy and quality of construction work when the engineer's professional status is recognized and his services engaged through a direct independent agreement with the owner.

When a combined engineering and construction contract is indicated, either because of the owner's preference or by the customs of the country in which the work is to be performed, the engineer should have it clearly understood that he will be guided by the principles of engineering practice and professional ethics in carrying out his duties under the contract. This should be done irrespective of whether he expects to work as a co-venturer or as an employee of the contractor.

Joint Venture Contracts. An extraordinarily large construction project to be accomplished under a general contract may require a greater concentration of financial, administrative, and technical resources than can be mobilized by one construction company. This has led to the development of the joint venture type of contract in which several firms combine their assets, plant, and personnel to undertake such a project. A joint venture is similar to an ordinary partnership or corporation in that each party in the combination shares in the work, risks, and profit or losses of the contract in accordance with the terms of the joint venture agreement. It differs from a partnership or corporation in that it is usually restricted to one contract and is not a continuing relationship.

A joint venture of several firms may enter into any of the usual types of construction contracts acceptable to the owner. As a prerequisite there should be an independent written agreement between the joint venturers which gives a clear arrangement for the financing and management of the work under the contract and the manner in which the risks and profits or losses are to be shared. The joint venture agreement should be subject to the approval of the owner and may be made a part of the construction contract if desired.

The joint venture type of contract came into prominence in the construction of large dam projects, notably Hoover Dam, and has been increasing in popularity. During World War II most of the large defense projects were accomplished under this form of agreement, and joint ventures are now standard practice in the construction industry in the United States.

Incentive Type Contracts for Work outside the United States. Many construction projects outside the continental limits of the United States are attractive to American contractors. There are many uncertainties in this class of work, however, and, therefore, the contracts are usually of the cost-plus type. Typical of the hazards to be encountered in overseas work are unstable foreign currencies, remote and frequently primitive living conditions for American personnel, limited and untrained native labor supply, and difficulties of transportation and communication. Work under these conditions is not susceptible of the efficient administrative control to be expected of contractors on similar work in the United States. Unless a strong incentive factor is provided in the contract in addition to the fee, construction costs are likely to be relatively high.

To give the contractor additional incentive to operate the project at maximum efficiency and minimum cost and to penalize him for inefficient work under these conditions, a cost-plus-base-fee contract with an incentive provision has been developed. Also, an element of competitive bidding has been incorporated which is helpful in selecting the most competent contractor for the work. It also determines the amount of the fee and establishes the basis of the incentive adjustment of the fee.

In practice, prospective bidders are invited to submit their qualifications to the engineer, and those found unsuitable for the work are eliminated from the bidding. Prequalification of bidders places a heavy responsibility on the engineer, and great care is necessary in this step of the procedure. Among the factors involved in establishing contractors' qualifications are experience in the type of construction required, experience in the locality where the work is to be performed, financial resources, quality of personnel and organization, and construction equipment available.

Contractors are instructed to prepare bids upon the basis that all costs will be reimbursed as for the ordinary cost-plus-fixed-fee contract. Each bidder prepares his own estimate of the cost of the work, and this is called *target estimate*. The bidder also proposes the base fee for which he offers to do the work.

If the actual cost of the work turns out to be less than the target estimate of the successful bidder, his fee is increased, usually by about 20 per cent of the saving. If the actual cost of the work exceeds the target estimate, the base fee of the contractor is reduced, usually by about 10 per cent of the excess, with the added provision that such a decrease shall not exceed 40 per cent of the base fee. This guarantees that the contractor will receive a minimum fee of 60 per cent of the base fee irrespective of the amount of the excess actual cost above the target estimate.

In analyzing the bids and awarding the contract, weighted consideration is given to the contractors' qualifications, job analysis, target estimate, and proposed base fee. The form of the contract is identical to the ordinary cost-plus-fixed-fee type excepting the target estimate, base fee, and fee adjustment clauses. In addition provision must be made for adjustment of the target estimate to recognize changes in wage rates, material prices, and other factors beyond the control of the contractor during the life of the contract.

This type of contract is unnecessarily complex and speculative for projects within the United States, and therefore it is not recommended for domestic work. It has been used satisfactorily for the construction of the 325-mile highway from Cochabamba to Santa Cruz in Bolivia and the Artibonite River dam and irrigation project in Haiti.

Selection of Type of Contract. Though the selection of the contract form is, strictly speaking, a function of the owner, it usually devolves on the engineer to furnish recommendations on which the owner may base his decision. In making a decision the fundamental differences between competitive-bid and negotiated contracts should be clearly understood and kept constantly in mind. A competitive-bid contract is a straightforward business transaction in which the owner takes his chances in an open market. He agrees to pay a specific price for a definite service, and constant inspection and supervision are required to make certain that the service is completed in the manner prescribed in the contract. In contrast, under the usual types of negotiated contracts the owner and the contractor are partners with a common objective, namely, to complete the work in accordance with the owner's requirements as quickly and as economically as possible. In this case emphasis is on the control of expenditures, which are reimbursable to the contractor, as well as on inspection and supervision of his workmanship.

It will be advantageous to the owner to adopt a competitive-bid contract when sufficient time is available to work out the plans and specifications in detail, thus minimizing costly changes and extra work orders at a later date; when activity in the construction industry is at a low ebb, thus insuring keen competition and lower bids; and when some control can be exercised on the quality and character of the bidders.

A negotiated contract will be indicated when it is desirable to begin work before the completion of detail plans and specifications, thus assuring earlier completion of the work; when the requirements of the project cannot be determined definitely in advance of the early phases of construction; or when the nature of the project is such that an accurate estimate cannot be made for bidding purposes.

When a decision has been made between a competitive-bid and a negotiated contract, an analysis of the unique features of the project will usually indicate the most suitable variation to adopt. This selection will be as significant as that determining the general type.

When a competitive-bid contract is selected, a decision must be made between the lump-sum and unit-price types. This will be based upon the type of construction involved. In general, the lump-sum contract is used for building construction, equipment, and machinery. The unit-price form is used for most kinds of engineering construction, such as bridges, dams, highways, and foundations. Local conditions may lead to exceptions to these rules, however.

When the owner decides upon the use of a negotiated contract with a selected contractor, the choice of the type of contract will usually be restricted to one of the variations of the cost-plus-fixed-fee form, in view of the disfavor in which the cost-plus-percentage type is held by most engineers. The particular variant to be adopted will depend on local conditions or personal preference. It should be noted, however, that practically all these variations introduce additional speculative elements into the contract as compared with the straightforward fixed-fee basis and in most instances complicate the payment provisions of the contract unnecessarily. An exception to this rule is found in construction work to be performed in remote and primitive localities. For these conditions an incentive type contract is indicated.

It should be noted that a contract may be negotiated on a lump-sum or unit-price basis as well as on a cost-plus basis. Ordinarily, however, this would require relatively complete plans and specifications, and in most instances it would be to the owner's advantage to obtain competitive bids.

Subcontracts. Generally all subcontracts for specialized phases of construction work are awarded on a competitive-bid basis irrespective of the type of the prime contract although some general contractors deal with favored subcontractors on a negotiated basis. Subcontracts may be of the lump-sum type, the unit-price type, or a combination of the two types as in the case of competitive-bid prime contracts. Subcontracts are the responsibility of the prime contractor, and the basis on which they are drawn is of no concern to the engineer and the owner under competitive-bid prime contracts, except that all subcontracts are subject to approval by the engineer. Under cost-plus prime contracts, however, all subcontracts are items of reimbursable cost, and therefore their amounts and terms should be closely scrutinized. Under these conditions negotiated cost-plus subcontracts should not be permitted, and usually a minimum of three competitive bids should be required for all subcontracts to make certain that the cost of the work is kept to a minimum.

Renegotiation of Contracts and Subcontracts. During World War II a congressional act provided for the renegotiation of any contract or subcontract on Government war projects amounting to more than $100,000 to determine whether the profits or costs were excessive. If found to be excessive the contractor was required to return the excess to the Government. Renegotiation proceedings were held before a board of experts and were initiated at the request of the Government. When renegotiation was initiated in connection with one war contract, costs and profits under all contracts between that contractor and other Government agencies were considered in combination, and the total profits from all work were considered in arriving at the amount of the excess. This procedure is contrary to the principles of contract law in which the court does not pass on the sufficiency of a bargain. However, it was enforceable during the emergency on the grounds that no one should be allowed to profit excessively because of the war. In 1950 the act was re-established on a continuing and more or less permanent basis. This law is restricted to defense contracts and does not apply in normal practice.

Equipment Rental Contracts. In addition to the foregoing, which are contracts for services, contracts for the rental of construction equipment are frequently necessary in the course of construction projects. This type of agreement is required when the contractor lacks equipment necessary to the accomplishment of the work. Under cost-plus types of prime contracts the terms of equipment rental contracts are a direct concern of the owner. Otherwise settlement of the matter pertains only to the contractor and the renter of the equipment, and the owner is not involved. Construction equipment may be rented on the basis of a lump sum for the duration of the job, a fixed rate per day or per hour, or on the basis of actual cost of ownership plus a predetermined profit for the owner of the equipment. The agreement may or may not include all operation costs and all personnel required in its use. Frequently a recapture clause is included whereby the renter may take over ownership of the equipment when the total rental reaches a specified amount or in case of a breach of the contract.

Questions

1. What are the advantages and disadvantages of competitive-bid contracts?

2. Explain the difference between the methods of payment under lump-sum and unit-price contracts.

3. The following projects are to be constructed under competitive-bid contracts. Make a selection between the lump-sum and unit-price types, and give the reasons for each choice.

    a. A sewer system.
    b. A small residence.
    c. A masonry dam.
    d. A steel bridge superstructure.
    e. A steam turbine.
    f. The foundations for a large building.

4. When would it be advisable to use a combination of the lump-sum and unit-price methods of payment?

5. What is the main criticism of the cost-plus-percentage contract?

6. Why is the cost-plus-fixed-fee type considered the most satisfactory form of negotiated contract?

7. What are the advantages of the cost-plus-a-sliding-scale-of-fees type of negotiated contract?

8. Under what conditions would the cost-plus type of contract be indicated in lieu of the competitive-bid form?

9. What are the objections to a contract for combined engineering and construction services?

10. Why should subcontracts always be awarded on the basis of competitive bids when the prime contract is of a cost-plus type?

8

BIDDING PROCEDURE

Of the documents required for competitive-bid contracts, three are related to the preparation of the bid. The *advertisement* invites bids or notifies contractors that bids are to be received. The *instructions to bidders* contain detailed information as to the various factors to be considered in arriving at the amount of the bid or the unit prices, as the case may be. The *bid,* or *proposal,* form prescribes the method to be followed in preparing the bid for submission. For public works these will be formal documents. For relatively unimportant private work the opposite extreme of informality would exist, and the necessary functions might be performed by letters or even telephone conversations. Nevertheless, in one form or another these three documents are required in connection with the award of any competitive-bid contract.

U.S. Government Requirements Relative to the Advertising and Opening of Bids. Before any contract with the U.S. Government for construction work can be awarded, it is required by law that the date for taking of bids be advertised and that bids be accepted from any qualified bidder who desires to bid on the work. During emergencies, such as a war, government contracting officers are permitted to negotiate directly with contractors without the formality of advertising. Under all normal conditions, however, the advertisement is required. State, county, and city public works are usually handled in a similar manner. The following federal statutes govern the procedure for the receiving of bids on government contracts.

Advertisements for Proposals for Purchases and Contracts for Supplies or Services for Departments of Government.

All purchases and contracts for supplies or services, in any of the departments of the Government, except for personal services, shall be made by advertising a sufficient time previously for proposals respecting the same, when the public exigencies do not require the immediate delivery of the articles, or performance of the service. When immediate delivery or performance is required by the public exigency, the articles or services required may be procured by open purchase or contract, at the places and in the manner in which such articles are usually bought and sold, or such services engaged between individuals.

(Revised Statutes Section 3709.)

Opening Bids.

Whenever proposals for supplies have been solicited, the parties responding to such solicitation shall be duly notified of the time and place of opening the bids, and be permitted to be present either in person or by attorney, and a record of each bid shall then and there be made.

(Revised Statutes Section 3710.)

The Advertisement. The purpose of the advertisement, or notice to contractors, is to inform prospective bidders that a contract is to be awarded and thus to obtain adequate competition for the work. The advertisement should be inserted in a newspaper or magazine which has a wide circulation, and it should be published sufficiently in advance of the date set for the opening of bids to allow contractors ample time to prepare their estimates, obtain prices and sub-bids for specialty work, and to make other arrangements necessary in arriving at the bid amounts. The time required to complete these operations will, of course, vary with the size and complexity of the project, but, ordinarily, three weeks should be considered about the minimum. When local competition is sufficient and satisfactory, the advertisement may be posted in a public place, such as a post office or city hall, or invitations to bid may be mailed to a number of contractors. The advertisement should be brief and written in a simple style.

The essential elements to be included are as follows.

a. Brief description of the work and its location. This should be written with the view of attracting the attention of only those interested and qualified to bid.

b. Name and address of the owner.

c. Name and address of person authorized to receive bids.

d. The place, date, and hour of the opening of bids and the restrictions relative to the submission, change, and withdrawal of bids.

e. Character of bids. This should state whether the bids are to be on a lump-sum or unit-price basis, and whether they are for the entire project or certain parts only. Note also should be made of any alternate bids requested.

f. Principal items of the work with approximate quantities involved. This informs the contractor immediately whether his equipment, organization, and experience are suitable for the work.

g. Bid surety. The amount of the surety and whether it is to be cash, a certified check, or a bid bond should be stated, as should be the provision for the return of the surety of unsuccessful bidders.

h. Contract surety. The amount and type of the performance and payment surety should be stated. If a certificate of surety is to be required with the bid, this also should be mentioned.

i. Conditions of payment. Required only if different from normal.

j. Information relative to the plans and specifications. It should be stated where the plans and specifications may be obtained or examined. This will usually be in the offices of the owner, the engineer, and sometimes in the offices of contractors' associations. Mention should also be made of any charge or deposit to be required for the plans and specifications when taken by contractors and provision for its recovery when the documents are returned.

k. Conditions of award of contract. Such as the reservation to accept the lowest responsible bid and to reject any or all bids.

l. Name of the engineer and the owner or their authorized representatives. In the latter case, the authorization should be stated.

EXAMPLES OF ADVERTISEMENTS

SOIL EXPLORATIONS—SPEC. SM-34

STATE DEPARTMENT OF PUBLIC WORKS, ALBANY, N. Y.— Pursuant to the provisions of Highway Law, sealed proposals will be received until 10:30 A.M., Eastern Standard Time, on April 2, 1952, by Henry A. Cohen, Director, Bureau of Contracts and Accounts, 14th Floor, The Governor Alfred E. Smith State Office Building, Albany, N. Y., for

SOIL EXPLORATIONS AND FOUNDATION INVESTIGATIONS AT VARIOUS SITES ON THE NEW YORK STATE THRUWAY, CATSKILL SECTION, SUBDIVISION NOS. 6 AND 7, RAVENA TO ROUTE 23A, GREENE COUNTY, SUBDIVISION 3A, DELAWARE AVENUE TO SOUTHERN BLVD., ALBANY COUNTY; WHICH CONSTITUTES CONTRACT NO. SPEC. SM-34.

Maps, plans and specifications may be examined and obtained at the following offices:

(a) Bureau of Contracts and Accounts, 14th Floor, The Governor Alfred E. Smith State Office Building, Albany 1, N. Y.

(b) G. L. Nickerson, District Engineer, 353 Broadway, Albany 1, N. Y.

(c) Charles Schaefer, Department of Public Works, 270 Broadway, New York City.

The deposit for a set of plans, specifications, and proposal forms is $5.00. A refund in full will be made to bidders for return of one set in good condition within 30 days of award or rejection of bids; refund for all other sets in good condition, similar period, will be 50% of deposit.

The Engineer's estimate of cost for this work is $27,525.00.

Proposal for this contract must be submitted in a separate sealed envelope with the name of the contract plainly endorsed on the outside of the envelope. Proposal must be accompanied by draft or certified check, payable to the order of the "State of New York, Commissioner of Taxation and Finance" for the sum of $1,377.00. The retention and disposal of the bidding check, the execution of the contract and bonds shall conform to the provisions of the Highway Law, as set forth in "Instruction to Bidders."

The right is reserved to reject any or all bids.

B. D. TALLAMY
Superintendent of Public Works

WELFARE ISLAND BRIDGE OVER EAST RIVER

New York, N. Y.

Sealed bids for the construction of the Welfare Island Bridge over the East River will be received by the Department of Public Works in Room 2200, Municipal Building, Manhattan, New York 7, N. Y., until 2:00 P.M. (EST) on Thursday, November 15, 1951, at which time they will be opened and read in public.

The work consists of the construction of a vertical-lift highway bridge over the east channel of the East River, with approaches on Welfare Island and in the Borough of Queens. The vertical-lift span is 418 feet long, and is flanked by two towers. The lift span superstructure and the towers are of steel construction and are founded on reinforced concrete granite faced piers protected by timber fenders. The bridge approaches consist of asphalt macadam paved embankment ramps and structural steel spans supported by reinforced concrete piers. The approach spans and bridge have a 34-foot roadway and one 6-foot sidewalk. Submarine cables carry control and power circuits across the channel from a transformer vault in the east abutment. The project will also include machinery and control houses, roadway lighting, miscellaneous bridge equipment such as warning and barrier gates, air buffers, electrical control and operation equipment, clearance indicator system, storm and sanitary sewers, and utility relocations. The estimated quantities of work for some of the principal items of construction are as follows:

Item of Construction	*Quantity*	*Unit*
Common Excavation-River Piers and Channel	6,000	C.Y.
Rock Excavation-River Piers and Channel	7,500	C.Y.
Concrete	16,000	C.Y.
Reinforcing Steel	725,000	Lbs.
Structural Carbon Steel	4,400,000	Lbs.
Structural Silicon Steel	1,275,000	Lbs.
Sheaves, Shafts, Bearings	271,000	Lbs.
Wire Ropes and Fittings	136,000	Lbs.
Embankment-Approaches	52,000	C.Y.
Asphalt Macadam Paving on Stone Base Course	14,000	S.Y.

Contract documents including plans and specifications may be examined at the office of the Department of Public Works, Municipal Building, and at the Office of Tippetts Abbett McCarthy Stratton, Engineers, 62 West 47 St., New York, N. Y. Contract documents may be obtained from the Department of Public Works upon deposit of fifty dollars ($50.00) per complete set. The full amount of the deposit will be refunded upon return of said documents in good condition within thirty (30) days after date of opening of bids.

A certified check or bid bond in the amount of one hundred thousand dollars ($100,000) must be furnished with each bid. Performance and payment bonds, each in the amount of forty per cent (40%) of the total contract price, will be required. Liquidated damages for delay will be one hundred fifty dollars ($150.00) per day. The Department of Public Works reserves the right to reject all bids and to waive informalities in a bid.

DEPARTMENT OF PUBLIC WORKS
NEW YORK CITY, N. Y.
FREDERICK H. ZURMUHLEN
Commissioner

Instructions to Bidders. The instructions to bidders, also known as information for bidders, is a document in which all bidders are furnished identical information on the unique features of the work and detailed instructions on the procedure to be followed in submitting bids. This is desirable in order that all bidders receive uniformly fair treatment and that all bids are prepared on a common basis. The information given is similar in character to that in the advertisement, but it is expanded in greater detail and stated more explicitly. In addition to the topics in the advertisement, the instructions to bidders should include the following.

a. Requirements as to bidder's experience record or prequalification data. These will be used in judging the bidder's ability to perform the work and may eliminate unqualified contractors.

b. Instructions relative to the procedure to be followed in writing and submitting the bid.

c. A list of the plans and specifications and a detailed estimate of quantities for unit-price contracts or an exact definition of the scope of the work if the contract is to be on a lump-sum basis.

d. Time of completion of the work. This statement should specify the date on which construction shall start and the number of calendar days to be allowed for its completion. It may be desirable to request estimates of the time of completion from bidders. These estimates may then become a consideration in the award of the contract.

e. Statement as to whether the bidder or the owner is to be responsible for the accuracy of the bidding information with particular reference to subsoil data, test borings, errors in the plans, and the like.

f. Details of any formalities required in the bid and provision for the rejection of any informalities or of the entire bid in which they are included.

g. Reference to the authority for the accomplishment of the work and any other legal considerations involved in the award of the contract.

h. Miscellaneous instructions relative to the unique features of the work.

EXAMPLES OF INSTRUCTIONS TO BIDDERS

Standard Form 22
Revised March 1953
General Services Administration
General Regulation No. 13

INSTRUCTIONS TO BIDDERS

(Construction Contracts)

(These instructions are not to be incorporated in the contract)

1. *Explanation to Bidders.* Any explanation desired by bidders regarding the meaning or interpretation of the drawings and specifications must be requested in writing and with sufficient time allowed for a reply to reach them before the submission of their bids. Oral explanations or instructions given before the award of the contract will not be binding. Any interpretation made will be in the form of an addendum to the specifications or drawings

and will be furnished to all bidders, and its receipt by the bidder shall be acknowledged.

2. *Conditions at Site of Work.* Bidders should visit the site to ascertain pertinent local conditions readily determined by inspection and inquiry, such as the location, accessibility, and general character of the site, labor conditions, the character and extent of existing work within or adjacent thereto, and any other work being performed thereon.

3. *Bidder's Qualifications.* Before a bid is considered for award, the bidder may be requested by the Government to submit a statement of facts in detail as to his previous experience in performing similar or comparable work, and of his business and technical organization and financial resources and plant available and to be used in performing the contemplated work.

4. *Bid Guaranty.* Where security is required, failure to submit the same with the bid may be cause for rejection. The bidder, at his option, may furnish a bid bond, postal money order, certified check, or cashier's check, or may deposit, in accordance with Treasury Department regulations, bonds or notes of the United States (at par value) as security in the amount required: *Provided,* That where the total amount of the bid is $2,000 or less, the contracting agency may declare a bid bond unacceptable by so stating in the specifications or Invitation for Bids.

In case security is in the form of postal money order, certified check, cashier's check, or bonds or notes of the United States, the Government may make such disposition of the same as will accomplish the purpose for which submitted. Checks may be held uncollected at the bidder's risk. Checks, or the amounts thereof, and bonds or notes of the United States deposited by unsuccessful bidders will be returned as soon as practicable after the opening.

5. *Preparation of Bids.* (*a*) Bids shall be submitted on the forms furnished, or copies thereof, and must be manually signed. If erasures or other changes appear on the forms, each such erasure or change must be initialed by the person signing the bid.

(*b*) The form of bid will provide for quotation of a price, or prices, for one or more items which may be lump sum bids, alternate prices, scheduled items resulting in a bid on a unit of construction or a combination thereof, etc. Where required on the bid form, bidders must quote on all items and *they are warned* that failure to do so may disqualify the bid. When quotations on all items are not required, bidders should insert the words "no bid" in the space provided for any item on which no quotation is made.

(*c*) Alternative bids will not be considered unless called for.

(*d*) Unless specifically called for, telegraphic bids will not be considered. Modification by telegraph of bids already submitted will be considered if received prior to the time fixed in the Invitation for Bids. Telegraphic modifications shall not reveal the amount of the original or revised bid.

6. *Submission of Bids.* Bids must be submitted as directed on the bid form.

7. *Receipt and Opening of Bids.* (*a*) Bids will be submitted prior to the time fixed in the Invitation for Bids. Bids received after the time so fixed are late bids; and the exact date and hour of mailing such bids, as shown by the cancellation stamp or by the stamp of an approved metering device will be recorded. Such late bids will be considered, *Provided,* They are

received before the award has been made, *And provided further,* The failure to arrive on time was due solely to a delay in the mails for which the bidder was not responsible; otherwise late bids will not be considered but will be held unopened until the time of award and then returned to the bidder, unless other disposition is requested or agreed to by the bidder.

(*b*) Subject to the provisions of paragraph 5(*d*) of these instructions, bids or bid modifications which were deposited for transmission by telegraph in time for receipt, by normal transmission procedure, prior to the time fixed in the Invitation for Bids and subsequently delayed by the telegraph company through no fault or neglect on the part of the bidder, will be considered if received prior to the award of the contract. The burden of proof of such abnormal delay will be upon the bidder, and the decision as to whether or not the delay was so caused will rest with the officer awarding the contract.

(*c*) No responsibility will attach to any officer for the premature opening of, or the failure to open, a bid not properly addressed and identified.

8. *Withdrawals of Bids.* Bids may be withdrawn on written or telegraphic request received from bidders prior to the time fixed for opening. Negligence on the part of the bidder in preparing the bid confers no right for the withdrawal of the bid after it has been opened.

9. *Bidders Present.* At the time fixed for the opening of bids, their contents will be made public for the information of bidders and others properly interested, who may be present either in person or by representative.

10. *Bidders Interested in More than One Bid.* If more than one bid be offered by any one party, by or in the name of his or their clerk, partner, or other person, all such bids will be rejected. A party who has quoted prices to a bidder is not thereby disqualified from quoting prices to other bidders or from submitting a bid directly for the work.

11. *Award of Contract.* (*a*) The contract will be awarded as soon as practicable to the lowest responsible bidder, price and other factors considered, provided his bid is reasonable and it is to the interest of the Government to accept it.

(*b*) The Government reserves the right to waive any informality in bids received when such waiver is in the interest of the Government. In case of error in the extension of prices, the unit price will govern.

(*c*) The Government further reserves the right to accept or reject any or all items of any bid, unless the bidder qualifies such bid by specific limitation; also to make an award to the bidder whose aggregate bid on any combination of bid items is low.

12. *Rejection of Bids.* The Government reserves the right to reject any and all bids when such rejection is in the interest of the Government; to reject the bid of a bidder who has previously failed to perform properly or complete on time contracts of a similar nature; and to reject the bid of a bidder who is not, in the opinion of the Contracting Officer, in a position to perform the contract.

13. *Contract and Bonds.* The bidder to whom award is made shall, within the time established in the bid and when required, enter into a written contract with the Government and furnish performance and payment bonds on Government Standard Forms. The bonds shall be in the amounts indicated in the specifications or the Invitation for Bids.

INFORMATION FOR BIDDERS

CONSTRUCTION
OF
WELFARE ISLAND BRIDGE

Over East Channel of East River

New York, N. Y.
October 1, 1952

1. *Receipt and Opening of Bids.* Sealed bids in duplicate for the construction of the Welfare Island Bridge will be received by the Department of Public Works in Room 2200, Municipal Building, Manhattan, New York 7, N. Y., until 2:00 P.M. (EST) on Thursday, November 15, 1952, at which time they will be opened and read in public. Any bids received after the time and date specified will not be considered.

2. *Drawings and Specifications.* Contract documents including drawings and specifications may be examined at the office of the Department of Public Works, Municipal Building, New York City, and the New York office of Tippetts Abbett McCarthy Stratton, Engineers. Contract documents may be obtained from the Department of Public Works upon deposit of fifty dollars ($50.00) per complete set. The full amount of the deposit will be refunded upon return of said documents in good condition within thirty (30) days after date of opening of bids.

3. *Examination of Site and Contract Documents.* Each bidder should carefully examine the contract documents and the site of the work, and fully acquaint himself with all conditions and matters which can in any way affect the work or the cost thereof.

4. *Preparation of Bid.* Bids must be submitted on the attached prescribed form which is divided into Schedule A—Substructure and Approaches, Schedule B—Superstructure, and Schedule C—Bridge and Approaches. A bid may be submitted on any or all of the schedules, provided any schedule for which prices are proposed is completely filled out. Bid prices must be filled in, in ink, in both words and figures. In case of discrepancy between unit prices and amounts, unit prices will govern. Erasures or other changes must be noted over the signature of the bidder.

5. *Signature to Bids.* Each bid must contain the name, residence, and place of business of the person or persons making the bid and must be signed by the bidder with his usual signature. Bids by partnerships must furnish the full names of all partners and must be signed with the partnership name by one of the members of the partnership or by an authorized representative, followed by the signature and designation of the person signing. Bids by corporations must be signed with the legal name of the corporation, followed by the name of the State of incorporation and by the signature and designation of the president, secretary, or other person authorized to bind it in the matter.

6. *Bid Guarantee.* Each bid must be accompanied by cash, certified check, or bid bond in the amount of one hundred thousand dollars ($100,000.00). If a bid bond is submitted, the approved form of bid bond attached herewith must be exectued by the bidder as principal and have as surety thereon, a

company satisfactory to the Department of Public Works. The bid guarantee will be returned to all except the three (3) lowest bidders within ten (10) days after the opening of bids, and the remaining cash, checks or bid bonds will be returned within forty-eight (48) hours after the Department of Public Works and the successful bidder have executed the contract for the work. If no such contract is executed within forty-five (45) days after the date of the opening of bids, or if all bids are rejected prior to such time, the guarantees will be returned at the time of rejection or at the end of the forty-five (45) day period. If the bidder to whom the contract is awarded refuses or neglects to execute it, or fails to furnish the required performance and payment bonds as required herein, the amount of his guarantee, or as much thereof as may be applicable to the amount of the award to him, shall be forfeited and shall be retained by the City of New York as liquidated damages.

7. *Qualification of Bidders.* Each bidder shall inclose with his bid a statement setting forth his financial condition, the personnel and qualifications of his working organization, prior experience and performance record. The bidder shall also state that he has available or under his control, plant of the character and in the amount required to complete the proposed work within the specified time. Upon request by the Department of Public Works, the bidder shall furnish a list of the plant proposed for use on the work.

8. *Modifications prior to Date of Bid Opening.* Modifications of the drawings and specifications will be made by addendum to the contract documents. Copies of each addendum will be furnished to each bidder.

9. *Submission of Bid.* Bids must be submitted in sealed envelopes bearing on the outside, the name of the bidder, his address, and the name of the project for which the bid is submitted, and the date of opening. If forwarded by mail, the sealed envelope containing the bid, and marked as directed above, must be enclosed in another envelope properly addressed.

10. *Withdrawal of Bid.* Bids may be withdrawn in writing prior to the date of opening of bids. After such date a bidder may not withdraw his bid until the expiration of forty-five (45) days, after which time a bid may be withdrawn only in writing and in advance of actual award of contract.

11. *Award of Contract.* Award of contract to separate contractors on the basis of Schedules A and B or to a single contractor on the basis of Schedule C will be made if at all, to the lowest responsible bidder or bidders complying with the requirements of the contract documents, provided the bid is reasonable and it is to the interest of the City of New York to accept it. The Department of Public Works, however, reserves the right to reject all bids and to waive any informality in bids received.

12. *Contract and Bond.* The bidder to whom award is made shall execute the standard form of Agreement of the Department of Public Works of the City of New York and furnish the performance and payment bonds required in the specifications, within ten (10) days after the agreement form is presented to him for signature.

DEPARTMENT OF PUBLIC WORKS
NEW YORK CITY, N. Y.
FREDERICK H. ZURMUHLEN
Commissioner

Responsibility for Accuracy of Bidding Information. In

many types of construction work the amounts of bids will depend on local conditions at the site some of which cannot be determined exactly in advance. This applies particularly to the construction of underground structures, such as foundations, sewers, and water mains, where the character of the subsoil determines the cost of the work. For the purposes of design, explorations are usually made by the owner by means of borings, test pits, test piles, and other methods, and information obtained in this manner is made available to bidders. At best, such tests cannot cover the entire site thoroughly. The information obtained thereby does not reveal all conditions, and sometimes it is misleading or subject to incorrect interpretation. This may have important effects on the actual cost of the work as compared with that anticipated in the bid, and the question then arises as to the responsibility for the discrepancy. Many lawsuits have originated from disputes caused in this way. Therefore, it is necessary that the contract documents state explicitly which party is to assume responsibility for the accuracy and interpretation of bidding information furnished by the owner.

Unless covered by a statement to the contrary, the owner will always be held responsible for the accuracy of information which he furnishes to bidders. The interpretation of such information is a risk that the owner, particularly in private work, usually should not assume on the grounds that the contractor is versed in such matters whereas the owner is not, and the risk is one of the normal hazards of the contracting business. Therefore, responsibility for the accuracy and interpretation of information furnished by the owner is usually placed on the contractor. It might be argued that the engineer is likewise familiar with such technical matters and should assume responsibility for their accuracy. It should be noted, however, that the engineer ordinarily acts as the agent of the owner, and his fee or salary usually does not contemplate such risks, whereas, when assumed by the contractor, it is considered that his bid will be adjusted adequately to insure himself against the most probable unfavorable conditions, consistent with the competitive nature of his business. In his agreement with the owner, the engineer should make certain that he does not inadvertently place himself in a position where he might be made responsible in questions of this kind.

Sufficient protection is usually secured for the owner by placing on the drawings or in the specifications a statement that all such information (boring data, driving records, and load tests of test piles, etc.) is furnished to bidders as the best information available, to be interpreted and used as the bidder sees fit and on the bidder's own responsibility, and that the owner disclaims all responsibility for its accuracy and sufficiency. The bidder is usually made responsible for informing himself on the character of the site and for making any additional investigations necessary to assure himself of the nature of conditions to be encountered in the work.

An alternative procedure permits a readjustment of the contract amount if actual subsurface or latent conditions should differ materially from those indicated in the bidding data or those inherent in the type of work covered by the contract. U.S. Government agencies, in particular, favor this procedure, which in effect places the responsibility for the accuracy of bidding information on the owner. Its interpretation, however, remains the responsibility of the contractor. This method reduces the risk in the contractor's bid and should reduce construction costs as well. Moreover, the owner pays for the actual work performed whether more or less than that indicated in the bidding information, which is consistent with ethical business practice. This method has the disadvantage that contractors can almost always show some discrepancy between the actual conditions and those indicated in the bidding information, and controversies result. This may not be serious in the case of Government contracts because of the great administrative and legal strength given to Government contracting officers, but it may be a constant source of annoyance, embarrassment, and litigation on private work.

Bid Form. In order that all bids may be prepared in a similar manner the appropriate form should be prescribed. This is desirable to facilitate the analysis and comparison of bids and to detect informalities, particularly in view of the fact that frequently large numbers of qualifications, reservations, and alternative bids may be included. Furthermore, the bid form is a convenience to contractors in that it tends to insure accuracy and prevent omissions. The bid form should contain the following elements.

a. Price for which the contractor offers to perform the specified work. If the contract is to be on a lump-sum basis, the price will be the total amount of the bid without a breakdown of quantities, or, if there is more than one bid item, a lump-sum amount will be given for each item. For a unit-price contract, there will be a list of the bid items, the engineer's estimate of the quantity of each item, and a unit price for each item. It should be noted that the quantities are approximate and may be varied during construction. All prices should be given in both words and figures, and if there are discrepancies, the amounts written out govern. In case of discrepancy between unit prices and the respective total for each item, unit prices govern, and the total amount is corrected accordingly.

b. Time of completion. This should state that the work will begin on a definite date, or within a specified number of days, after the receipt of a notice from the owner to proceed with the work. It should also be stated that the work will be completed within a specified number of calendar days after receipt of the notice to proceed.

c. Bid surety. It should be stated that a bid bond, certified check, or other form of guarantee accompanies the bid as required.

d. Agreement to furnish contract surety. By this statement the bidder agrees to furnish the required contract surety if the contract is awarded to him. The bidder is sometimes required to furnish a certificate of surety properly executed by the guarantor. This certifies that if the bidder is awarded the contract the guarantor will furnish the required surety.

e. List of addenda to the plans and specifications which were considered when the bid was prepared. Frequently changes are incorporated in the plans and specifications after the advertisement has been published, and prospective bidders are so notified by means of addenda. It is necessary to note the inclusion in the bid of such changes, particularly in the case of lump-sum contracts.

f. List of subcontractors. Under some conditions it may be desirable to require the bidder to specify the subcontractors to be employed on specialty work. This is usually not necessary if the contract provides for the approval of all subcontractors by the owner and the engineer prior to their employment.

g. Experience record, financial statement, and plant and equipment questionnaire, when required.

h. Declaration that no fraud or collusion exists, with particular reference to illegal relationships between the bidder and representatives of the owner, pooling of bids by several bidders, straw bids submitted by an employee or other representative of the bidder, and similar illegal acts.

i. Statement that the site has been examined and that the plans and specifications are understood by the bidder.

j. Signature and witnesses.

Selection of Bid Items for Unit-Price Contracts. The bid form for unit-price contracts contains a tabulation of the various items of work, the unit of measurement, and the estimated quantity of each item. In arriving at the items to be included the work is broken down into the basic types of construction, such as excavation, concrete and structural steel. Each type and each classification of a type of construction comprises one bid item. This procedure necessarily conforms with the methods of estimating quantities and costs.

Each item of work must be capable of being measured accurately and must be such that all elements of its unit cost will vary in direct proportion with any changes in the total number of units. For example, the cost of 1 cubic yard of concrete consists of the sum of the costs of the various ingredients and the cost of mixing, placing, and curing. Within reasonable limits the magnitudes of these elements relative to each other remain approximately constant and independent of the number of cubic yards poured. Therefore, it is convenient to consider "concrete in place" as one bid item rather than to set up separate items for cement, sand, gravel, mixing, placing, etc. When the character of the concrete work is such that the relative amounts of form work and reinforcing steel are also constant throughout the various parts of the structure it would be convenient to consider "reinforced concrete in place" as one item. If the work includes heavy, massive, unreinforced concrete members requiring minimum form work in one part of the structure and thin heavily reinforced sections requiring expensive form work in another part, the form work and reinforcing steel should be made separate bid items. Likewise, each class of concrete mixture should be made a separate bid item. Under some conditions of competition and supply in the cement industry it is desirable to set up cement as a separate bid item, and this is done frequently on large projects. When practicable, work that may be performed by different trades or different subcontractors should be segregated into separate bid items. The list of bid items should be carefully

TABLE III

TYPICAL ESTIMATING UNITS FOR UNIT-PRICE CONTRACTS

Clearing and grubbing	Acre
Excavation, various classes	Cubic yard
Piling	Linear foot
Fine grading	Square yard
Pavement	Square yard
Sidewalks	Square foot
Curb and gutter	Linear foot
Sewers, each diameter size	Linear foot for various intervals of depth *
Water mains, each diameter size	Linear foot for various intervals of depth *
Roofing	Square (10 ft. by 10 ft. = 100 sq. ft.)
Cement	Barrel (4 bags = 3.8 cu. ft.)
Sand and gravel	Cubic yard
Concrete, various classes	Cubic yard
Lumber	Thousand feet, board measure (M.B.M.)
Concrete forms	Square foot of contact surface
Brick, face and common	Thousand
Masonry, including mortar	Cubic yard
Plaster	Square foot
Reinforcing steel	Pound or ton
Structural steel	Pound or ton
Steel rail	Pound or ton
Metal castings	Pound (or piece)
Painting	Square (10 ft. by 10 ft. = 100 sq. ft.)
Motors, pumps, etc.	Each (lump sum)

* An alternate method provides for trench excavation and backfill as separate items on a cubic-yard basis. When this is used, sewer and water pipe units include the cost of furnishing and placing each size of pipe exclusive of excavation and backfill.

examined and analyzed to make certain that all phases of the work are included in one item or another and that none of the items overlaps so that some element of the work is included more than once, resulting in duplication in payments.

Within the foregoing requirements it will be convenient to keep the number of bid items to a minimum. These are more or less standardized for the common types of construction, and unless there are good reasons to the contrary it is usually advisable to conform with standard practice in this respect.

Units for Measurement and Payment under Unit-Price Contracts. The units to be adopted for bidding purposes and for measurement and payment under unit-price contracts conform closely with standard estimating practice. In general, units are made as large as practicable for accurate measurement with the provision that the unit adopted shall be a true measure of the cost of the work when the quantity varies.

When the plans and specifications for any portion of the work are complete in all details and no variation whatsoever in the quantities is contemplated, it will be convenient to adopt a lump sum as the bid unit for that portion of the work.   Likewise, a lump sum may be used for such items as machinery, equipment, and special castings.

Typical units for the common types of construction are shown in Table III.

EXAMPLES OF BID FORMS

Standard Form 21 Revised March 1953 General Services Administration General Regulation No. 13 **BID FORM** **(Construction Contract)**	Reference
Read the Instructions to Bidders (Standard Form 22) *This form to be submitted in*	Date of Invitation

Name and Location of Project

_____

(Date)

TO:

In compliance with your invitation for bids of the above date, the undersigned hereby proposes to furnish all labor, equipment, and materials and perform all work for

at

in strict accordance with the specifications, schedules, drawings, and conditions for the consideration of the following amount(s)

and agrees that, upon written acceptance of this bid, mailed, or otherwise furnished, within .... calendar days (............ calendar days unless a shorter period be inserted by the bidder) after the date of opening of bids, he will within .... calendar days (unless a longer period is allowed) after receipt of the prescribed forms, execute Standard Form 23, Construction Contract, and give performance bond and payment bond on Government standard forms, if these forms are required, with good and sufficient surety or sureties.

The undersigned agrees that if awarded the contract, he will commence the work .... after the date of receipt of notice to proceed, and that he will complete the work within .... calendar days after the date of receipt of notice to proceed.

The undersigned acknowledges receipt of the following addenda to the drawings and/or specifications (*Give number and date of each*) :

The undersigned represents (*check appropriate boxes*) : (1) that the aggregate number of employees of the bidder and its affiliates is ☐ 500 or more, ☐ less than 500; (2) (*a*) that he ☐ has ☐ has not, employed or retained any company or person (*other than a full-time bona fide employee working solely for the bidder*) to solicit or secure this contract; and (*b*) that he ☐ has, ☐ has not, paid or agreed to pay to any company or person (*other than a full-time bona fide employee working solely for the bidder*) any fee, commission, percentage or brokerage fee, contingent upon or resulting from the award of this contract, and agrees to furnish information relating thereto as requested by the contracting officer. (*Note: For interpretation of the representation, including the term "bona fide employee," see General Services Administration Regulations, Title 44, secs. 150.7 and 150.5 (d) Fed. Reg., Dec. 31, 1952, Vol. 17, No. 253.*)

Enclosed is bid guarantee, consisting of ....

in the amount of ....

Name of Firm or Individual (*Type or print*)	Full Name of All Partners (*Type or print*)
Business Address (*Type or print*)	
By (*Signature in ink. Type or print name under signature*)	
Title (*Type or print*)	
State of Incorporation (*Type or print*)	

DIRECTIONS FOR SUBMITTING BIDS

Envelopes containing bids, guarantee, etc., must be sealed, marked, and addressed as follows:

CAUTION: Do not include in the envelope any bids for other work. Bids should not be qualified by exceptions to the bidding conditions.

BID (Unit Price) *

FOR CONTRACT NO. 3A

To the Niagara Falls Bridge Commission:

FOR MATERIAL AND LABOR NECESSARY FOR THE CONSTRUCTION OF THE SUPERSTRUCTURE OF ARCH SPAN OF THE RAINBOW BRIDGE OVER THE NIAGARA RIVER, CONCRETE APPROACH LAYOUT.

The undersigned agrees to furnish all labor, materials, plant and other facilities, and to perform all work necessary or proper for or incidental to the construction of the work herein called for, complete in every respect, in strict accordance with the Plans and Specifications and any future changes made therein as provided in the Contract and Specifications, and to perform all other obligations and assume all liability imposed upon the Contractor by the Contract; and further agrees to accept in full compensation therefor the prices named in the following schedule and, except as otherwise provided in the Contract and Specifications, such prices only:

* Reproduced by permission of Waddell & Hardesty, Consulting Engineers, New York City.

SCHEDULE OF PRICES—IN AMERICAN MONEY

Item No.	Approximate Quantities	Items with Unit Prices Written in Words	Unit Prices in Figures		Amounts	
			Dollars	Cents	Dollars	Cents
1	400	Cubic Yards of 1:2:3½ Concrete in Arch Abutments, for.................... ....per cubic yard.				
2	1,420	Cubic Yards of 1:1¾:3½ Concrete in Slabs and Curbs on Arch Span, for.................... ....per cubic yard.				
3	200,000	Pounds of Bar Reinforcement for Structures, for.................... ....per pound.				
4	340,000	Pounds of Reinforcing Trusses for Roadway Slabs, for............ ....per pound.				
5	2,440,000	Pounds of Structural Carbon Steel in Floorsystem of Arch Span, for.................... ....per pound.				
6	440,000	Pounds of Structural Silicon Steel in Floorsystem of Arch Span, for.................... ....per pound.				
7	1,140,000	Pounds of Structural Carbon Steel in Spandrel Columns of Arch Span, for.................... ....per pound.				
8	940,000	Pounds of Structural Carbon Steel in Rib Laterals, for ....per pound.				
9	6,150,000	Pounds of Structural Silicon Steel in Arch Ribs, for............ ....per pound.				
10		Provision for Utilities, for the Lump-Sum Price of.......... ....dollars.				
		Total or Gross Sum Bid Written in Words........				

It is understood that the approximate quantities shown in the foregoing Schedule are solely for the purpose of facilitating the comparison of Bids, and that the Contractor's compensation will be computed upon the basis of the actual quantities in the completed work, whether they be more or less than those shown herein.

The undersigned agrees to complete the work in every detail by August 1, 1941. It is agreed that no progress payments will be made after August 1, 1941, until the final completion of the work, and that a deduction of $500.00 per day will be made for every calendar day after the above completion date that the Contract is not completed in every detail.

To induce the acceptance of this Bid, the undersigned hereby makes each and every representation and warranty made by the Contractor in the Contract and the other papers bound herewith.

Said Information for Bidders, all papers required by it and submitted herewith, said Contract and all papers made part hereof by its terms, are hereby made part of this Bid.

The undersigned bidder hereby represents as follows:

(a) that no Commissioner, officer, agent, or employee of the Commission is personally interested, directly or indirectly, in this Contract, or the compensation to be paid hereunder; and that no representation, statement or statements, oral or in writing, of the Commission, its Commissioners, officers, agents or employees, has induced him to enter into this Contract excepting only those contained in this form of Contract and the papers made a part hereof by its terms;

(b) that this Bid is made without connection with any person, firm or corporation making a bid for the same work, and is in all respects fair, and without collusion or fraud;

(c) that said bidder has visited and examined the site of the work, and has carefully examined the drawings and the Contract which includes the Information for Bidders, Bid, Contract, Bond, and Specifications, and will execute the Contract and perform all its items, covenants and conditions, all in strict conformity to the requirements of the Specifications and Drawings.

The undersigned hereby designates..

....

as his office to which notices may be delivered or mailed.

Dated ...., 1940

....
Name of Bidder
Corporation, Firm or Individual.

By ....

....
Title

....
Business Address of Corporation,
Firm or Individual.

Irregularities in the Preparation and Submission of Bids. Usually bids are completed at the last moment before the date and hour set forth for the bid opening. This allows the bidder maximum time to analyze the job requirements and enables him to take advantage of changing prices. As a consequence, however, the final preparation of the bid is likely to be hurried, and there is a danger of errors and other inadvertent irregularities which may void the bid or result in a serious loss to the bidder if he should be awarded the contract on the basis of an incorrect amount.

Any deviation in the bid from the requirements of the plans and specifications or any violation of the instructions to bidders constitutes an informality which the owner may waive if he so desires. In general, however, an informal bid may be rejected. On the other hand, an error in the bid amount may be binding upon the bidder.

Past experience has shown that certain types of irregularities in bidding occur repeatedly, and these, among others, should be carefully checked before the bid is submitted. Typical of such irregularities are the following.

1. Bid mailed but not received prior to time of bid opening.

2. Modification of previously submitted bid mailed but not received prior to time of bid opening.

3. Telegraphed bids garbled or otherwise not clear.

4. Bid not dated.

5. Bid not signed.

6. Receipt of changes in the drawings not acknowledged.

7. Receipt of addenda to the specifications not acknowledged.

8. Required list of available plant and equipment not given.

9. Required description of work previously executed not furnished.

10. Bid not submitted with required number of copies.

11. Required plan of operations not furnished.

12. Total amount of bid incorrect (error in addition).

13. Extended amounts for certain items incorrect (error in multiplying quantities of item by unit price).

14. Error in unit price by giving price for incorrect unit (for example, per cubic foot rather than per cubic yard).

15. Error in modifying bid by telegram changing item cost and not following through by revising total accordingly.

16. Price not quoted for all items of a lot when specifications indicate bids will be considered only for all items of lot.

17. Prices not quoted for alternates when such are called for by the specifications.

18. Price submitted for an alternate in lieu of an item called for by the specifications.

Receiving and Opening Bids. Contractors may submit bids at any time prior to the hour designated for their opening, and bids submitted may be withdrawn or changed at any time before the official opening. No changes are permitted after the opening of bids. For submission, the bid, with accompanying papers, should be placed in a sealed envelope addressed to the person authorized to receive bids and endorsed with the bidder's name and the title of the project. At the proper time all bids are opened and read aloud publicly so that all bidders and others interested may be present as witnesses or to tabulate amounts.

Analysis and Comparison of Bids. Ordinary lump-sum bids require only a simple comparison of amounts and a survey to detect informalities. Frequently, though, there may be large numbers of alternative bids requested to cover substitute types of construction or, perhaps, important variations in the entire project. Such bids will require careful analysis in order to determine the low bidder for the various combinations. Important engineering and financial decisions must also be made at this time inasmuch as the character and scope of the work are not definitely determined until after the bids are opened, and a determination is made as to the alternatives to be adopted.

For unit-price contracts the bid consists of a unit price for each item in the engineer's estimate. To compare bids the unit prices are multiplied by the quantities for each item, respectively, and the total thus obtained for the entire estimate is considered as a lump sum. If the bid contains an error in the multiplication of the unit prices by the quantities, the unit price should govern, and, if there is an error in addition, the correct sum governs.

It is understood and should be noted in the instructions to bidders that the engineer's estimate is approximate and that the actual quantities may vary somewhat from those estimated. Therefore the bid comparison is based on approximate costs. The actual cost of the contract can be determined exactly only after the work is completed and the various quantities are accurately measured in place. This sometimes discloses that the apparent low bidder actually was not low, and another contractor would have received the award if the estimate had been more accurate.

The estimated quantities are given solely for the purposes of indicating the scope of the work and comparing bids. The unit prices in the bids are binding on the contractor unless there is a gross error in the estimated quantities. It has been held that the engineer's estimate is a representation which, if grossly in error, results in a mistake and may provide sufficient grounds for the contractor to obtain a change in unit price. For this reason some contracts provide for a readjustment of the contract unit prices if the actual quantities of work performed should vary from the estimated quantities by more than a stipulated percentage, usually 25 per cent. Such a clause may call for a reconsideration of the entire contract or, more frequently, for a negotiation of new unit prices for the differential quantities.

If a gross error should be detected in the low bid, it usually will be in the owner's best interest to reject the bid and award the contract to the second lowest bidder. A question will then arise as to the assessment of a penalty against the bidder who submitted the erroneous bid. This may be in the amount of the bid bond or certified check submitted with the erroneous bid, or an amount equal to the difference between the erroneous low bid and the second lowest bid. In the case of a lawsuit on this question the court usually will attempt to determine the actual damages suffered by the owner because of the erroneous bid and assess that amount against the contractor who made the mistake.

Awarding the Contract. The selection of the contractor requires two independent operations, namely, a study of the qualifications of the bidders to eliminate irresponsible and undesirable contractors and an analysis and comparison of bids to determine the lowest responsible bidder.

Proving irresponsibility on the part of a bidder is a difficult and delicate matter except in the most obvious cases. The default of a previous contract is considered sufficient grounds, and in some quarters litigation in connection with previous contract work will disqualify the bidder. A review of the bidder's experience record to determine his normal capacity for work and a statement of the amount of work he has currently on hand will reveal his available capacity for new work. The contractor's previous record for completing work on time should also be considered. If there appears to be danger of the contractor obligating himself beyond his capacity, he should be disqualified unless he can show evidence of his ability to expand to the necessary extent. Dishonesty, evidence of lack of ability, insufficient working capital, and inadequate surety are likewise grounds for disqualification.

For private works the contract may be awarded to any bidder at the discrimination of the owner. Usually, however, it will go to the lowest responsible bidder unless special considerations outweigh the saving in cost, in which case one of the higher bids may be accepted. Formerly, almost all government statutes required that public works contracts be let to the lowest bidder without reservation. This led to so much difficulty and expense through poor quality of workmanship and defaulted contracts that, except in a relatively few local governments, it is now permitted to award such contracts to the lowest responsible bidder. This allows some control over the qualifications of the contractor to be selected, but if the award is made to other than the lowest responsible bidder it may easily result in responsible bidders refusing to bid on future work for this owner or engineer. The submission of a bid by a contractor entails considerable expense for which he receives no compensation unless he is awarded the contract, and, if he is invited to bid, he has a right to assume that the owner and his engineer will act in good faith in considering his bid.

To establish his qualifications for the contract it may be advisable to require each bidder to furnish a qualification statement with his bid, preferably in the form of a questionnaire, which should include the following information.

a. A complete and detailed financial statement.

b. A record of his experience going back at least five years and giving references in connection with each contract.

c. The particulars in connection with any failures to complete previous contracts.

d. Names and experience records of the principals and key personnel of his firm.

e. Explanation of the manner in which he has inspected the proposed work.

f. A detailed plan for performing the proposed work.

g. List of equipment available for the proposed work.

h. List of equipment to be rented and from whom.

i. List of equipment to be purchased.

j. Items of work to be sublet and names and addresses of proposed subcontractors if known.

k. Name of the individual who would have direct personal supervision over the work.

l. When time of completion is important it may be desirable to require the bidder to submit a preliminary progress schedule with his bid.

Notice of Award and Notice to Proceed. After the analysis of bids and selection of the contractor, a formal *notice of award* should be issued to notify the contractor of his selection and should contain instructions as to time and place designated for the signing of the contract. The letter should state specifically the conditions of the award. After the acceptance of the bid and the signing of the contract a formal *notice to proceed* should be forwarded to the contractor authorizing him to begin work. This should be explicit inasmuch as the contract time for completion will be dated in accordance with the conditions therein.

Letter of Intent. After bids are opened, considerable time may be required to select the contractor, and additional time will be necessary to prepare the contract documents for execution. When the work is urgent, it may be desirable to avoid some of the delay by authorizing the start of work in advance of the signing of the contract. This may be accomplished by combining a *letter of intent* and the notice to proceed in one communication. It is then necessary that the letter state the character and scope of the work to be done and the compensation the contractor is to receive for his services. It is also necessary that the letter obligate both parties to enter into the contract at a later date, allowing the work to proceed in the meantime. Such a letter is, in effect, a contract to enter a contract, and when accepted it is binding on both parties.

Unbalanced Bids. For unit-price contracts, a balanced bid is one in which each bid item is priced to carry its share of the cost of the work and also its share of the contractor's profit. Occasionally contractors will raise the prices on certain items and make corresponding reductions of the prices on others with the total amount of the bid unchanged. The result is an unbalanced bid. In general, unbalanced bids are undesirable and should not be permitted when detected although the practice is sometimes justified from the contractor's viewpoint. Some of the purposes of unbalanced bidding are as follows.

a. To discourage certain types of construction and to encourage others which may be more favorable to the contractor's organization and equipment.

b. When the contractor believes the engineer's estimate for the quantities of certain items is low, by unbalancing his bid in favor of such items, he can secure an increased profit in the payment for the increased actual quantities of the work without increasing the apparent total amount of his bid proportionately.

c. Unreliable contractors may increase the bid prices for the first items of work to be completed with corresponding reductions elsewhere in the estimate with the intention of securing excessive payments on the unbalanced bids and then defaulting the contract. This leaves the surety to complete the work which was underpriced.

d. By unbalancing his bid in favor of the items which will be completed early in the progress of the work, the contractor can build up his working capital for the remainder of the work.

Of the foregoing reasons for unbalanced bidding, the last has some justification when dealing with reliable contractors. The expenses of mobilizing the construction plant, bringing equipment and materials to the site, and the general costs of getting the work started are appreciable. These items usually do not appear in the bid and, therefore, are liquidated only as the work on the bid items progresses. This causes a hardship to the contractor in that his work capital is tied up in the work to the advantage of the owner. There appears to be no way to overcome this imposition on the contractor without exposing the owner to undue risks. For some specialized types of work, notably dredging, it is common practice to permit a payment item in the contract to cover plant mobilization costs, but this does not appear to be feasible for general construction work.

The prevention of unbalanced bids requires a knowledge of construction costs in order that unreasonable bids on individual items may be detected. An obvious case of unbalanced bidding should be considered sufficient grounds for rejecting the entire bid.

EXAMPLE OF AN UNBALANCED BID

For a foundation contract, the engineer's estimate for the purpose of comparing bids indicated a total of 11,000 cubic yards of excavation of which 1,000 cubic yards were estimated to be rock and the remainder earth excavation. A balanced bid of $5.00 per cubic yard and $1.00 per cubic yard, respectively, gives a total cost for excavation of $15,000. Suppose the contractor by investigations of his own concluded that the engineer's estimate for the quantity of rock excavation was too low, and he unbalanced his bid by increasing the unit price for rock excavation and making a corresponding decrease in the unit price for earth excavation. He could then maintain the same total of $15,000.

		BALANCED BID		UNBALANCED BID	
ITEM *Excavation*	*Engineer's Estimate*	*Unit Price*	*Amount*	*Unit Price*	*Amount*
Rock	1,000 cu. yd.	$5.00	$ 5,000	$10.00	$10,000
Earth	10,000 cu. yd.	1.00	10,000	0.50	5,000
Total			$15,000		$15,000

Upon completion of the work suppose it was found by measurement that the actual quantities were: rock excavation, 2,000 cubic yards, and earth excavation, 9,000 cubic yards. The amounts that would be paid to the contractor under each of the two bid items are as follows.

		BALANCED BID		UNBALANCED BID	
ITEM *Excavation*	*Actual Quantities*	*Unit Price*	*Amount*	*Unit Price*	*Amount*
Rock	2,000 cu. yd.	$5.00	$10,000	$10.00	$20,000
Earth	9,000 cu. yd.	1.00	9,000	0.50	4,500
Total			$19,000		$24,500

By unbalancing his bid as shown, the contractor would have obtained an additional payment of $5,500 as compared with the balanced bid. This would not have been readily evident at the time of the award of the contract.

Bidding Practice for U.S. Government Contracts. The general principles of bidding procedure are applicable, in most instances, for contracts with the U.S. Government. There are some notable exceptions, however. Whereas for private works the contractor may change his bid at any time before the official opening, or withdraw it at any time before its formal acceptance, ordinarily for government work his bid cannot be withdrawn after it has once been opened, during the interval between the opening of bids and the award of the contract. The standard bid form permits the Government sixty days after the bid is opened in which to accept it, but the bidder may specify a shorter period if he so desires. Unless he is willing to assume the risk of changing prices or other conditions during this period, the bidder should specify a shorter interval.

The Davis-Bacon Act of March 3, 1931, amended August 30, 1935, requires the acceptance in the bid prices of a scale of minimum wages to be paid all classes of laborers and mechanics. The wage scale is predetermined by the Department of Labor and is included in the specifications for the work.

A mistake in the bid is binding on the bidder unless discovered and changed before the bids are opened. If the low bidder discovers, after the opening of bids, that an error has been made, he may appeal for relief, through the contracting officer, to the Comptroller General. The amount of the claim may not exceed the difference between the accepted bid and the next low bid, and the Comptroller General's decision in the matter is final. Similarly, if the contracting officer makes a mistake in the acceptance of a bid, the government is bound by the amount accepted.

Separate competitive bids may not be submitted by two corporations when one is a subsidiary, owned and controlled by the other. Consideration will be given to only one bid from such a combination.

When identical bids are received and none appears to have superior advantages to the Government, the award may be made by the drawing of lots or by the arbitrary selection of one of the bidders.

Suggestions for Obtaining Lower Bids. The character of the bidding information which is furnished to contractors by the engineer will have important effects on the amount of the bid. The following suggestions are offered as a means for eliminating some of the causes of high bids.

a. Avoid incomplete plans and specifications. All details, dimensions, types of construction, and the like should be available to the contractor for bidding purposes.

b. Do not place the responsibility for incomplete or uncertain data on the contractor. In this event he will have to protect himself against the worst conditions to be expected.

c. Whenever practicable allow the contractor sufficient time to complete the work at an efficient rate of production. Emergency conditions may require speedy construction, but it should be understood that the cost of the work will almost always be increased thereby.

d. Place the fewest possible restrictions on the contractor, consistent with obtaining the results desired.

e. Do not impose unreasonable or unfair conditions on the contractor.

f. Do not impose restrictions as to construction methods on the contractor. Specify results.

g. In so far as possible use stock sizes and standard products in the design. Do not specify the product of a particular manufacturer, however.

h. Review plans and specifications to eliminate all non-essentials.

i. Allow bidders sufficient time between the advertisement and the opening of bids for a careful analysis and estimate of the cost of the work and the conditions under which it will be performed.

j. If a special type of contract is to be used, differing from the usual forms, a copy should be furnished to each bidder.

k. No addenda to the plans or specifications should be issued later than four days before the time for receipt of bids.

l. Establish a reputation among contractors for fair and reasonable requirements during construction and honest, impartial decisions on controversial questions.

Awarding Subcontracts. Bids on subcontracts are usually on an informal basis and, strictly speaking, are a matter between the prime contractor and the subcontractor inasmuch as the prime contractor assumes full responsibility for the work so accomplished. Nevertheless the owner should exercise some control over the award of subcontracts by specifying that all subcontractors are subject to the approval of the engineer. This is desirable in order to assure competent workmanship and to avoid delays and other difficulties which might occur with the acceptance of unqualified subcontractors. Whether the selection of the subcontractors should be subject to approval before or after the award of the prime contract is a controversial question. If the prime contractor is required to submit a list of his subcontractors with his bid, some protection is afforded the subcontractors in that the prime contractor is not permitted to bargain with their competitors for better prices after the award and thereby increase his profit over that contemplated by his bid. On the other hand, it is sometimes held that the prime contractor is entitled to receive the benefits of the best bargain obtainable inasmuch as the owner's interests are not involved.

Competitive Bidding under Incentive Type Contracts. Although incentive type contracts are of the general form of the cost-plus-fixed-fee agreement with a bonus and penalty clause, an element of competitive bidding may be obtained by requiring the bidder to submit his own estimate of the cost of the work, called the target estimate, as in the normal competitive-bid arrangement. In heavy engineering projects this will usually be on a unit-price basis. The bidder also proposes the base fee for which he will undertake the work. The incentive adjustments to the base fee are determined from the difference between the contractor's own target estimate and the actual cost of the work at the end of the job, it being understood that all defined costs are reimbursable to the contractor. All bids are analyzed and evaluated from a consideration of the following factors which are weighted in a manner appropriate to the magnitude and character of the work.

1. Bidders' qualifications (about 50 per cent).
2. Job analysis and plan of operations (about 20 per cent).
3. Target estimate (about 20 per cent).
4. Base fee (about 10 per cent).

The relative rating of the bidders which results from the foregoing analysis will indicate the most advantageous selection from the owner's viewpoint. In appraising the bidders' target estimates, the engineer's estimate should be used as the base rather than the lowest estimated cost. Similarly, in appraising the base fee proposals, the engineer's estimate of a fair and reasonable fee should serve as the basis of comparison.

Questions

1. Why is it desirable to announce by means of an advertisement the taking of bids in connection with the award of a public works contract?

2. What information should the advertisement contain?

3. Write an advertisement for bids for the construction of a water treatment plant by a city department of water supply.

4. What is the purpose of the instructions to bidders?

5. What information should be included in the instructions to bidders?

6. Write the instructions to bidders for the water treatment plant in question 3.

7. Select the bid items and units for a unit-price contract for the construction of a highway. The required construction operations consist of clearing and grubbing, excavation, embankments, concrete box culverts, fine grading, concrete pavement, guard railing, and fencing.

8. Prepare a bid form for the highway contract in question 7.

9. Explain the procedure to determine whether a bidder is qualified to enter into a construction contract.

10. What is an unbalanced bid?

11. Read and report on "Legal Warranty of Engineer's Boring Data" in *Civil Engineering,* June, 1945, page 296.

12. Under a contract for the foundation work and excavation for a public school built for the Board of Education of the City of New York, there was no indication in the record of borings or in the plans and specifications that difficulties would be encountered in the excavation work. When the excavation progressed massive old foundation walls were encountered. The inspectors insisted on the removal of the walls and the resultant extra work increased the cost of the contract to about twice the amount of the bid. The action of the contractor for additional compensation was resisted by the owner on the ground of provisions formulated in the specifications to the effect that "because of the well-understood difficulties under which borings and subsoil explorations are made the owner shall not be held responsible for any differences between indications shown in the plans or other subsoil data and the actual conditions disclosed by excavation," and "no compensation shall be paid the contractor for any differences in the character of the earth to be excavated or for any pumping that may be required." The contractor on the other hand contended that "a difference in the character of the earth to be excavated" applied only to conditions that would be found in natural earth. Who should pay the extra excavation cost? (See *Engineering News-Record,* January 30, 1941, page 60.)

13. Through overlooking a sheet of computations upon which a bid for the repair of concrete substructures for a city was based, a contractor made a bid of $143,171, instead of $193,421. The mistake was discovered the day after the bids were opened. The next lowest bid was $205,395, and the highest was $334,604.

Before the contract was awarded and promptly after the mistake was discovered, the bidding company notified the city and asked to be relieved from the bid. But the city awarded the contract to the company and the company forfeited the $5,000 deposit that accompanied the bid when it refused to enter into a contract on the basis of the erroneous bid. Was the company entitled to compel the city by suit to cancel the bid and refund the deposit? (See *Contractors and Engineers Monthly,* January, 1950, page 95.)

9

COMPETITIVE-BID CONTRACTS

The purpose of a contract is to define, with exactness, the rights and responsibilities of both parties to the transaction. In construction contracts, legal, financial, and engineering considerations are involved, and the relationships between the contracting parties are complex. Therefore, discerning judgment is a prerequisite in arriving at the appropriate terms to be included. It is significant to note that for competitive-bid contracts the terms, except for the contract amount, are dictated by one party, the owner, without consulting the other because ordinarily he is not known at the time the contract is written. This is a necessary condition since contractors cannot be expected to prepare bids without information as to the contract conditions under which the work is to be performed, but it does have important implications relative to the amounts of bids and the conduct of the work.

Inasmuch as a simple oral agreement or an offer and acceptance by letter is just as binding as a lengthy document setting forth all the conditions in detail, the decision as to the form and content of the contract is a matter of supplementing the essential elements of the agreement to a degree commensurate with the importance and complexity of the work. Though brief contracts are desirable, it is always advisable to sacrifice brevity in favor of clarity in order that the obligations of both parties may be defined explicitly.

The formal contract documents to be accepted and agreed to by both parties consist of the contract, the specifications, and the drawings.* Of these the contract is the strongest from the legal viewpoint, it being the document which contains the basic promises or covenants and which bears the signatures of the contracting

---

* The advertisement, instructions to bidders, and the bid, although generally classed as contract documents, are not strictly speaking parts of the contract but preliminary to it. They may be incorporated in the contract by reference, however, if this is considered necessary or desirable. Normally

parties. If any statements of contractual importance are placed in the advertisement or instructions to bidders, these documents must be made parts of the contract by reference, or the statements must be repeated in one of the formal contract documents.

Language of the Contract. All contract documents should be written in a simple style and in so far as possible they should be self-explanatory. Unusual technical terms should be defined where they occur and it is sometimes advisable to devote one article in the contract to definitions of general terms including the various contract documents, the parties to the contract, the engineer, contractor, subcontractor, and the like. The purpose of these definitions is to limit the meaning of significant terms to that intended throughout the contract documents rather than to give general definitions of words which may have several meanings, depending on how they are used.

The parties to the contract are mentioned throughout as if each were singular in number. The designations "Party of the First Part" and "Party of the Second Part" which were formerly used are now seldom encountered in construction contracts. Instead, direct reference is made to the Owner, the Contractor, or the Engineer. Acceptable alternatives to the Owner are the Company, City, State or Government as the case may be. In government contracts the authorized representative of the government agency having cognizance over the work will be a contracting officer who acts for the government. He in turn will be represented by an officer in charge in lieu of the engineer. Each of these terms should be defined.

In the past it has frequently happened that construction contracts have been written in such complex legal language as to obscure the meaning intended. In general, the trend is away from legal phraseology and other esoteric terms and toward simple, direct language which is in common use. Most lawyers as well as engineers and contractors favor this practice which results in documents which are clear and understandable to those without technical and legal training.

these documents do not contain matter of contractual importance, and their inclusion would be pointless except as evidence of the intentions of the contracting parties. An exception would exist when no formal contract is executed. In this case, the bid, together with the acceptance, would comprise the agreement.

Form of the Contract. Contracts for construction work follow legal practice and are encountered in many different forms. In general, all contain an introduction or preamble which states the date of the agreement and designates the parties thereto, as in the following.

> This Agreement, entered into the .... day of ...., 19........, by ...., hereinafter called the Owner and ...., hereinafter called the Contractor, Witnesseth that the parties hereto do mutually agree as follows:

After the preamble, the articles or covenants are written. These may be limited to the basic essentials for a valid contract, namely, the conditions precedent to the agreement, a statement of the work to be done, the time in which it is to be done, the compensation to be paid for its performance, and the signatures of the contracting parties, affirming their agreement as to the conditions imposed by the contract. Such a brief contract is seldom adequate for construction work, however, and supplemental articles are usually required to define in more particularity the intended rights and responsibilities of the signing parties. These should include a statement that the specifications and drawings are each made part of the contract and that their requirements are equally enforceable. Also, such matters as changes in the work, liquidated damages, and control of the work during construction should be covered. Any special statutes or ordinances relating to or governing the conduct of the work should be incorporated by reference, and the contractor's attention should be directed to them specifically.

It is recognized that it is impossible to write a single contract which would be suitable for all phases of engineering construction. However, certain basic articles are required for all contracts, and these can be more or less standardized. Supplemental articles will depend upon the character and magnitude of individual projects, but many of these are also approaching standardization after repeated use.

The form of the contract is influenced considerably by the methods to be used in its preparation. Some organizations develop a standard form which they consider suitable for all work. This is printed with blank spaces in which are written special clauses pertaining to an individual job, such as a description of the work to be done and the amount of the compensation to the contractor. Other articles are standardized in accordance with the owner's general policy and require no individual treatment. When preparing contracts for government agencies, federal, state, county, city, or village, detailed inquiries should be made of the agency counsel to determine special requirements to be incorporated in the contract. Detailed requirements of the work not covered in the contract are incorporated into the specifications which are made a part of the agreement by reference.

The contract is sometimes divided into parts. The first part contains the basic articles relative to the scope of the work, compensation to the contractor, time of completion, and the like and is called the *agreement*. The second part is called the *general conditions of the contract,* and it contains all the supplemental articles. Both the agreement and the general conditions may be standardized and printed for use on all jobs. A third document called the *special conditions* is then written specifically for each job to cover special contractual provisions not included in the standard agreement or the general conditions. These three documents, together with the technical specifications and drawings, make up the contract documents. This is the standard procedure recommended by the American Institute of Architects, and it is used for practically all building construction in the United States. For engineering construction there appears to be no advantage in dividing the contract into parts in this manner. In fact, the contract is complicated unnecessarily by the addition of two extra documents. The contract procedure is simplified considerably by placing all the legal and general contractual requirements in the contract proper, and this document can be standardized and printed if desired. Specific requirements for a particular job and exceptions or amendments to the standard articles can be placed in the general provisions of the specifications, thereby eliminating the general conditions and the special conditions, as separate documents, from the contract papers. The discussions of contracts and specifications throughout this book are based on the latter procedure because of its simplicity in outlining and classifying the contents of the contract documents.

The articles of the contract should be followed by a final or concluding paragraph affirming the agreement as in the following.

> In witness whereof, the parties hereto have executed this Agreement as of the day and year first above written.

This is followed by the signatures of the contracting parties, the signatures of the witnesses, and the seal if required.

Articles of the Contract. In general, the articles of the contract are legal in nature and deal with the contractual promises of the signing parties. Inasmuch as the specifications and drawings are made part of the Agreement by reference, it might be said that every requirement, whether in the contract, specifications, or drawings, is contractual and that the complete assembly makes up the contract. For convenience, however, distinction is made between these documents, and the requirements of the plans and specifications are valid through a contractual promise in the agreement that the work will be performed in accordance with their provisions. This distinction has legal significance. For instance, when it is necessary to produce the original contract in a lawsuit a copy of the original specifications and drawings will suffice for reference purposes. In all respects, however, the requirements of the contract, the specifications, and the drawings are equally enforceable.

Inasmuch as the articles of the contract and some of the general provisions of the specifications may be of similar character and the boundaries between the documents are not rigidly defined, there will always be a question as to the allocation of some topics. This question can be settled best on the basis of clarity and consistency in composition. Many of the requirements could be placed in either document appropriately. In this connection practice varies from the thought that the contract should be as brief as possible, covering only the essential provisions and allocating all supplemental articles to the other documents; to the opposite extreme in which all legal and business matters are placed in the contract, restricting the specifications to the technical requirements. Probably the best solution is some place between the two. A good rule to follow is that the contract being the document of greatest strength should include, in addition to the basic requirements, all provisions requiring special emphasis and those likely to lead to conflicts. If the contract is being prepared for repeated use on a number of jobs, its subject matter must be kept general with the understanding that the specifications will cover detailed contract requirements although some of these may be similar to the articles of the contract. On the other hand a contract being written for a particular project may contain specific references to that project which would not be feasible in a standard contract form. In either event it is important to note the relationship between the documents in order to prevent overlapping, omissions, and conflicts.

When a condition is specified completely in one document it should not be duplicated in the others. In this connection it is advisable to cover a topic completely when it is first mentioned rather than to specify it partially in one place and refer to an additional treatment elsewhere in the documents.

The articles of the contract and the specifications as well should be confined to the phases of the contract procedure related to the work after the award of the contract, all matters pertaining to the bidding being covered in the advertisement and the instructions to bidders. The word "bidder" should not appear in the contract.

There is no fixed rule as to the order in which the articles of the contract should appear in the composition and the extent to which they are covered will depend on the terms to be included and also the contents of the specifications. Usually the essential articles are placed directly after the preamble, and they are followed by the supplemental clauses. This sequence is used in the discussion of the various contract articles which follows.

Statement of the Work. In this article the services to be performed by the Contractor are defined. These will usually be the furnishing of all materials and labor and the performing of the work in accordance with the requirements of the contract, drawings, and specifications, all of which should be designated specifically by title, number, and date and made a part of the contract. If the contract is divided into parts, such as the agreement, general conditions of the contract, and special conditions, as recommended by the American Institute of Architects, mention should also be made of these supplemental documents. The description of the project should be concise but sufficiently complete to define the scope of the work covered by the contract. The contractor will be required to perform only the work described in this article and that defined in the other contract documents designated, and he will be entitled to extra compensation in addition to the contract amount when any omitted item is recalled.

Time of Completion. This article states the date when the work is to begin and the time in which it is to be completed. Usually time is measured from the date of the signing of the contract. The contractor is allowed a reasonable interval after this date, usually ten days, in which to get the work started. The time of completion may be set at a specific calendar date, or the contractor may be allowed a specific number of days in which to complete the work. In the latter method, it is always advisable to specify *calendar* days which includes Sundays and holidays; otherwise there will always be a question as to what time is intended. The use of *working* days in this connection is indefinite. In fixing the time of completion, a reasonable period should be allowed for the contractor to complete his work. A time schedule requiring extra equipment and workmen and overtime work, as compared with normal progress, will always result in increased construction costs. These extra cost factors must be weighed against the urgency for the completed work. Frequently the benefits to the owner of being able to place the completed work into service at an earlier date will justify the extra cost required, but otherwise the amount of the contract will be increased unnecessarily.

The Contract Amount. This article states the amount the contractor is to receive in consideration for the accomplishment of the work. The consideration will be a lump sum or unit prices or possibly both, depending on the basis of the bid. Excepting this article the contract form may be the same for both lump-sum and unit-price contracts.* For unit-price contracts a statement should be included permitting an equitable adjustment in unit prices if the quantities originally contemplated should be so changed that the application of the contract unit prices would create a hardship to either the owner or the contractor. This qualification is sometime left indefinite to be negotiated at the time of the change in the quantities. It is preferable, however, to state in the contract the limit of variations from the estimated quantities, which will be allowed with no change in the contract unit prices. This is usually set at 20 per cent to 25 per cent of the original estimated quantities. If it is desired to fix in advance the changes in unit prices to cover variations from the original estimated quantities the bid may be worded to provide unit-price quotations for various increments more or less than the base estimate. The payment article of the contract would then contain corresponding provisions. The statement of the work, time

---

* Methods of measurement and payment for bid items under unit-price contracts normally are covered in the specifications.

of completion, and the contract amount are frequently combined in one article of the contract.

Liquidated Damages. Closely related to the contract amount, the article on liquidated damages when used covers the situation in which the contractor fails to complete the work within the time specified and thereby causes the owner to suffer losses and other hardships. In order to be valid this article should declare that time is of the essence of the contract; that is to say, time is part of the consideration of the agreement. Actual damages are difficult to determine exactly and in lieu thereof both parties agree to the payment to the owner by the contractor of a predetermined lump sum for each day of delay until the work is completed. This lump sum is not to be considered as a penalty but as the liquidated actual damages which the owner would suffer as a result of the failure of the contractor to complete the work as agreed, and the contract should so state. Penalties are difficult to enforce under contract law.

In arriving at the amount to be assessed as liquidated damages, it must be kept in mind that this amount is intended to be a measure of the actual damages suffered. Although agreed to in the contract, liquidated damages may be challenged at any time by the contractor if it appears that the amount is unreasonable as compared with the actual loss suffered by the owner on account of the delay. Under these conditions the owner must be able to justify the amount, or it may be revised or voided by the court.

Provision should also be made to grant extensions of the contract time of completion for any delays resulting from causes beyond the contractor's control, which are not to be considered normal hazards of the contract. Delays of this classification, such as those due to acts of God, disasters, strikes, and the like, are deducted from the time for which the contractor is liable for liquidated damages.

Bonus Clauses. As an incentive for the contractor to complete the work before the date specified, a bonus clause is sometimes written into the contract providing for the payment to the contractor of an additional lump-sum amount for each day the completed work is available for use by the owner prior to the time of completion specified in the contract. Ordinarily the amount of the bonus payment should be consistent with the provision for the payment by the contractor of liquidated damages for delays.

Bonus provisions have the effect of strengthening liquidated damage clauses because of the tendency of the courts to view liquidated damages alone as a penalty without a corresponding consideration. Although liquidated damage clauses are valid without bonus provisions proponents of the latter hold that it is only equitable to permit the contractor to profit by performance better than that anticipated if he is to be penalized for failure to do so. On the other hand opponents object to bonus clauses because of the additional speculative element introduced into the contract. Bonus clauses are not permissible in U.S. Government contracts.

Supplemental Articles. When accepted and signed by the contracting parties, the foregoing articles satisfy the essential legal requirements for a valid contract, but because of the complexity of most construction projects they are seldom adequate to define the intentions of the parties and additional covenants are needed. The topics to be covered will depend on the magnitude and character of the work, but the usual practice is to include only such contractual provisions as will apply to any project. This permits the contract to be standardized and printed for general usage. All unique features of a particular project may then be covered in the specifications and made part of the contract by reference.

The supplemental articles most frequently used in current practice may be classified broadly under the following headings.

a. Plans and specifications.
b. Responsibilities and rights of the owner during construction.
c. Authority and duties of the engineer.
d. Responsibilities and rights of the contractor.
e. Progress and control of the work.
f. Payments to the contractor.
g. Insurance, safety, and sanitation.

Each of the foregoing classifications may require several articles in the contract. When the contract is to be divided into parts these supplemental articles normally would be placed in the part termed general conditions of the contract.

Plans and Specifications. The articles under this classification give the contract requirements relative to the intent of the plans and specifications, conflicts, omissions and errors, and the like. Some contracts also contain a general article pertaining to shop drawings and other drawings to be prepared by the contractor. It is advisable to require a minimum of this type of work by the contractor, however, as such requirements usually result in a disproportionate increase in the cost. For this reason, it is usually preferable to cover shop drawings in detail in the specifications.

Some states have laws which require that all contracts, plans, and specifications for work, the cost of which exceeds a specified amount, shall be registered. When this is required the contract should contain an article placing responsibility for the registration either on the owner or the contractor. Unless the requirements of the law are complied with in this respect, the validity of the contract may be endangered.

The article covering the intent of the plans and specifications deserves special attention. This article usually states that all work called for in the specifications and not shown on the plans or shown on the plans and not called for in the specifications shall be of like effect as if shown or mentioned in both. If a conflict should exist between the provisions of the specifications and those of the drawings the specifications usually govern. Omissions from the drawings or specifications, or the misdescription of details of the work, which are evidently necessary to carry out the intent of the plans and specifications normally do not relieve the contractor from performing such omitted or misdescribed work. Under this article he is required to perform all work necessary to provide a complete structure as if it were fully and correctly described in the plans and specifications. The fairness of this requirement is open to question, but it is used almost universally and is enforceable when the plans and specifications are sufficient to indicate clearly the general intentions in connection with the work. A contract to perform work according to specific plans and specifications implies that they are understood.

The contractor sometimes is required to "guarantee the sufficiency of the plans," making him responsible for the adequacy of the design and this provision also is enforceable when agreed to by the contractor. Under this condition, the contractor may be held liable for a failure even though the work was performed in accordance with the plans and specifications.

With regard to the interpretation of the plans and specifications, the engineer should be given authority to decide all questions, disputes, or conflicts, and all work should be performed in accordance with his directions. In this connection the contractor should be required to check all drawings and report any discrepancies to the engineer for his determination. When dimensions are affected by existing conditions the contractor should be required to take measurements notwithstanding any scale or figure dimensions on the drawings and report such information to the engineer.

No changes or deviations from the drawings and the dimensions given therein should be permitted without authority from the engineer regardless of whether or not error is believed to exist.

Responsibilities and Rights of the Owner during Construction. In general, very few direct contacts are necessary between the owner and the contractor after the contract is signed, all details of the work being handled for the owner by the engineer. This procedure should be rigidly adhered to in order to prevent divided or conflicting authority on the job. Therefore, one of the owner's first responsibilities is to designate in the contract a representative, usually the engineer, who shall have complete charge of the work and the authority to direct and supervise all operations in so far as the owner's interests are concerned. He is also obliged to furnish the land on which the work is to be performed, all the property surveys required to establish the boundaries of the site, and the funds for prompt payment to the contractor of all monies due under the contract, but these items ordinarily are not matters to be covered in the contract.

With regard to the rights of the owner, reservation should be made in the contract for him to perform work with his own forces or to let other contracts in connection with the work and to require the contractor to connect and coordinate his work with that of others without interference to either. Unless provisions are made to the contrary the contractor may be entitled to extra compensation if he is delayed or otherwise suffers damages because of interference by the owner's forces or those of other contractors on the job.

On most construction projects it is desirable for the owner to take possession and use any completed portions of the work irrespective of the time of completion of the entire contract. Use of portions of the work in this manner should not constitute acceptance which should be considered only when the entire project is completed. It should be noted that use of portions of the work by the owner may delay or interfere with the contractor's operations in the remainder, and if so he is entitled to extra compensation unless provisions are made to the contrary.

The owner may assert the right to suspend work under the contract temporarily if it appears to be in his interest to do so. The contractor should then be required to resume work upon receipt of written notice from the owner. This article should provide for reimbursement to the contractor to cover any expense incurred by him as a result of the suspension of the work. It is unreasonable to expect a contractor to stand by indefinitely under a temporary suspension order, and the contract should permit him to abandon the work completely if the suspension is not lifted within a specified time, say ninety days. He is then entitled to payment for all work done on the abandoned portions of the project.

Changes in the Work. Frequently in the course of construction work it becomes necessary to make changes in the plans and to introduce additional work which was not contemplated at the time of the award of the contract. In a similar manner, it may become desirable to omit certain parts of the project which were included in the contractor's bid. In either case it is necessary to revise the contract by the issuance of a change order. This should always be done in writing, and no changes should ever be undertaken on the basis of verbal orders, except in an emergency. The change order may be in letter form and should state explicitly the work required and the price to be paid to the contractor, if the change involves any additional cost, or the credit to be allowed the owner, if the change involves a reduction in cost.

The method of arriving at the cost of changes should be given careful consideration by the engineer, and an agreement with the contractor, as to the amount, should be made before he is authorized to proceed, in order to avoid disputes after the work is completed. The following are the usual methods of payment for changes.

a. The actual cost of the change plus a percentage of the cost, usually about 10 per cent for overhead and 6 to 10 per cent for profit.

b. The actual cost of the change plus a fixed fee determined in advance of the work on the basis of the estimated cost.

c. A lump-sum amount to cover all costs and profit to the contractor.

d. Unit prices to cover all costs and profit to the contractor.

It is essential that every construction contract shall permit the owner to order extra work or make changes by altering, adding to, or deducting from the work at any time without invalidating the contract or the contractor's bond. All such changes should be executed under the conditions of the original contract except that the cost of the extra work or changes and any claims for extension of time caused thereby should be adjusted at the time the change is ordered. The cost of changes may be established in the contract on the basis of cost plus a specific percentage or methods for arriving at the cost by negotiation may be specified as described above. When changes are anticipated unit prices for additions and deductions may be obtained in the original bid.

The amounts involved in changes and extra work are contractual matters to be negotiated between the owner and the contractor, when not provided for in the original contract, inasmuch as it is usually impracticable to request bids from others. Ordinarily the engineer does not have the right to order changes in the work unless he is expressly authorized to do so by the owner.

Termination of the Contract. The owner may reserve the right to terminate the contract by written notice without fault of the contractor at any time when it is in his interest to do so. When this is done the contractor is entitled to compensation for all work done by him prior to the time of annulment and for all expenses not otherwise paid for and as are required in abandoning the work. The engineer should be authorized to determine the amounts of such expenses.

The contract also may be terminated by the owner because of the failure of the contractor to perform in accordance with the requirements of the contract. The conditions which should be considered sufficient causes for annulment because of the contractor's breach of the contract are as follows.

a. If the contractor should be adjudged bankrupt.

b. If he should make a general assignment for the benefit of his creditors.

c. If he should persistently or repeatedly fail or refuse to supply enough properly skilled workmen or proper materials.

d. If he should fail to make prompt payment for materials or labor or to subcontractors.

e. If he should persistently disregard laws, ordinances or the instructions of the engineer.

When the engineer determines that sufficient cause exists the contractor should be given written notice that his employment is terminated, and the owner may take possession of all materials, tools, and equipment on the premises and finish the work by the most expedient method at the expense of the contractor or his surety. The expense of finishing the work, including all additional costs for managerial and administrative services, should be determined by the engineer.

The Authority and Duties of the Engineer. During construction the engineer serves in a dual capacity. He is the authorized representative of the owner for the purpose of supervising and inspecting the work to make certain that all operations are performed in accordance with the requirements of the contract. Also he is the interpreter of the plans and specifications and the arbiter of disputes. The owner is fully within his rights in designating a representative to act for him, but questions of interpretation and arbitration can be assigned to the engineer for decision only on the mutual assent of both parties to the contract. Careful distinction must be made between designation and mutual agreement in this respect. The anomalous situation in which the authorized representative of one party to a contract may also act as the unbiased referee of disputes between the two parties is acceptable only because of the professional integrity of the engineer.

As the representative of the owner it is the engineer's duty to establish all lines and elevations; to make all field measurements necessary to determine the amounts of payments to the contractor; to inspect all materials and workmanship furnished by the contractor; and to see that the work conforms to the requirements of the design; and, in exceptional cases, to establish the order or precedence of operations and construction methods. The last prerogative should be exercised only when required by conditions of the design or when the owner's interests are involved. It is desirable that the contractor be given a free hand, in so far as is practicable, in matters of construction procedure and methods. Under an independent contractor relationship, control of the methods for accomplishing the work is vested in the contractor since he is obligated to produce specific results. This relationship is destroyed if the owner or the owner's representative assumes excessive control of the work, and the laws of agency may be invoked. However, it should be noted that the contractor assumes full responsibility when he acts contrary to the specific requirements of the contract.

As the arbiter of disputes it is the engineer's duty to interpret the requirements of the contract and, in case of errors or omissions, to define the intentions of the plans and specifications. His decisions should be based on professional judgment and should be impartial in all respects. Actually the jurisdiction of the engineer is restricted to the settlement of questions of fact. In the last analysis questions of the law of the contract can be decided only by a board of arbitration or a court of law. Therefore, the usual stipulation that "the engineer's decisions shall be final and binding on both parties" cannot be enforced literally. Even on questions of fact it is desirable that provision be made for either party to appeal to a higher authority. Under government contracts such appeal may be made to the contracting officer and, beyond him, to the head of the department having cognizance over the work. If the engineer's decision is not acceptable on private work a board of arbitration is the only alternative to settlement by litigation.

The status of the engineer during construction should be defined in a few carefully worded articles in the contract, the first of which should state that the work shall be subject at all times to his supervision and direction. He should be given authority to determine the amount or quantity of the various kinds of work, and the quality of materials and workmanship to be paid for under the contract and to decide all questions which may arise relative to the performance of the work. This article should also provide that any question as to meaning of the plans and specifications and any obscurity or discrepancy in their wording and intent shall be decided by the engineer and his decision shall be binding on both parties to the contract. Errors or ommissions in the plans and specifications should be amended or corrected by the engineer consistent with a reasonable interpretation of the intent of the plans and specifications.

Construction contracts frequently provide that all lines and grades and other construction surveys will be furnished by the engineer. However, practice in this respect varies. For the construction of buildings, bridges, and similar structures the contractor is sometimes required to perform all layout and survey work with the provision that base points and elevations only will be furnished by the engineer. Therefore, if the contract is to be used for several kinds of work, it is advisable to cover layouts and construction surveys in the specifications rather than in the contract.

The article on inspection and examination of the work should be thorough and complete. It should provide that all materials and workmanship shall be subject to inspection and test by the engineer at any and all times, and, when found by the engineer to be in variance with the requirements of the plans and specifications, all defective work shall be removed and replaced at the contractor's expense. Failure to remove such defective work should be considered sufficient grounds to terminate the contract. Reservation should be made, however, that defective work may be retained if desired by the owner, but when this is done there should be a reasonable adjustment in the contract price.

Inspectors, if mentioned at all, should be established as representatives and assistants of the engineer with authority to reject materials until the question can be referred to the engineer but with no authority to obligate the engineer in any way. Ordinarily inspectors are not authorized to revoke or alter any of the requirements of the plans and specifications or to approve or accept any portion of the work.

Responsibilities and Rights of the Contractor. In prescribing the responsibilities of the contractor, it should be kept in mind that the main objective of the contract method is that the work shall be done by a skilled constructor in order to relieve the owner of all responsibilities in connection with management, purchase of materials, employment of labor, and construction operations. The contract price is assumed to include a reasonable profit to the contractor for the accomplishment of this objective.

Therefore, the contract should establish a clear understanding as to the scope of the contractor's obligations, with the detailed instructions set forth in the plans and specifications.

The contractor should be required to designate a representative who is authorized to act for him in all matters pertaining to the contract when he himself is not available. The contractor's representative will usually be the construction superintendent. This designation is necessary in order that there may be some one on the job at all times to whom directions and instructions may be issued with the same effect as if delivered to the contractor in person.

Along with changes and extra work orders, the matter of local conditions at the site of the work is the source of more controversies and disputes than any other condition of the contract, and the owner's intentions in this respect should be clearly stated. Latent or subsurface conditions frequently cannot be determined definitely before the beginning of construction. On private work the contractor is usually obliged to accept full responsibility for the conditions encountered, it being stated that the information given relative to these conditions is the best obtainable, but the owner accepts no responsibility as to its accuracy, and the contractor is to use it at his own risk. U.S. Government contracts, on the other hand, allow a reasonable adjustment in the contract amount if subsurface conditions are later found to be materially different from those indicated on the plans.

Traditionally it was customary to preface each section of the specifications with a statement to the effect that, "The contractor is to furnish all materials, equipment, and labor necessary to construct or install the following work. . . ." Much needless repetition can be avoided by including one such statement in the contract to cover all work contemplated by the plans and specifications with such exceptions as may be noted in the specifications. The exceptions will usually be any materials or labor scheduled to be furnished by the owner or other contractors. Note also should be made of these exceptions in the instructions to bidders.

In connection with the materials, equipment, and processes required, it should be stipulated, unless otherwise provided for, that all royalties and license fees are to be paid by the contractor and that he shall be responsible in all cases of suits or claims for infringement of patent rights. Similarly, the contractor is usually required to provide all building permits.

The conditions under which the contractor has the right to terminate the contract should be defined. In general, the contractor is entitled to terminate the contract if the owner fails to make payment as prescribed in the contract or if the work should be suspended by the owner or by order of a court for a period longer than a specified time, usually three months. A suspension of longer duration may introduce changed conditions with respect to materials prices and wage rates, and the contractor should not have to bear any excess expenses caused thereby.

The contract should be explicit as to the responsibility for delays to the contractor caused by acts of the owner or by other contractors employed on the work. In current practice it is sometimes considered that such delays constitute a normal hazard of the contract and therefore are not sufficient grounds for an extension of the time of completion when liquidated damages are concerned. Many contractors protest that this is an unfair interpretation and claim that a contractor is entitled to an extension of time and possibly the payment of damages when he is delayed or otherwise suffers loss because another contractor's work is behind schedule. There is much to be said for this argument, and it appears that the most equitable solution is to permit extensions of time when the contractor suffers unreasonable delay because of acts of the owner or other contractors as determined by the engineer. The accurate determination of damages under these conditions is very difficult and in case of disagreement arbitration is recommended for the settlement of any questions along this line. The contractor is also entitled to an extension of time to cover delays to the work caused by unavoidable delays in transportation, fire, strikes, floods, and like casualties. Bad weather normally does not justify an extension of time unless it can be shown conclusively that such weather is unprecedented and could not have been anticipated as a normal hazard of the contract.

Progress and Control of the Work. The purpose of the articles under this heading is to establish general requirements regarding the quality of materials and workmanship, schedules of progress of the work, approval of subcontracts, employees of the contractor, and similar items dealing with the conduct of the work. These should be stated in general terms so as not to conflict with detailed requirements in the specifications. It is intended that the control of the work should be vested in the owner, through the engineer, in such a manner that the owner's rights and interests will be protected without limiting the contractor's freedom to plan and administer the conduct of the work along the lines best suited to his organization and equipment.

Before the work is started and materials ordered the contractor should be required to consult with the engineer relative to materials, equipment and methods for prosecuting the work. The work should be carried on at such time and in such parts of the project and with such forces of workmen, materials, and equipment as may be ordered by the engineer to complete the work in accordance with the plans, specifications, and contract. The contractor should be allowed to work at night or outside regular working hours only on the written permission of the engineer and then only after suitable arrangements have been made to inspect the work in progress and after the contractor has provided satisfactory lighting and has complied with other requirements and regulations for night work.

One article should be devoted to the quality of workmanship and materials. Unless otherwise specified, it is intended that all materials, workmanship, equipment, and articles incorporated in the work are to be of the most suitable grade of their respective kinds. Where such items are referred to in the specifications as "equal to" any particular standard the engineer shall decide the question of equality. Information concerning materials and equipment including samples should be submitted to the engineer for approval when required.

Most contracts authorize the engineer to remove from the work employees of the contractor considered incompetent, careless, insubordinate, or otherwise objectionable. This prerogative must be used with the utmost discretion, or the contractor's status as an independent contractor may be converted to that of an agent of the owner thereby relieving him of many responsibilities under the contract.

The contract should also require the contractor to submit the names of all subcontractors to the engineer for approval as soon as practicable after the contract is signed. No subcontractor shall be employed until such approval is obtained. Full responsibility for the acts and omissions of subcontractors is placed upon the contractor, and nothing contained in the contract should create any contractual relation between any subcontractor and the owner.

A general article is sometimes included requiring the contractor to furnish a schedule of expected progress which shows the dates on which the various parts of the work are expected to be begun and completed. On important work, however, there will usually be special requirements as to the sequence of operations and work to be furnished by others. When this is contemplated the requirements relative to progress schedules can be covered more effectively in the specifications.

Upon the completion of the work the contractor should be required to remove from the owner's property and from all public and private property at his own expense all temporary structures, equipment, rubbish, and waste materials resulting from his operations.

Payments to the Contractor. As the work progresses, the contractor is entitled to partial payments which are made on the basis of the engineer's estimates of the amount of the work completed. On lump-sum contracts the estimate will be expressed as a percentage of the total work to be done. For unit-price contracts the engineer will measure the quantities of work completed under each item, and these will indicate the degree of completion. In arriving at the amount of the partial payment, consideration may be given to the amount of material delivered to the site and work done by the contractor at his assembly plant, as well as to completed work. Partial payments are usually made by the owner once each month, although on large contracts it may be desirable to pay semi-monthly in order to reduce the amount of working capital required by the contractor.

Partial payments usually cover 90 per cent of the work completed during the payment period, 10 per cent being retained by the owner as protection against claims or charges against him because of some act or omission on the part of the contractor. Possible grounds for such claims or charges are as follows.

a. Failure of contractor to reimburse subcontractors, material dealers, or employees.

b. Damage to adjacent property or to another contractor.

c Defective work not corrected.

d. Evidence that the work cannot be completed for the unpaid balance.

e. Evidence that the contractor may default.

The final payment to the contractor will include the retained percentage as well as the amount due for the work completed during the last payment period. Before the final payment is made, however, the engineer will make the final inspection of the work as a condition of its formal acceptance by the owner. The final inspection will determine any omitted portions of the work, all defective work to be corrected, and any deductions to be made in the contract amount for defective work to be accepted in lieu of its removal and replacement. The contractor should be required to deliver to the owner a release of all claims or liens arising out of the contract before the final payment is made. If any claims or liens are outstanding against the contractor, the amount claimed should be deducted from the final payment and retained by the owner until the matter is settled. If the contractor has any unsatisfied claims against the owner for damages, extra work, or other reasons, the release should be required notwithstanding, specific exceptions being included in the release to cover the unsatisfied claims.

Insurance, Safety, and Sanitary Requirements. The hazards and risks incident to construction operations are innumerable, and careful control is necessary to reduce the probability of accidents to a minimum. It is understood that the contractor accepts the responsibility for damages to persons or property due to acts, omissions, or accidents during construction. This, however, will not be sufficient to provide full protection to the owner since suits or claims for damages may name him, as well as the contractor, as a defendant. Furthermore, there is the matter of possible loss due to fire, storms, floods, thefts, and similar occurrences during construction. The most positive method to protect the owner fully is by means of insurance, and the requirements in this connection should be specified in the contract. A construction contract is not one of insurance also unless it is so stipulated.

The common risks in construction work are injury to the contractor's employees, injury to other persons or to property other than that owned by the contractor, and damage to the owner's or the contractor's property due to unforeseen occurrences. Injury to workmen on the job should always be covered by workmen's compensation and employer's liability insurance at the contractor's expense. To avoid loss from damage claims arising out of injuries to persons not connected with the work or to property other than that owned by the contractor, public liability insurance should also be required. Practice varies with respect to insurance to cover loss by fire, wind, theft, floods, and the like. Comprehensive insurance of the builder's risk type is frequently specified. If the contractor is required to deliver a completed structure before the owner assumes title to the work, the responsibility rests with the contractor as to whether he will protect the completed portions of the work with insurance. On the other hand, the interests of the owner may indicate the advisability of complete insurance coverage as the work progresses, and he may elect to obtain the insurance himself at his own expense; or he may require the contractor to obtain it under the contract. In any event, the types and amounts of insurance desired should be determined and specified before the contract is let, inasmuch as the bid prices will be affected.

Closely allied to risks and insurance are the matters of safety precautions and sanitation. Specific instructions should be included relative to barricades around openings, ladders, warning signs, and the like; also provisions should be made for the location and construction of toilets and other sanitation facilities. These matters are the responsibility of the contractor, as is the removal of such facilities when the work is completed.

Standard Contracts. In this chapter the writing of competitive-bid construction contracts has been discussed in general. For a guide in the composition of the detailed requirements and the wording and arrangement of the articles, a study of the standard documents of the various technical societies and government agencies is recommended. As examples of typical practice, the U.S. Government standard form for construction contracts, the uniform contract form of the American Railway Engineering Association, and the standard agreement and general conditions of the contract of the American Institute of Architects are reproduced here for reference.

The United States standard form was prepared by a joint committee of representatives of the various governmental agencies having cognizance over construction work and is used for practically all government contracts except during war emergencies. It differs from contracts for private construction in that it contains a number of articles required by statute law to be included in all U.S. Government contracts.

The uniform contract of the American Railway Engineering Association is one of the oldest of the standard contracts and is representative of current practice for private engineering construction work.

The agreement and general conditions recommended by the American Institute of Architects are used for practically all private building construction in the United States. They also illustrate the method of dividing the contract into two parts, whereas in the American Railway Engineering Association form all contractual promises or covenants are consolidated into one document.

Each of the standard forms reproduced herein has been developed and used over a long period and has covered construction work costing many millions of dollars. Moreover, they have been tested many times in court and are believed to represent sound current practice for their respective purposes in the United States.

EXAMPLES OF COMPETITIVE-BID CONTRACTS

STANDARD FORM OF CONSTRUCTION CONTRACT
USED BY THE U.S. GOVERNMENT

Standard Form 23 Revised March 1953 General Services Administration General Regulation No. 13 **CONSTRUCTION CONTRACT**	Contract No.
	Date of Contract
Name and Address of Contractor	Check Appropriate Box ☐ Individual ☐ Partnership ☐ Incorporated in the State of

Department or Agency

Contract for (*Work to be performed*)

Place

Amount of Contract (*Express in words and figures*)

Administrative Data (*Optional*)

THIS CONTRACT, entered into this date by The United States of America, hereinafter called the Government, represented by the Contracting Officer executing this contract, and the individual, partnership, or corporation named above, hereinafter called the Contractor, witnesseth that the parties hereto do mutually agree as follows:

Statement of Work. The Contractor shall furnish all labor, equipment, and materials and perform the work above described for the amount stated above in strict accordance with the General Provisions (Standard Form 23A), specifications, schedules, drawings, and conditions all of which are made a part hereof and designated as follows:

Work Shall Be Started	Work Shall Be Completed

Alterations. The following changes were made in this contract before it was signed by the parties hereto:

In witness whereof, the parties hereto have executed this contract as of the date entered on the first page hereof.

THE UNITED STATES   CONTRACTOR
OF AMERICA

By _____

_____   *(Name of Contractor)*

_____   By _____
(Official title)   *(Signature)*

_____
(Title)

INSTRUCTIONS

1. This form shall be used, as required by GSA regulations, for contracts for the construction, alteration, or repair of public buildings or works.

2. The full name and business address of the Contractor must be inserted in the space provided on the face of the form. The Contractor shall sign in the space provided above with his usual signature and typewrite or print name under all signatures to the contract and bonds.

3. An officer of a corporation, a member of a partnership, or an agent signing for the Contractor shall place his signature and title after the word "By" under the name of the Contractor. A contract executed by an attorney or agent on behalf of the Contractor shall be accompanied by two authenticated copies of his power of attorney, or other evidence of his authority to act on behalf of the Contractor.

Standard Form 23A
March 1953
Prescribed by General
Services Administration
General Regulation No. 13

GENERAL PROVISIONS

(Construction Contracts)

1. DEFINITIONS

(a) The term "head of the department" as used herein shall mean the head or any assistant head of the executive department or independent establishment involved, and the term "his duly authorized representative" shall mean any person authorized to act for him other than the Contracting Officer.

(b) The term "Contracting Officer" as used herein, shall include his duly appointed successor or his authorized representative.

2. SPECIFICATIONS AND DRAWINGS

The Contractor shall keep on the work a copy of the drawings and specifications and shall at all times give the Contracting Officer access thereto. Anything mentioned in the specifications and not shown on the drawings, or shown on the drawings and not mentioned in the specifications, shall be of like effect as if shown or mentioned in both. In case of difference between drawings and specifications, the specifications shall govern. In any case of discrepancy either in the figures, in the drawings, or in the specifications, the matter shall be promptly submitted to the Contracting Officer, who shall promptly make a determination in writing. Any adjustment by the Contractor without this determination shall be at his own risk and expense. The Contracting Officer shall furnish from time to time such detail drawings and other information as he may consider necessary, unless otherwise provided.

3. CHANGES

The Contracting Officer may at any time, by a written order, and without notice to the sureties, make changes in the drawings and/or specifications of this contract and within the general scope thereof. If such changes cause an increase or decrease in the amount due under this contract, or in the time required for its performance, an equitable adjustment shall be made and the contract shall be modified in writing accordingly. Any claim of the Contractor for adjustment under this clause must be asserted in writing within 30 days from the date of receipt by the Contractor of the notification of change: *Provided, however,* That the Contracting Officer, if he determines that the facts justify such action, may receive and consider, and adjust any such claim asserted at any time prior to the date of final settlement of the contract. If the parties fail to agree upon the adjustment to be made the dispute shall be determined as provided in Clause 6 hereof. But nothing provided in this clause shall excuse the Contractor from proceeding with the prosecution of the work as changed. Except as otherwise herein provided, no charge for any extra work or material will be allowed.

4. CHANGED CONDITIONS

The Contractor shall promptly, and before such conditions are disturbed, notify the Contracting Officer in writing of: (1) subsurface or latent physical conditions at the site differing materially from those indicated in this contract, or (2) unknown physical conditions at the site, of an unusual nature, differing materially from those ordinarily encountered and generally recognized as inhering in work of the character provided for in this contract. The Contracting Officer shall promptly investigate the conditions, and if he finds that such conditions do so materially differ and cause an increase or decrease in the cost of, or the time required for, performance of this contract, an equitable adjustment shall be made and the contract modified in writing accordingly. Any claim of the Contractor for adjustment hereunder shall not be allowed unless he has given notice as above required; provided that the Contracting Officer may, if he determines the facts so justify, consider and adjust any such claim asserted before the date of final settlement of the contract. If the parties fail to agree upon the adjustment to be made, the dispute shall be determined as provided in Clause 6 hereof.

5. TERMINATION FOR DEFAULT—DAMAGES FOR DELAY— TIME EXTENSIONS

(*a*) If the Contractor refuses or fails to prosecute the work, or any separable part thereof, with such diligence as will insure its completion within the time specified in this contract, or any extension thereof, or fails to complete said work within such time, the Government may, by written notice to the Contractor, terminate his right to proceed with the work or such part of the work as to which there has been delay. In such event the Government may take over the work and prosecute the same to completion, by contract or otherwise, and the Contractor and his sureties shall be liable to the Government for any excess cost occasioned the Government thereby, and for liquidated damages for delay, as fixed in the specifications or accompanying papers, until such reasonable time as may be required for the final completion of the work, or if liquidated damages are not so fixed, any actual damages occasioned by such delay. If the Contractor's right to proceed is so terminated, the Government may take possession of and utilize in completing the work such materials, appliances, and plant as may be on the site of the work and necessary therefor.

(*b*) If the Government does not terminate the right of the Contractor to proceed, as provided in paragraph (*a*) hereof, the Contractor shall continue the work, in which event he and his sureties shall be liable to the Government, in the amount set forth in the specifications or accompanying papers, for fixed, agreed, and liquidated damages for each calendar day of delay until the work is completed or accepted, or if liquidated damages are not so fixed, any actual damages occasioned by such delay.

(*c*) The right of the Contractor to proceed shall not be terminated, as provided in paragraph (*a*) hereof, nor the Contractor charged with liquidated or actual damages, as provided in paragraph (*b*) hereof because of any delays in the completion of the work due to unforeseeable causes beyond the control and without the fault or negligence of the Contractor, including, but not restricted to, acts of God, or of the public enemy, acts of the Government, in either its sovereign or contractual capacity, acts of another contractor in the performance of a contract with the Government, fires, floods, epidemics, quarantine restrictions, strikes, freight embargoes, and unusually severe weather, or delays of subcontractors or suppliers due to such causes: *Provided,* That the Contractor shall within 10 days from the beginning of any such delay, unless the Contracting Officer shall grant a further period of time prior to the date of final settlement of the contract, notify the Contracting Officer in writing of the causes of delay. The Contracting Officer shall ascertain the facts and the extent of the delay and extend the time for completing the work when in his judgment the findings of fact justify such an extension, and his findings of fact thereon shall be final and conclusive on the parties hereto, subject only to appeal as provided in Clause 6 hereof.

6. DISPUTES

Except as otherwise provided in this contract, any dispute concerning a question of fact arising under this contract which is not disposed of by agreement shall be decided by the Contracting Officer, who shall reduce his decision to writing and mail or otherwise furnish a copy thereof to the Contractor. Within 30 days from the date of receipt of such copy, the Contractor may appeal by mailing or otherwise furnishing to the Contracting Officer a written appeal addressed to the head of the department, and the decision of the head of the department or his duly authorized representatives for the hearings of such appeals shall, unless determined by a court of competent jurisdiction to have been fraudulent, arbitrary, capricious, or so grossly erroneous as necessarily to imply bad faith, be final and conclusive: *Provided,* That, if no such appeal to the head of the department is taken, the decision of the Contracting Officer shall be final and conclusive. In connection with any appeal proceeding under this clause, the Contractor shall be afforded an opportunity to be heard and to offer evidence in support of its appeal. Pending final decision of a dispute hereunder, the Contractor shall proceed diligently with the performance of the contract and in accordance with the Contracting Officer's decision.

7. PAYMENTS TO CONTRACTORS

(*a*) Unless otherwise provided in the specifications, partial payments will be made as the work progresses at the end of each calendar month, or as soon thereafter as practicable, or at more frequent intervals as determined by the Contracting Officer, on estimates made and approved by the Contracting Officer. In preparing estimates the material delivered on the site and preparatory work done may be taken into consideration.

(*b*) In making such partial payments there shall be retained 10 per cent on the estimated amount until final completion and acceptance of all work covered by the contract: *Provided, however,* That the Contracting Officer, at any time after 50 per cent of the work has been completed, if he finds that satisfactory progress is being made, may make any of the remaining partial payments in full: *And provided further,* That on completion and acceptance of each separate building, public work, or other division of the contract, on which the price is stated separately in the contract, payment may be made in full, including retained percentage thereon, less authorized deductions.

(*c*) All material and work covered by partial payments made shall thereupon become the sole property of the Government, but this provision shall not be construed as relieving the Contractor from the sole responsibility for all materials and work upon which payments have been made or the restoration of any damaged work, or as a waiver of the right of the Government to require the fulfillment of all of the terms of the contract.

(*d*) Upon completion and acceptance of all work required hereunder, the amount due the Contractor under this contract will be paid upon the presentation of a properly executed and duly certified voucher therefor, after the Contractor shall have furnished the Government with a release, if required, of all claims against the Government arising under and by virtue of this contract, other than such claims, if any, as may be specifically excepted by the Contractor from the operation of the release in stated amounts to be set forth therein.   If the Contractor's claim to amounts payable under the contract has been assigned under the Assignment of Claims Act of 1940, as amended (41 U.S.C. 15), a release may also be required of the assignee at the option of the Contracting Officer.

8. MATERIALS AND WORKMANSHIP

Unless otherwise specifically provided for in the specifications, all equipment, materials, and articles incorporated in the work covered by this contract are to be new and of the most suitable grade of their respective kinds for the purpose and all workmanship shall be first class.   Where equipment, materials, or articles are referred to in the specifications as "equal to" any particular standard, the Contracting Officer shall decide the question of equality. The Contractor shall furnish to the Contracting Officer for his approval the name of the manufacturer of machinery, mechanical and other equipment which he contemplates incorporating in the work, together with their performance capacities and other pertinent information.   When required by the specifications, or when called for by the Contracting Officer, the Contractor shall furnish the Contracting Officer for approval full information concerning the materials or articles which he contemplates incorporating in the work. Samples of materials shall be submitted for approval when so directed. Machinery, equipment, materials, and articles installed or used without such approval shall be at the risk of subsequent rejection.   The Contracting Officer may in writing require the Contractor to remove from the work such employee as the Contracting Officer deems incompetent, careless, insubordinate, or otherwise objectionable, or whose continued employment on the work is deemed by the Contracting Officer to be contrary to the public interest.

9. INSPECTION

(*a*) Except as otherwise provided in paragraph (*d*) hereof all material and workmanship, if not otherwise designated by the specifications, shall be subject to inspection, examination, and test by the Contracting Officer at any and all times during manufacture and/or construction and at any and all places where such manufacture and/or construction are carried on.   The Government shall have the right to reject defective material and workmanship or require its correction.   Rejected workmanship shall be satisfactorily corrected and rejected material shall be satisfactorily replaced with proper material without charge therefor, and the Contractor shall promptly segregate and remove the rejected material from the premises.   If the Contractor fails to proceed at once with the replacement of rejected material and/or the correction of defective workmanship the Government may, by contract or otherwise, replace such material and/or correct such workmanship and charge the cost thereof to the Contractor, or may terminate the right of the Contractor to proceed as provided in Clause 5 of this contract, the Contractor and surety being liable for any damage to the same extent as provided in said Clause 5 for terminations thereunder.

(b) The Contractor shall furnish promptly without additional charge, all reasonable facilities, labor, and materials necessary for the safe and convenient inspection and test that may be required by the Contracting Officer. All inspection and tests by the Government shall be performed in such manner as not unnecessarily to delay the work. Special, full size, and performance tests shall be as described in the specifications. The Contractor shall be charged with any additional cost of inspection when material and workmanship are not ready at the time inspection is requested by the Contractor.

(c) Should it be considered necessary or advisable by the Government at any time before final acceptance of the entire work to make an examination of work already completed, by removing or tearing out same, the Contractor shall on request promptly furnish all necessary facilities, labor, and material. If such work is found to be defective or nonconforming in any material respect, due to fault of the Contractor or his subcontractors, he shall defray all the expenses of such examination and of satisfactory reconstruction. If, however, such work is found to meet the requirements of the contract, the actual direct cost of labor and material necessarily involved in the examination and replacement, plus 15 per cent, shall be allowed the Contractor, and he shall, in addition, if completion of the work has been delayed thereby, be granted a suitable extension of time on account of the additional work involved.

(d) Inspection of material and finished articles to be incorporated in the work at the site shall be made at the place of production, manufacture, or shipment, whenever the quantity justifies it, unless otherwise stated in the specifications; and such inspection and written or other formal acceptance, unless otherwise stated in the specifications, shall be final, except as regards latent defects, departures from specific requirements of the contract, damage or loss in transit, fraud, or such gross mistakes as amount to fraud. Subject to the requirements contained in the preceding sentence, the inspection of material and workmanship for final acceptance as a whole or in part shall be made at the site. Nothing contained in this paragraph (d) shall in any way restrict the Government's rights under any warranty or guarantee.

10. SUPERINTENDENCE BY CONTRACTOR

The Contractor shall give his personal superintendence to the work or have a competent foreman or superintendent, satisfactory to the Contracting Officer, on the work at all times during progress, with authority to act for him.

11. PERMITS AND RESPONSIBILITY FOR WORK, ETC.

The Contractor shall, without additional expense to the Government, obtain all licenses and permits required for the prosecution of the work. He shall be responsible for all damages to persons or property that occur as a result of his fault or negligence in connection with the prosecution of the work. He shall also be responsible for all materials delivered and work performed until completion and final acceptance, except for any completed unit thereof which theretofore may have been finally accepted.

12. OTHER CONTRACTS

The Government may undertake or award other contracts for additional work, and the Contractor shall fully cooperate with such other contractors and Government employees and carefully fit his own work to such additional work as may be directed by the Contracting Officer. The Contractor shall not commit or permit any act which will interfere with the performance of work by any other contractor or by Government employees.

13. PATENT INDEMNITY

Except as otherwise provided, the Contractor agrees to indemnify the Government and its officers, agents, and employees against liability, including costs and expenses, for infringement upon any Letters Patent of the United States (except Letters Patent issued upon an application which is now or may hereafter be, for reasons of national security, ordered by the Government to be kept secret or otherwise withheld from issue) arising out of the performance of this contract or out of the use or disposal by or for the account of the Government of supplies furnished or construction work performed hereunder.

14. ADDITIONAL BOND SECURITY

If any surety upon any bond furnished in connection with this contract becomes unacceptable to the Government, or if any such surety fails to furnish reports as to his financial condition from time to time as requested by the Government, the Contractor shall promptly furnish such additional security as may be required from time to time to protect the interests of the Government and of persons supplying labor or materials in the prosecution of the work contemplated by this contract.

15. COVENANT AGAINST CONTINGENT FEES

The Contractor warrants that no person or selling agency has been employed or retained to solicit or secure this contract upon an agreement or understanding for a commission, percentage, brokerage, or contingent fee, excepting bona fide employees or bona fide established commercial or selling agencies maintained by the Contractor for the purpose of securing business. For breach or violation of this warranty the Government shall have the right to annul this contract without liability or in its discretion to deduct from the contract price or consideration the full amount of such commission, percentage, brokerage, or contingent fee.

16. OFFICIALS NOT TO BENEFIT

No member of or Delegate to Congress, or Resident Commissioner, shall be admitted to any share or part of this contract, or to any benefit that may arise therefrom; but this provision shall not be construed to extend to this contract if made with a corporation for its general benefit.

17. BUY AMERICAN ACT

The Contractor agrees that in the performance of the work under this contract the Contractor, subcontractors, material men, and suppliers shall use only such unmanufactured articles, materials, and supplies (which term "articles, materials, and supplies" is hereinafter referred to in this clause as "Supplies") as have been mined or produced in the United States, and only such manufactured supplies as have been manufactured in the United States substantially all from supplies mined, produced, or manufactured, as the case may be, in the United States. Pursuant to the Buy American Act (41 U.S.C. 10a–d), the foregoing provisions shall not apply (i) with respect to supplies excepted by the head of the department from the application of that Act, (ii) with respect to supplies for use outside the United States, or (iii) with respect to the supplies to be used in the performance of work under this contract which are of a class or kind determined by the head of the department or his duly authorized representative not to be mined, produced, or manufactured, as the case may be, in the United States in sufficient and reasonably available commercial quantities and of a satisfactory quality, or (iv) with respect to such supplies, from which the supplies to be used in the performance of work under this contract are manufactured, as are of a class or kind determined by the head of the department or his duly authorized representative not to be mined, produced, or manufactured, as the case may be, in the United States in sufficient and reasonably available commercial quantities and of a satisfactory quality, provided that this exception (iv) shall not permit the use in the performance of work under this contract of supplies manufactured outside the United States if such supplies are manufactured in the United States in sufficient and reasonably available commercial quantities and of a satisfactory quality.

18. CONVICT LABOR

In connection with the performance of work under this contract, the Contractor agrees not to employ any person undergoing sentence of imprisonment at hard labor.

19. NONDISCRIMINATION IN EMPLOYMENT

In connection with the performance of work under this contract, the Contractor agrees not to discriminate against any employee or applicant for employment because of race, creed, color, or national origin; and further agrees to insert the foregoing provision in all subcontracts hereunder except subcontracts for standard commercial supplies or for raw materials.

20. DAVIS-BACON ACT (40 U.S.C. 276a–a(7))

(a) All mechanics and laborers employed or working directly upon the site of the work will be paid unconditionally and not less often than once a week, and without subsequent deduction or rebate on any account (except such payroll deductions as are permitted by the Copeland Act (Anti-Kickback) Regulations (29 CFR, Part 3)) the full amounts due at time of payment, computed at wage rates not less than those contained in the wage determination decision of the Secretary of Labor which is attached hereto and made a part hereof, regardless of any contractual relationship which may be alleged to exist between the Contractor or subcontractor and such laborers and mechanics; and a copy of the wage determination decision shall be kept posted by the Contractor at the site of the work in a prominent place where it can be easily seen by the workers.

(b) In the event it is found by the Contracting Officer that any laborer or mechanic employed by the Contractor or any subcontractor directly on the site of the work covered by this contract has been or is being paid at a rate of wages less than the rate of wages required by paragraph (*a*) of this clause, the Contracting Officer may (1) by written notice to the Government Prime Contractor terminate his right to proceed with the work, or such part of the work as to which there has been a failure to pay said required wages, and (2) prosecute the work to completion by contract or otherwise, whereupon such Contractor and his sureties shall be liable to the Government for any excess costs occasioned the Government thereby.

(*c*) Paragraphs (*a*) and (*b*) of this clause shall apply to this contract to the extent that it is (1) a prime contract with the Government subject to the Davis-Bacon Act or (2) a subcontract under such prime contract.

21. EIGHT-HOUR LAWS—OVERTIME COMPENSATION

No laborer or mechanic doing any part of the work contemplated by this contract, in the employ of the Contractor or any subcontractor contracting for any part of said work contemplated, shall be required or permitted to work more than eight hours in any one calendar day upon such work, except upon the condition that compensation is paid to such laborer or mechanic in accordance with the provisions of this clause. The wages of every laborer and mechanic employed by the Contractor or any subcontractor engaged in the performance of this contract shall be computed on a basic day rate of eight hours per day and work in excess of eight hours per day is permitted only upon the condition that every such laborer and mechanic shall be compensated for all hours worked in excess of eight hours per day at not less than one and one-half times the basic rate of pay. For each violation of the requirements of this clause a penalty of five dollars shall be imposed for each laborer or mechanic for every calendar day in which such employee is required or permitted to labor more than eight hours upon said work without receiving compensation computed in accordance with this clause, and all penalties thus imposed shall be withheld for the use and benefit of the Government: *Provided,* That this stipulation shall be subject in all respects to the exceptions and provisions of the Eight-Hour Laws as set forth in 40 U.S.C. 321, 324, 325, 325a, and 326, which relate to hours of labor and compensation for overtime.

22. APPRENTICES

Apprentices will be permitted to work only under a bona fide apprenticeship program registered with a State Apprenticeship Council which is recognized by the Federal Committee on Apprenticeship, U.S. Department of Labor; or if no such recognized Council exists in a State, under a program registered with the Bureau of Apprenticeship, U.S. Department of Labor.

23. PAYROLL RECORDS AND PAYROLLS

(*a*) Payroll records will be maintained during the course of the work and preserved for a period of three years thereafter for all laborers and mechanics working at the site of the work. Such records will contain the name and address of each such employee, his correct classification, rate of pay, daily and weekly number of hours worked, deductions made, and actual wages paid. The Contractor will make his employment records available for inspection by authorized representatives of the Contracting Officer and the U.S. Department of Labor, and will permit such representatives to interview employees during working hours on the job.

(*b*) A certified copy of all payrolls will be submitted weekly to the Contracting Officer. The Government Prime Contractor will be responsible for the submission of certified copies of the payrolls of all subcontractors. The certification will affirm that the payrolls are correct and complete, that the wage rates contained therein are not less than the applicable rates contained in the wage determination decision of the Secretary of Labor attached to this contract, and that the classifications set forth for each laborer or mechanic conform with the work he performed.

24. COPELAND (ANTI-KICKBACK) ACT—NONREBATE OF WAGES

The regulations of the Secretary of Labor applicable to Contractors and subcontractors (29 CFR, Part 3), made pursuant to the Copeland Act, as amended (40 U.S.C. 276c) and to aid in the enforcement of the Anti-Kickback Act (18 U.S.C. 874) are made a part of this contract by reference. The Contractor will comply with these regulations and any amendments or modifications thereof, and the Government Prime Contractor will be responsible for the submission of affidavits required of subcontractors thereunder. The foregoing shall apply except as the Secretary of Labor may specifically provide for reasonable limitations, variations, tolerances, and exemptions.

25. WITHHOLDING OF FUNDS TO ASSURE WAGE PAYMENT

There may be withheld from the Contractor so much of the accrued payments or advances as may be considered necessary to pay laborers and mechanics employed by the Contractor or any subcontractor the full amount of wages required by this contract. In the event of failure to pay any laborer or mechanic all or part of the wages required by this contract, the Contracting Officer may take such action as may be necessary to cause the suspension, until such violations have ceased, of any further payment, advance, or guarantee of funds to or for the Government Prime Contractor.

26. SUBCONTRACTS—TERMINATION

The Contractor agrees to insert Clauses 20 through 26 hereof in all subcontracts and further agrees that a breach of any of the requirements of these clauses may be grounds for termination of this contract. The term "Contractor" as used in such clauses in any subcontract shall be deemed to refer to the subcontractor except in the phrase "Government Prime Contractor."

THE STANDARD FORM OF CONSTRUCTION CONTRACT ISSUED BY THE AMERICAN RAILWAY ENGINEERING ASSOCIATION FOR LUMP-SUM AND UNIT-PRICE CONTRACTS
1947

THIS CONTRACT, made this .... day of ...., 19........, by and between .... a corporation organized and existing under the laws of the State of ...., hereinafter called the .... Company, and ...., hereinafter called the Contractor.

WITNESSETH: That, in consideration of the covenants and agreements herein contained, to be performed by the parties hereto, and of the payments hereinafter agreed to be made, it is mutually agreed as follows:

The Contractor shall furnish all of the materials, superintendence, labor, tools, equipment and transportation, except as hereinafter specified, and shall execute, construct and finish, in an expeditious, substantial and workmanlike manner, to the satisfaction and acceptance of the Chief Engineer of the Company, all of the work required for ...., in accordance with the plans identified by the signatures of the Contractor and the Chief Engineer, dated ...., 19........, and described ...., and the Specifications hereto attached, all of which are made a part of this contract.

The work covered by this contract shall be commenced on or before the .... day of ...., 19........, and shall be completed on or before the .... day of ...., 19.........

[For those desiring a liquidating damage clause the following reading is suggested for insertion at this point, as given in Note 1.]

[For those who believe that a so-called "bonus" cause not only strengthens the "liquidated damages" clause, but is also an urge to the Contractor for early completion, as well as being fair and equitable, the following reading is suggested at this point, as given in Note 2.]

And in consideration of the completion of the work described herein, and the fulfillment of all stipulations of this contract to the satisfaction and acceptance of the Chief Engineer of the Company, the Company shall pay, or cause to be paid, to the Contractor, the amount due to the Contractor, based on the following prices:

[Here insert a schedule of Items and Prices.]

1. TERMS OF EMPLOYMENT

The Company reserves no control whatsoever over the employment, discharge, compensation of or services rendered by the Contractor's employees, and it is the intention of the parties to this agreement that the Contractor shall be and remain an Independent Contractor, and that nothing in this agreement contained shall be construed as inconsistent with that status. The Contractor agrees to pay the contributions measured by the wages of his (its) employees required to be made under the Unemployment Compensation Insurance, Social Security and Retirement Laws or similar laws, State and Federal, applicable to the work hereunder undertaken by the Contractor or his (its) subcontractors, and to accept exclusive liability for said contributions; the Contractor further promises and agrees to indemnify and hold harmless the Company, its successors and assigns, from any and all liability arising therefrom.

2. INTENT OF PLANS AND SPECIFICATIONS

All work that may be called for in the specifications and not shown on the plans, or shown on the plans and not called for in the specifications, shall be executed and furnished by the Contractor as if described in both these ways; and should any work or material be required which is not detailed in the specifications or plans, either directly or indirectly but which is nevertheless necessary for the proper carrying out of the intent thereof, the Contractor is to understand the same to be implied and required, and shall perform all such work and furnish any such material as fully as if they were particularly delineated or described.

3. CONTRACTOR'S UNDERSTANDING

It is understood and agreed that the Contractor has, by careful examination, satisfied himself as to the nature and location of the work, the conformation of the ground, the character, quality and quantity of the materials to be encountered, the character of equipment and facilities needed preliminary to and during the prosecution of the work, the general and local conditions, and all other matters which can in any way affect the work under this contract. No verbal agreement or conversation with any officer, agent or employee of the Company, either before or after the execution of this contract, shall affect or modify any of the terms or obligations herein contained.

4. LAND OF COMPANY, USE OF, BY CONTRACTOR

The Company shall provide the land upon which the work under this contract is to be done, and will, so far as it can conveniently do so, permit the Contractor to use so much of its land as is required for the erection of temporary construction facilities and storage of materials, together with the right of access to same, but beyond this, the Contractor shall provide, at his cost and expense, any additional land required.

5. CONSENT TO TRANSFER

The Contractor shall not let, transfer or assign this contract as a whole, without the consent in writing from the Company. Parts or portions of the work and the furnishing of and delivery of materials may be subcontracted with the consent of the Chief Engineer, but such consent does not relieve the Contractor from any of his obligations and liabilities under the contract.

6. RISK

The work under this contract in every respect shall be at the risk of the Contractor until finished and accepted, except damage or injury caused directly by Company's agents or employees.

7. ENGINEER AND CHIEF ENGINEER

Wherever in this contract the word Engineer is used, it shall be understood as referring to the Chief Engineer of the Company, acting personally or through an assistant duly authorized in writing for such act by the Chief Engineer, and wherever the words Chief Engineer are used it shall be understood as referring to the Chief Engineer in person, and not to any assistant engineer.

8. WAIVER

It is expressly understood and agreed that any waiver on the part of the Company or the Engineer, of any term, provision or covenant of this contract, shall not constitute a precedent, nor bind the Company or the Engineer, to a waiver of any succeeding breach of the same or any other of the terms, provisions or covenants of this contract.

9. ADJUSTMENT OF DISPUTE

All questions or controversies which may arise between the Contractor and the Company, under or in reference to this contract, shall be subject to the decision of the Chief Engineer, and his decision shall be final and conclusive upon both parties.

10. BOND

The Contractor unless notified to the contrary, shall, at the time of the execution and delivery of this contract, and before the taking effect of the same in other respects, furnish and deliver to the Company a written bond of indemnity to the amount of .... dollars in form and substance and with surety thereon satisfactory and acceptable to the Company, to insure the faithful performance by the Contractor of all the covenants and agreements on the part of the Contractor contained in this contract.

This bond shall remain in force and effect for the full amount or such smaller sum as may at any time be specified by the Chief Engineer.

11. PERMITS AND INSURANCE

Before doing any work under this contract the Contractor at his (its) own expense shall secure and keep in effect until the entire work is completed and accepted the following items:

(a) All permits, licenses and authorizations of a temporary nature required by Federal, State, County or Municipal authorities and assume payment of salaries and expenses of city and other public inspectors, policemen or officers which may be required in connection therewith. (Permits for permanent structures will be secured and paid for by the Company.)

(b) Fire insurance in the name of and payable to the Company for the benefit of the Contractor or the Company as the Chief Engineer shall find their interests to appear. Such policies to cover such structures or structural material and supplies on hand and subject to damage as a result of fire, in such amount as necessary to cover their value.

(c) Workmen's Compensation insurance covering employees of the Contractor or any subcontractor, as required by the State or States in which the work is to be performed, so that the Company shall be duly protected from any liability or claims for damages for personal injury, including death, which may arise while engaged upon the work covered by this contract. The Contractor shall submit to the Chief Engineer evidence by certificate that such insurance in satisfactory amount is in force.

(d) Public Liability Insurance Standard, Intermediate or Full Coverage, whichever form of coverage the Chief Engineer designates, shall be furnished in amounts of $...................., $...................., and Property Damage insurance in amount of $.....................

(e) Contractor's Contingent Public Liability and Property Damage insurance, if there are one or more subcontractors, in the same amounts as required in (d) above.

Policies required in (d) and (e) shall be endorsed to cover the collapse and explosion hazard in addition to all other hazards unless endorsement be waived by the Chief Engineer, and shall expressly provide that if during the course of this work watchmen, flagmen or other employees of the Company are loaned or assigned to the Contractor to perform work in connection with the work under this contract, such employees, while so engaged, shall be considered employees of the Contractor for the purpose of this insurance.

All the policies must be written by reliable and well-rated companies acceptable to the Chief Engineer. The policies shall be so written that they protect both the Contractor and the Company against any action which may be instituted against either of them. Certified copies of all policies shall be submitted to the Chief Engineer for approval. When approved they shall be retained by the Company and the Contractor notified of their approval.

12. INDEMNITY

The Contractor shall indemnify and save harmless the Company from and against all losses and all claims, demands, payments, suits, actions, recoveries and judgments of every nature and description made, brought or recovered against the Company by reason of any act or omission of the Contractor, his agents or employees, in the execution of the work or in guarding the same.

In case no bond is furnished the Company may require indemnity insurance in amount, form and substance, satisfactory and acceptable to the Company, which insurance shall provide for the protection of the Company against failure of the Contractor to comply with the conditions of this covenant. The Contractor shall take out and pay the premiums on such insurance.

13. SUPERINTENDENCE

The Contractor shall constantly superintend all of the work embraced in this contract, in person or by a duly authorized representative acceptable to the Company.

14. NOTICE—HOW SERVED

Any notice to be given by the Company to the Contractor under this contract shall be deemed to be served if the same be delivered to the person in charge of the office used by the Contractor, or to his representative at or near the work, or deposited in the postoffice, postpaid, addressed to the Contractor at his last known place of business.

15. PROTECTION

Whenever the local conditions, laws or ordinances require, the Contractor shall furnish and maintain, at his own cost and expense, necessary passageways, guard fences and lights and such other facilities and means of protection as may be required.

16. TIMELY DEMAND FOR POINTS AND INSTRUCTIONS

The Contractor shall provide reasonable and necessary opportunities and facilities for setting points and making measurements. He shall not proceed until he has made timely demand upon the Engineer for, and has received from him, such points and instructions as may be necessary as the work progresses. The work shall be done in strict conformity with such points and instructions.

17. PRESERVATION OF STAKES

The Contractor shall carefully preserve bench marks, reference points and stakes, and in case of wilful or careless destruction, he will be charged with the resulting expense and shall be responsible for any mistakes that may be caused by their unnecessary loss or disturbance.

18. REPORT ERRORS AND DISCREPANCIES

If the Contractor, in the course of the work, finds any discrepancy between the plans and the physical conditions of the locality, or any errors or omissions in plans or in the layout as given by points and instructions, it shall be his duty to immediately inform the Engineer, in writing, and the Engineer shall promptly verify the same. Any work done after such discovery, until authorized, will be done at the Contractor's risk.

19. INSPECTION

All work and material shall be at all times open to the inspection, acceptance or rejection of the Engineer or his authorized representative. The Contractor shall give the Engineer reasonable notice of starting any new work and shall provide reasonable and necessary facilities for inspection even to the extent of taking out portions of finished work; in case the work is found satisfactory, the cost of taking out and replacement shall be paid by the Company. No work shall be done at night without the previous approval of the Engineer.

20. DEFECTIVE WORK OR MATERIAL

The Engineer shall have power to reject or condemn all work or material which does not conform to this contract and any omission or failure on the part of the Engineer to disapprove or reject any defective work or material shall not be construed to be an acceptance thereof.

The Contractor shall remove, at his own expense, any work or material condemned by the Engineer, and shall rebuild and replace the same without extra charge, and in default thereof the same may be done by the Company at the Contractor's expense, or in case the Chief Engineer shall not consider the defect of sufficient importance to require the Contractor to rebuild or replace any imperfect work or material, he shall have power and is hereby authorized to make an equitable deduction from the stipulated price established under the terms of this agreement.

21. CHANGE OF FACILITIES OF OTHER UTILITIES

If in the conduct of the contemplated work any temporary changes or alterations in water, oil or gas pipe lines, sewers, drains, conduits, fences, trolley tracks, electric line or power lines, telephone or telegraph or other wires, poles, etc., are necessary, either for the convenience of the Contractor or for the performance of the work, the responsibility for making such changes will rest with the Contractor unless otherwise provided elsewhere in this agreement; and he shall arrange for such changes to be made at his own expense.

If such changes are of a permanent character and made necessary by the improvement itself and not incident to the performance of the work, then in that case such changes will be arranged for by the Company or others without cost to the Contractor.

22. WORK ADJACENT TO RAILWAY OR OTHER PROPERTY

Wherever the work embraced in this contract is near the tracks, structures or buildings of the Company or of other railways, or persons, the Contractor shall use proper care and vigilance to avoid injury to persons or property. The work shall be so conducted as not to interfere with the movement of trains or other operations of the railway; or, if in any case such interference be necessary, the Contractor shall not proceed until he has first obtained specific authority and directions therefor from the proper designated officer of the Company and has the approval of the Engineer.

23. RIGHTS OF VARIOUS INTERESTS

Wherever work being done by Company forces or by other contractors is contiguous to work covered by this contract, the respective rights of the various interests involved shall be established by the Engineer, to secure the completion of the various portions of the work in general harmony.

24. CONTRACTOR NOT TO HIRE COMPANY'S EMPLOYEES

The Contractor shall not employ or hire any of the Company's employees without the permission of the Engineer.

25. ORDER OF COMPLETION; USE OF COMPLETED PORTIONS

The Contractor shall complete any portion or portions of the work in such order of time as the Engineer may require. The Company shall have the right to take possession of and use any completed or partially completed portions of the work, notwithstanding the time for completing the entire work or such portions may not have expired; but such taking possession and use shall not be deemed an acceptance of the work so taken or used or any part thereof. If such prior use increases the cost of or delays the work, the Contractor shall be entitled to such extra compensation, or extension of time, or both, as the Chief Engineer may determine.

26. CHANGES

The Company shall have the right to make any changes that may be hereafter determined upon, in the nature or dimensions of the work, either before or after its commencement, and such changes shall in no way affect or void the obligations of this contract. If such changes make any change in the cost of the work, an equitable adjustment shall be made by the Chief Engineer to cover the same, but the Contractor shall not claim compensation for anticipated profits.

27. EXTRA WORK

(a) No bill or claim for extra work or material shall be allowed or paid unless the doing of such extra work or the furnishing of such extra material shall have been authorized in writing by the .... Engineer.

(b) The price for such work shall be determined by the Chief Engineer, who may either fix a unit price or a lump sum price, or may, if he so elects, provide that the price shall be determined by the actual cost, to which shall be added .... percent to cover general expenses and superintendence, profits, contingencies, use of tools, Contractor's risk and liability. If the Contractor shall perform any work or furnish any material which is not provided in this contract, or which was not authorized in writing by the Engineer, said Contractor shall receive no compensation for such work or materials so furnished, and does hereby release and discharge the Company from any liability therefor.

(c) If the Contractor shall proceed with such extra work or the furnishing of such extra material after receiving the written authority therefor, as hereinbefore provided, then such work or material, stated in the written authority of the Engineer, shall be covered, governed and controlled by all the terms and provisions of this contract, subject to such prices as may be agreed upon or fixed by the Chief Engineer.

(d) If the Contractor shall decline or fail to perform such work or furnish such extra material as authorized by the Engineer in writing, as aforesaid, the Company may then arrange for the performance of the work in any manner it may see fit, the same as if this contract had not been executed, and the Contractor shall not interfere with such performance of the work.

28. UNAVOIDABLE DELAYS: EXTENSION OF TIME ON PARTS OF WORK

If the Contractor shall be delayed in the performance of the work from any cause beyond his control, he may, upon written application to the Chief Engineer within three days of such delay, be granted such extension of time as the Chief Engineer shall deem equitable and just.

29. SUSPENSION OF WORK

The Company may at any time stop the work, or any part thereof, by giving .... days' notice to the Contractor in writing. The work shall be resumed by the Contractor within ten days after the date fixed in the written notice from the Company to the Contractor so to do. The Company shall not be held liable for any damages or anticipated profits on account of the work being stopped, or for any work done during the interval of suspension. It will, however, pay the Contractor for expense of men and teams necessarily retained during the interval of suspension; provided the Contractor can show that it was not reasonably practical to move these men and teams to other points at which they could have been employed. The Company will further pay the Contractor for time necessarily lost during such suspension at the rate of .... percent per annum on the estimated value of materials, equipment and fixtures furnished by the Contractor on the work which is necessarily idle during such suspension, said rate of .... percent per annum being understood to include depreciation, interest and insurance. But if the work, or any part thereof, shall be stopped by the notice in writing aforesaid, and if the Company does not give notice in writing to the Contractor to resume work at a date within .... of the date fixed in the written notice to suspend, then the Contractor may abandon that portion of the work so suspended and he will be entitled to the estimates and payments for work done on such portion so abandoned, as provided in Section 38 of this contract.

30. FAILURE OF PERFORMANCE BY CONTRACTOR

(a) If the Chief Engineer of the Company shall at any time be of the opinion that the Contractor is neglecting to remedy any imperfection in the work, or is not progressing with the work as fast as necessary to insure its completion within the time and as required by the contract, or is otherwise violating any of the provisions of this contract, the Chief Engineer, in behalf of the Company, shall have the power, and it shall be his duty to notify the Contractor in writing to remedy such imperfections, proceed more rapidly with said work, or otherwise comply with the provisions of this contract.

(b) If on the expiration of ten days after the serving of such written notice upon the Contractor, the Contractor shall continue to neglect the work and shall fail to satisfy the Engineer of his efforts, ability and intentions, to remedy the specified deficiencies, the Company may terminate the employment of the Contractor and may take possession of the work and of all materials, tools and appliances thereon, and employ such means as may be, in the Engineer's judgment, necessary to finish the work. In such case the Contractor shall receive no further payment until the work shall be finished, when, if the unpaid balance that would be due under this contract exceeds the cost to the Company of finishing the work, such excess shall be paid to the Contractor; but if such cost exceeds such unpaid balance, the Contractor shall pay the difference to the Company.

(c) Upon failure of the Contractor to comply with any notice given in accordance with the provisions thereof, the Company shall have the alternative right, instead of assuming charge of the entire work, to place additional forces, tools, equipment and materials on parts of the work for the purpose of carrying on such parts of the work, and the Contractor shall be allowed therefor the contract price. The Company may retain the amount of the cost of such work, with .... percent added, from any sum or sums due or to become due the Contractor under this contract.

31. ANNULMENT WITHOUT FAULT OF CONTRACTOR

The Company shall have the right at any time, for reasons which appear good to it, to annul this contract upon giving notice in writing to the Contractor, in which event the Contractor shall be entitled to the full amount of the estimate for the work done by him under this contract up to the time of such annulment, including the retained percentage. The Contractor shall be reimbursed by the Company for such expenditures as in the judgment of the Chief Engineer are not otherwise compensated for, and as are required in preparing for and moving to and from the work; the intent being that an equitable settlement shall be made with the Contractor.

32. REMOVAL OF EQUIPMENT

In case of annulment of this contract before completion from any cause whatever, the Contractor, if notified to do so by the Company, shall promptly remove any part or all of his equipment and supplies from the property of the Company, failing which the Company shall have the right to move such equipment and supplies at the expense of the Contractor.

33. FAILURE TO MAKE PAYMENTS

Failure by the Company to make payments at the times provided by this contract shall give the Contractor the right to suspend work until payment is made, or at his option, after thirty days' notice in writing, should the Company continue to default, to terminate this contract and recover the price of all work done and materials provided and all damages sustained, and such failure to make payments at the time provided shall be a bar to any claim by the Company against the Contractor for delay in completion of the work, due to such suspension or failure to pay.

34. LIENS

If at any time there shall be evidence of any lien or claim for which the Company might become liable, and which is chargeable to the Contractor, the Company shall have the right to retain out of any payment then due or thereafter to become due, an amount sufficient to completely indemnify the Company against such lien or claim, and if such lien or claim be valid, the Company may pay and discharge the same, and deduct the amount so paid from any moneys which may be or become due and payable to the Contractor.

35. MONTHLY ESTIMATE

So long as the work herein contracted for is prosecuted in accordance with the provisions of this contract, and with such progress as may be satisfactory to the Chief Engineer, the said Chief Engineer will, on or about the first day of each month, make an approximate estimate of the proportionate value of the work done and of acceptable material furnished or delivered upon the Company's property at the site of the work, up to and including the last day of the previous month. The amount of said estimate after deducting .... percent and all previous payments, shall be due and payable to the Contractor at the office of the Treasurer of the Company on or before the .... day of the current month.

It is understood and agreed that the monthly estimates and certificates on unfinished work shall, in no case, be taken as an acceptance of the work, or a release of the Contractor from responsibility therefor, and that in computing the final estimate the Chief Engineer need not be bound by the preceding estimates and certificates.

36. CLEANING UP

The Contractor shall, as directed by the Engineer, remove from the Company's property and from all public and private property, at his own expense, all temporary structures, rubbish and waste materials, resulting from his operations.

37. ACCEPTANCE

The work shall be inspected for acceptance by the Company promptly upon receipt of notice in writing that the work is ready for such inspection.

38. FINAL ESTIMATE

Upon the completion and acceptance of the work, the Chief Engineer shall issue a final estimate over his signature, covering work provided for in this contract, completed and accepted by him, under the terms and conditions thereof, whereupon the balance found to be due the Contractor, including the retained percentage, shall be paid to the Contractor at the office of the Treasurer of the Company within .... days after the date of said final estimate, provided that, before the payment of said final estimate, the Contractor shall submit evidence satisfactory to the Chief Engineer that all payrolls, material bills and outstanding indebtedness in connection with the work have been paid.

THIS CONTRACT shall inure to the benefit of and be binding upon the legal representatives and successors of the parties respectively.

IN WITNESS WHEREOF, the parties hereto have executed this contract, in ...., the day and year first above written.

.... Company,

Attest: By ....

....
Secretary.

....
Contractor.

Witness: By ....

....

Note 1. And, whereas, time being the essence of this contract and it is agreed that the Company would suffer loss by the failure of the Contractor to have said work completed in all its parts on said day, and as it might be difficult and expensive to accurately compute the amount of such loss; in order to avoid such expense and difficulty, the Contractor hereby expressly covenants and agrees to pay to the Company the sum of .... Dollars per day for each and every day, Sundays and legal holidays only excepted, after said .... day of ...., 19........, during or upon which the said work or any part thereof remains incomplete and unfinished, not as a penalty, but as the liquidated actual losses which the Company will suffer on account of any failure on the part of the Contractor to have said work completed in all of its parts on said day; and that any sum which may be due the Company for such losses may be deducted and retained by the Company from any balance which may be due the Contractor when the said work shall have been finished and accepted as hereinafter provided. It is, however, agreed that in case any failure to complete the said work or some part thereof on said day shall be due to any cause beyond the Contractor's control and for which the Chief Engineer has granted an extension of time, such delays form no part of the number of days for each and every one of which the Contractor is to pay the Company the sum of .... Dollars as as herein specified.

Note 2. If the work is completed by the Contractor prior to the date herein given, then the Company agrees to pay the Contractor an equal amount as given for each day said completed work is available for use by the Company as determined by its Engineer.

THE STANDARD FORM OF AGREEMENT BETWEEN CONTRACTOR AND OWNER FOR CONSTRUCTION OF BUILDINGS

ISSUED BY THE AMERICAN INSTITUTE OF ARCHITECTS FOR USE WHEN A STIPULATED SUM FORMS THE BASIS OF PAYMENT

This form of agreement, fifth edition, has received the approval of the associated general contractors of America; the Contracting Plasterers' International Association; the Heating, Piping and Air Conditioning Contractors National Association; the National Building Granite Quarries Association, Inc.; the National Electrical Contractors Association; the Painting and Decorating Contractors of America, and the Producers' Council, Inc.

Copyright 1915–1918–1925–1937 by the American Institute of Architects, the Octagon, Washington, D. C. Reproduction of the material herein or substantial quotation of its provisions without permission of the American Institute of Architects violates the copyright laws of the United States and will be subject to legal prosecution.

This form is to be used only with the standard general conditions of the contract for construction of buildings.

THIS AGREEMENT made the .... day of .... in the

year Nineteen Hundred and .... by and between ....

.... hereinafter called the Contractor, and ....

.... hereinafter called the Owner, WITNESSETH, that the Contractor and the Owner for the considerations hereinafter named agree as follows:

ARTICLE 1. *Scope of the Work.* The Contractor shall furnish all of the materials and perform all of the work shown on the Drawings and described in the Specifications entitled

----

[Here insert the caption descriptive of the work as used on the Drawings and in the other Contract Documents.]

prepared by ....
acting as and in these Contract Documents entitled the Architect; and shall do everything required by this Agreement, the General Conditions of the Contract, the Specifications and the Drawings.

ARTICLE 2. *Time of Completion.* The work to be performed under this Contract shall be commenced .... and shall be substantially

Reprinted with permission of The American Institute of Architects.

completed ....

[Here insert stipulation as to liquidated damages, if any.]

ARTICLE 3. *The Contract Sum.* The Owner shall pay the Contractor for the performance of the Contract, subject to additions and deductions provided therein, in current funds as follows: ....

[State here the lump sum amount, unit prices, or both, as desired in individual cases.]

Where the quantities originally contemplated are so changed that application of the agreed unit price to the quantity of work performed is shown to create a hardship to the Owner or the Contractor, there shall be an equitable adjustment of the Contract to prevent such hardship.

ARTICLE 4. *Progress Payments.* The Owner shall make payments on account of the Contract as provided therein, as follows:

On or about the .... day of each month .... per cent of the value, based on the Contract prices, of labor and materials incorporated in the work and of materials suitably stored at the site thereof up to the .... day of that month, as estimated by the Architect, less the aggregate of previous payments; and upon substantial completion of the entire work, a sum sufficient to increase the total payments to .... per cent of the Contract price ....

[Insert here any provision made for limiting or reducing the amount retained after the work reaches a certain stage of completion.]

ARTICLE 5. *Acceptance and Final Payment.* Final payment shall be due

.... days after substantial completion of the work provided the work be then fully completed and the Contract fully performed.

Upon receipt of written notice that the work is ready for final inspection and acceptance, the Architect shall promptly make such inspection, and when he finds the work acceptable under the Contract and the Contract fully performed he shall promptly issue a final certificate, over his own signature, stating that the work provided for in this Contract has been completed and is accepted by him under the terms and conditions thereof, and that the entire balance found to be due the Contractor, and noted in said final certificate, is due and payable.

Before issuance of final certificate the Contractor shall submit evidence satisfactory to the Architect that all payrolls, material bills, and other indebtedness connected with the work have been paid.

If after the work has been substantially completed, full completion thereof is materially delayed through no fault of the Contractor, and the Architect so certifies, the Owner shall, upon certificate of the Architect, and without terminating the Contract, make payment of the balance due for that portion of the work fully completed and accepted. Such payment shall be made under the terms and conditions governing final payment, except that it shall not constitute a waiver of claims.

ARTICLE 6. *The Contract Documents.* The General Conditions of the Contract, the Specifications and the Drawings, together with this Agreement, form the Contract, and they are as fully a part of the Contract as if hereto attached or herein repeated. The following is an enumeration of the Specifications and Drawings:

IN WITNESS WHEREOF the parties hereto have executed this Agreement, the day and year first above written.

[Signatures and Affidavits]

THE GENERAL CONDITIONS OF THE CONTRACT FOR THE CONSTRUCTION OF BUILDINGS

STANDARD FORM OF THE AMERICAN INSTITUTE OF ARCHITECTS

Reprinted with permission of The American Institute of Architects.

INDEX TO THE ARTICLES OF THE GENERAL CONDITIONS

ART. 1. *Definitions.*

(*a*) The Contract Documents consist of the Agreement, the General Conditions of the Contract, the Drawings and Specifications, including all modifications thereof incorporated in the documents before their execution. These form the Contract.

(*b*) The Owner, the Contractor, and the Architect are those mentioned as such in the Agreement. They are treated throughout the Contract Documents as if each were of the singular number and masculine gender.

(*c*) The term Subcontractor, as employed herein, includes only those having a direct contract with the Contractor and it includes one who furnishes material worked to a special design according to the plans or specifications of this work, but does not include one who merely furnishes material not so worked.

(*d*) Written notice shall be deemed to have been duly served if delivered in person to the individual or to a member of the firm or to an officer of the corporation for whom it is intended, or if delivered at or sent by registered mail to the last business address known to him who gives the notice.

(*e*) The term "work" of the Contractor or Subcontractor includes labor or materials or both.

(*f*) All time limits stated in the Contract Documents are of the essence of the Contract.

(*g*) The law of the place of building shall govern the construction of this Contract.

ART. 2. *Execution, Correlation, and Intent of Documents.* The Contract Documents shall be signed in duplicate by the Owner and the Contractor. In case the Owner and the Contractor fail to sign the General Conditions, Drawings, or Specifications, the Architect shall identify them.

The Contract Documents are complementary, and what is called for by any one shall be as binding as if called for by all. The intention of the documents is to include all labor and materials, equipment, and transportation necessary for the proper execution of the work. It is not intended, however, that materials or work not covered by or properly inferable from any heading, branch, class, or trade of the specifications shall be supplied unless distinctly so noted on the drawings. Materials or work described in words which so applied have a well-known technical or trade meaning shall be held to refer to such recognized standards.

ART. 3. *Detail Drawings and Instructions.* The Architect shall furnish, with reasonable promptness, additional instructions, by means of drawings or otherwise, necessary for the proper execution of the work. All such drawings and instructions shall be consistent with the Contract Documents, true developments thereof, and reasonably inferable therefrom.

The work shall be executed in conformity therewith and the Contractor shall do no work without proper drawings and instructions.

The Contractor and the Architect, if either so requests, shall jointly prepare a schedule, subject to change from time to time in accordance with the progress of the work, fixing the dates at which the various detail drawings will be required, and the Architect shall furnish them in accordance with that schedule. Under like conditions, a schedule shall be prepared, fixing the dates for the submission of shop drawings, for the beginning of manufacture and installation of materials, and for the completion of the various parts of the work.

ART. 4. *Copies Furnished.* Unless otherwise provided in the Contract Documents the Architect will furnish to the Contractor, free of charge, all copies of drawings and specifications reasonably necessary for the execution of the work.

ART. 5. *Shop Drawings.* The Contractor shall submit, with such promptness as to cause no delay in his own work or in that of any other Contractor, two copies of all shop or setting drawings and schedules required for the work of the various trades, and the Architect shall pass upon them with reasonable promptness, making desired corrections, including all necessary corrections relating to artistic effect. The Contractor shall make any corrections required by the Architect, file with him two corrected copies, and furnish such other copies as may be needed. The Architect's approval of such drawings or schedules shall not relieve the Contractor from responsibility for deviations from drawings or specifications, unless he has in writing called the Architect's attention to such deviations at the time of submission, nor shall it relieve him from responsibility for errors of any sort in shop drawings or schedules.

ART. 6. *Drawings and Specifications on the Work.* The Contractor shall keep one copy of all drawings and specifications on the work, in good order, available to the Architect and to his representatives.

ART. 7. *Ownership of Drawings and Models.* All drawings, specifications, and copies thereof furnished by the Architect are his property. They are not to be used on other work and, with the exception of the signed Contract set, are to be returned to him on request, at the completion of the work. All models are the property of the Owner.

ART. 8. *Samples.* The Contractor shall furnish for approval all samples as directed. The work shall be in accordance with approved samples.

ART. 9. *Materials, Appliances, Employees.* Unless otherwise stipulated, the Contractor shall provide and pay for all materials, labor, water, tools, equipment, light, power, transportation, and other facilities necessary for the execution and completion of the work.

Unless otherwise specified, all materials shall be new and both workmanship and materials shall be of good quality. The Contractor shall, if required, furnish satisfactory evidence as to the kind and quality of materials.

The Contractor shall at all times enforce strict discipline and good order among his employees, and shall not employ on the work any unfit person or anyone not skilled in the work assigned to him.

ART. 10. *Royalties and Patents.* The Contractor shall pay all royalties and license fees. He shall defend all suits or claims for infringement of any patent rights and shall save the Owner harmless from loss on account thereof, except that the Owner shall be responsible for all such loss when a particular process or the product of a particular manufacturer or manufacturers is specified, but if the Contractor has information that the process or article specified is an infringement of a patent he shall be responsible for such loss unless he promptly gives such information to the Architect or Owner.

ART. 11. *Surveys, Permits, and Regulations.* The Owner shall furnish all surveys unless otherwise specified. Permits and licenses of a temporary nature necessary for the prosecution of the work shall be secured and paid for by the Contractor. Permits, licenses, and easements for permanent structures or permanent changes in existing facilities shall be secured and paid for by the Owner, unless otherwise specified.

The Contractor shall give all notices and comply with all laws, ordinances, rules, and regulations bearing on the conduct of the work as drawn and specified. If the Contractor observes that the drawings and specifications are at variance therewith, he shall promptly notify the Architect in writing, and any necessary changes shall be adjusted as provided in the Contract for changes in the work. If the Contractor performs any work knowing it to be contrary to such laws, ordinances, rules, and regulations, and without such notice to the Architect, he shall bear all costs arising therefrom.

ART. 12. *Protection of Work and Property.* The Contractor shall continuously maintain adequate protection of all his work from damage and shall protect the Owner's property from injury or loss arising in connection with this Contract. He shall make good any such damage, injury, or loss, except such as may be directly due to errors in the Contract Documents or caused by agents or employees of the Owner. He shall adequately protect adjacent property as provided by law and the Contract Documents.

The Contractor shall take all necessary precautions for the safety of employees on the work, and shall comply with all applicable provisions of Federal, State, and Municipal safety laws and building codes to prevent accidents or injury to person on, about, or adjacent to the premises where the work is being performed. He shall erect and properly maintain at all times, as required by the conditions and progress of the work, all necessary safeguards for the protection of workmen and the public and shall post danger signs warning against the hazards created by such features of construction as protruding nails, hod hoists, well holes, elevator hatchways, scaffolding, window openings, stairways, and falling materials; and he shall designate a responsible member of his organization on the work, whose duty shall be the prevention of accidents. The name and position of the person so designated shall be reported to the Architect by the Contractor.

In an emergency affecting the safety of life or of the work or of adjoining property, the Contractor, without special instruction or authorization from the Architect or Owner, is hereby permitted to act, at his discretion, to prevent such threatened loss or injury, and he shall so act, without appeal, if so instructed or authorized. Any compensation, claimed by the Contractor on account of emergency work, shall be determined by agreement or Arbitration.

Art. 13. *Inspection of Work.* The Architect and his representatives shall at all times have access to the work wherever it is in preparation or progress, and the Contractor shall provide proper facilities for such access and for inspection.

If the specifications, the Architect's instructions, laws, ordinances, or any public authority require any work to be specially tested or approved, the Contractor shall give the Architect timely notice of its readiness for inspection, and if the inspection is by another authority than the Architect, of the date fixed for such inspection. Inspections by the Architect shall be promptly made, and where practicable at the source of supply. If any work should be covered up without approval or consent of the Architect, it must, if required by the Architect, be uncovered for examination at the Contractor's expense.

Re-examination of questioned work may be ordered by the Architect, and if so ordered the work must be uncovered by the Contractor. If such work be found in accordance with the Contract Documents the Owner shall pay the cost of re-examination and replacement. If such work be found not in accordance with the Contract Documents the Contractor shall pay such cost, unless he shall show that the defect in the work was caused by another Contractor, and in that event the Owner shall pay such cost.

Art. 14. *Superintendence: Supervision.* The Contractor shall keep on his work, during its progress, a competent superintendent and any necessary assistants, all satisfactory to the Architect. The superintendent shall not be changed except with the consent of the Architect, unless the superintendent proves to be unsatisfactory to the Contractor and ceases to be in his employ. The superintendent shall represent the Contractor in his absence and all directions given to him shall be as binding as if given to the Contractor. Important directions shall be confirmed in writing to the Contractor. Other directions shall be so confirmed on written request in each case.

The Contractor shall give efficient supervision to the work, using his best skill and attention. He shall carefully study and compare all drawings, specifications, and other instructions and shall at once report to the Architect any error, inconsistency, or omission which he may discover, but he shall not be held responsible for their existence or discovery.

Art. 15. *Changes in the Work.* The Owner, without invalidating the Contract, may order extra work or make changes by altering, adding to, or deducting from the work, the Contract Sum being adjusted accordingly. All such work shall be executed under the conditions of the original contract except that any claim for extension of time caused thereby shall be adjusted at the time of ordering such change.

In giving instructions, the Architect shall have authority to make minor changes in the work, not involving extra cost, and not inconsistent with the purposes of the building, but otherwise, except in an emergency endangering life or property, no extra work or change shall be made unless in pursuance of a written order from the Owner signed or countersigned by the Architect, or a written order from the Architect stating that the Owner has authorized the extra work or change, and no claim for an addition to the contract sum shall be valid unless so ordered.

The value of any such extra work or change shall be determined in one or more of the following ways:

(a) By estimate and acceptance in a lump sum.

(b) By unit prices named in the contract or subsequently agreed upon.

(c) By cost and percentage or by cost and a fixed fee.

If none of the above methods is agreed upon, the Contractor, provided he receives an order as above, shall proceed with the work. In such case and also under case (c), he shall keep and present, in such form as the Architect may direct, a correct amount of the cost, together with vouchers. In any case, the Architect shall certify to the amount, including reasonable allowance for overhead and profit, due to the Contractor. Pending final determination of value, payments on account of changes shall be made on the Architect's certificate.

Should conditions encountered below the surface of the ground be at variance with the conditions indicated by the drawings and specifications the contract sum shall be equitably adjusted upon claim by either party made within a reasonable time after the first observance of the conditions.

Art. 16. *Claims for Extra Cost.* If the Contractor claims that any instructions by drawings or otherwise involve extra cost under this contract, he shall give the Architect written notice thereof within a reasonable time after the receipt of such instructions, and in any event before proceeding to execute the work, except in emergency endangering life or property, and the procedure shall then be as provided for changes in the work. No such claim shall be valid unless so made.

Art. 17. *Deductions for Uncorrected Work.* If the Architect and Owner deem it inexpedient to correct work injured or done not in accordance with the Contract, an equitable deduction from the contract price shall be made therefor.

Art. 18. *Delays and Extension of Time.* If the Contractor be delayed at any time in the progress of the work by any act or neglect of the Owner or the Architect, or of any employee of either, or by any separate Contractor employed by the Owner, or by changes ordered in the work, or by strikes, lockouts, fire, unusual delay in transportation, unavoidable casualties, or any causes beyond the Contractor's control, or by delay authorized by the Architect pending arbitration, or by any cause which the Architect shall decide to justify the delay, then the time of completion shall be extended for such reasonable time as the Architect may decide.

No such extension shall be made for delay occurring more than seven days before claim therefor is made in writing to the Architect. In the case of a continuing cause of delay, only one claim is necessary.

If no schedule or agreement stating the dates upon which drawings shall be furnished is made, then no claim for delay shall be allowed on account of failure to furnish drawings until two weeks after demand for such drawings and not then unless such claim be reasonable.

This article does not exclude the recovery of damages for delay by either party under other provisions in the contract documents.

ART. 19. *Correction of Work before Final Payment.* The Contractor shall promptly remove from the premises all work condemned by the Architect as failing to conform to the Contract, whether incorporated or not, and the Contractor shall promptly replace and re-execute his own work in accordance with the Contract and without expense to the Owner and shall bear the expense of making good all work of other contractors destroyed or damaged by such removal or replacement.

If the Contractor does not remove such condemned work within a reasonable time, fixed by written notice, the Owner may remove it and may store the material at the expense of the Contractor. If the Contractor does not pay the expenses of such removal within ten days' time thereafter, the Owner may, upon ten days' written notice, sell such materials at auction or at private sale and shall account for the net proceeds thereof, after deducting all the costs and expenses that should have been borne by the Contractor.

ART. 20. *Correction of Work after Final Payment.* Neither the final certificate nor payment nor any provision in the Contract Documents shall relieve the Contractor of responsibility for faulty materials or workmanship, and, unless otherwise specified, he shall remedy any defects due thereto, and pay for any damage to other work resulting therefrom, which shall appear within a period of one year from the date of substantial completion. The Owner shall give notice of observed defects with reasonable promptness. All questions arising under this article shall be decided by the Architect subject to arbitration.

ART. 21. *The Owner's Right to Do Work.* If the Contractor should neglect to prosecute the work properly or fail to perform any provision of this contract, the Owner, after three days' written notice to the Contractor may, without prejudice to any other remedy he may have, make good such deficiencies and may deduct the cost thereof from the payment then or thereafter due the Contractor, provided, however, that the Architect shall approve both such action and the amount charged to the Contractor.

ART. 22. *Owner's Right to Terminate Contract.* If the Contractor should be adjudged a bankrupt, or if he should make a general assignment for the benefit of his creditors, or if a receiver should be appointed on account of his insolvency, or if he should persistently or repeatedly refuse or should fail, except in cases for which extension of time is provided, to supply enough properly skilled workmen or proper materials, or if he should fail to make prompt payment to subcontractors or for material or labor, or persistently disregard laws, ordinances, or the instructions of the Architect, or otherwise be guilty of a substantial violation of any provision of the contract, then the Owner, upon the certificate of the Architect that sufficient cause exists to justify such action, may, without prejudice to any other right or remedy and after giving the Contractor seven days' written notice, terminate the employment of the Contractor and take possession of the premises and of all materials, tools, and appliances thereon and finish the work by whatever method he may deem expedient. In such case the Contractor shall not be entitled to receive any further payment until the work is finished. If the unpaid balance of the contract price shall exceed the expense of finishing the work including compensation for additional managerial and administrative services, such excess shall be paid to the Contractor. If such expense shall exceed such unpaid balance, the Contractor shall pay the difference to the

Owner. The expense incurred by the Owner as herein provided, and the damage incurred through the Contractor's default, shall be certified by the Architect.

Art. 23. *Contractor's Right to Stop Work or Terminate Contract.* If the work should be stopped under an order of any court, or other public authority, for a period of three months, through no act or fault of the Contractor or of anyone employed by him, or if the Architect should fail to issue any certificate for payment within seven days after it is due, or if the Owner should fail to pay to the Contractor within seven days of its maturity and presentation, any sum certified by the Architect or awarded by arbitrators, then the Contractor may, upon seven days' written notice to the Owner and the Architect, stop work or terminate this contract and recover from the Owner payment for all work executed and any loss sustained upon any plant or materials and reasonable profit and damages.

Art. 24. *Applications for Payments.* The Contractor shall submit to the Architect an application for each payment, and, if required, receipts or other vouchers, showing his payments for materials and labor, including payments to subcontractors as required by Article 37.

If payments are made on valuation of work done, such application shall be submitted at least ten days before each payment falls due, and, if required, the Contractor shall, before the first application, submit to the Architect a schedule of values of the various parts of the work, including quantities, aggregating the total sum of the contract, divided so as to facilitate payments to subcontractors in accordance with Article 37 (*e*), made out in such form as the Architect and the Contractor may agree upon, and, if required, supported by such evidence as to its correctness as the Architect may direct. This schedule, when approved by the Architect, shall be used as a basis for certificates of payment, unless it be found to be in error. In applying for payments, the Contractor shall submit a statement based upon this schedule, and, if required, itemized in such form and supported by such evidence as the Architect may direct, showing his right to the payment claimed.

If payments are made on account of materials delivered and suitably stored at the site but not incorporated in the work, they shall, if required by the Architect, be conditional upon submission by the Contractor of bills of sale or such other procedure as will establish the Owner's title to such material or otherwise adequately protect the Owner's interest.

Art. 25. *Certificates of Payments.* If the Contractor has made application as above, the Architect shall, not later than the date when each payment falls due, issue to the Contractor a certificate for such amount as he decides to be properly due, or state in writing his reasons for withholding a certificate.

No certificate issued nor payment made to the Contractor, nor partial or entire use or occupancy of the work by the Owner, shall be an acceptance of any work or materials not in accordance with this contract. The making and acceptance of the final payment shall constitute a waiver of all claims by the Owner, other than those arising from unsettled liens, from faulty work appearing after final payment or from requirement of the specifications, and of all claims by the Contractor, except those previously made and still unsettled.

Should the Owner fail to pay the sum named in any certificate of the Architect or in any award by arbitration, upon demand when due, the Contractor shall receive, in addition to the sum named in the certificate, interest thereon at the legal rate in force at the place of building.

ART. 26. *Payments Withheld.* The Architect may withhold or, on account of subsequently discovered evidence, nullify the whole or a part of any certificate to such extent as may be necessary to protect the Owner from loss on account of:

(*a*) Defective work not remedied.

(*b*) Claims filed or reasonable evidence indicating probable filing of claims.

(*c*) Failure of the Contractor to make payments properly to subcontractors or for material or labor.

(*d*) A reasonable doubt that the contract can be completed for the balance then unpaid.

(*e*) Damage to another Contractor.

When the above grounds are removed payment shall be made for amounts withheld because of them.

ART. 27. *Contractor's Liability Insurance.* The Contractor shall maintain such insurance as will protect him from claims under workmen's compensation acts and from claims for damages because of bodily injury, including death, which may arise from and during operations under this Contract, whether such operations be by himself or by any subcontractor or anyone directly or indirectly employed by either of them. This insurance shall be written for not less than any limits of liability specified as part of this contract. This insurance need not cover any liability imposed by Article 31 of these General Conditions. Certificates of such insurance shall be filed with the Owner if he so require.

ART. 28. *Owner's Liability Insurance.* The Owner shall be responsible for and at his option may maintain such insurance as will protect him from his contingent liability to others for damages because of bodily injury, including death, which may arise from operations under this contract, and any other liability for damages which the Contractor is required to insure under any provision of this contract.

ART. 29. *Fire Insurance.* The Owner shall effect and maintain fire insurance upon the entire structure on which the work of this contract is to be done to 100 per cent of the insurable value thereof, including items of labor and materials connected therewith whether in or adjacent to the structure insured, materials in place or to be used as part of the permanent construction including surplus materials, shanties, protective fences, bridges, or temporary structures, miscellaneous materials and supplies incident to the work, and such scaffoldings, stagings, towers, forms, and equipment as are not owned or rented by the contractor, the cost of which is included in the cost of the work. EXCLUSIONS: This insurance does not cover any tools owned by mechanics, any tools, equipment, scaffoldings, stagings, towers, and forms owned or rented by the Contractor, the capital value of which is not included in the cost of the work, or any cook shanties, bunk houses, or other structures erected for housing the workmen. The loss, if any, is to be made adjustable with and payable to the Owner as Trustee for the insureds as their interests may appear, except in such cases as may require payment of all or a proportion of said insurance to be made to a mortgagee as his interests may appear.

The Contractor and all subcontractors shall be named or designated in such capacity as insured jointly with the Owner in all policies, all of which shall be open to the Contractor's inspection. Certificates of such insurance shall be filed with the Contractor if he so requires. If the Owner fails to effect or maintain insurance as above and so notifies the Contractor, the Contractor may insure his own interest and that of the subcontractors and charge the cost thereof to the Owner. If the Contractor is damaged by failure of the Owner to maintain such insurance or to so notify the Contractor, he may recover as stipulated in the contract for recovery of damages. If extended coverage or other special insurance not herein provided for is required by the Contractor, the Owner shall effect such insurance at the Contractor's expense by appropriate riders to his fire insurance policy.

If required in writing by any party in interest, the Owner as Trustee shall, upon the occurrence of loss, give bond for the proper performance of his duties. He shall deposit any money received from insurance in an account separate from all his other funds and he shall distribute it in accordance with such agreement as the parties in interest may reach, or under an award of arbitrators appointed, one by the Owner, another by joint action of the other parties in interest, all other procedure being as provided elsewhere in the contract for Arbitration. If after loss no special agreement is made, replacement of injured work shall be ordered and executed as provide for changes in the work.

The Trustee shall have power to adjust and settle any loss with the insurers unless one of the Contractors interested shall object in writing within three working days of the occurrence of loss, and thereupon arbitrators shall be chosen as above. The Trustee shall in that case make settlement with the insurers in accordance with the directions of such arbitrators, who shall also, if distribution by arbitration is required, direct such distribution.

ART. 30. *Guaranty Bonds.* The Owner shall have the right, prior to the signing of the Contract, to require the Contractor to furnish bond covering the faithful performance of the Contract and the payment of all obligations arising thereunder, in such form as the Owner may prescribe and with such sureties as he may approve. If such bond is required by instructions given previous to the submission of bids, the premium shall be paid by the Contractor; if subsequent thereto, it shall be paid by the Owner.

ART. 31. *Damages.* If either party to this Contract should suffer damage in any manner because of any wrongful act or neglect of the other party or of anyone employed by him, then he shall be reimbursed by the other party for such damage, provided the Owner shall be responsible for and at his option insure against loss of use of any of his existing property, due to fire or otherwise, however caused.

Claims under this clause shall be made in writing to the party liable within a reasonable time of the first observance of such damage and not later than the time of final payment, except as expressly stipulated otherwise in the case of faulty work or materials, and shall be adjusted by agreement or arbitration.

The Contractor is relieved of responsibility for damages to the work due to causes beyond the control of and without fault or negligence of the Contractor.

ART. 32. *Liens.* Neither the final payment nor any part of the retained percentage shall become due until the Contractor, if required, shall deliver to the Owner a complete release of all liens arising out of this Contract, or receipts in full in lieu thereof, and, if required in either case, an affidavit that so far as he has knowledge or information the releases and receipts include all the labor and material for which a lien could be filed; but the Contractor may, if any subcontractor refuses to furnish a release or receipt in full, furnish a bond satisfactory to the Owner, to indemnify him against any lien. If any lien remain unsatisfied after all payments are made, the Contractor shall refund to the Owner all moneys that the latter may be compelled to pay in discharging such a lien, including all costs and a reasonable attorney's fee.

ART. 33. *Assignment.* Neither party to the Contract shall assign the Contract or sublet it as a whole without the written consent of the other, nor shall the Contractor assign any moneys due or to become due to him hereunder, without the previous written consent of the Owner.

ART. 34. *Mutual Responsibility of Contractors.* Should the Contractor cause damage to any separate contractor on the work the Contractor agrees, upon due notice, to settle with such contractor by agreement or arbitration, if he will so settle. If such separate contractor sues the Owner on account of any damage alleged to have been so sustained, the Owner shall notify the Contractor, who shall defend such proceedings at the Owner's expense, and, if any judgment against the Owner arise therefrom, the Contractor shall pay or satisfy it and pay all costs incurred by the Owner.

ART. 35. *Separate Contracts.* The Owner reserves the right to let other contracts in connection with this work. The Contractor shall afford other contractors reasonable opportunity for the introduction and storage of their materials and the execution of their work, and shall properly connect and coordinate his work with theirs.

If any part of the Contractor's work depends for proper execution or results upon the work of any other contractor, the Contractor shall inspect and promptly report to the Architect any defects in such work that render it unsuitable for such proper execution and results. His failure so to inspect and report shall constitute an acceptance of the other contractor's work as fit and proper for the reception of his work, except as to defects which may develop in the other contractor's work after the execution of his work.

To insure the proper execution of his subsequent work the Contractor shall measure work already in place and shall at once report to the Architect any discrepancy between the executed work and the drawings.

ART. 36. *Subcontracts.* The Contractor shall, as soon as practicable after the execution of the contract, notify the Architect in writing of the names of subcontractors proposed for the principal parts of the work and for such others as the Architect may direct and shall not employ any that the Architect may within a reasonable time object to as incompetent or unfit.

If the Contractor has submitted before execution of the contract a list of subcontractors and the change of any name on such list is required in writing by the Owner after such execution, the contract price shall be increased or diminished by the difference in cost occasioned by such change.

The Architect shall, on request, furnish to any subcontractor, wherever practicable, evidence of the amounts certified on his account.

The Contractor agrees that he is as fully responsible to the Owner for the acts and omissions of his subcontractors and of persons either directly or indirectly employed by them, as he is for the acts and omissions of persons directly employed by him.

Nothing contained in the contract documents shall create any contractual relation between any subcontractor and the Owner.

ART. 37. *Relations of Contractor and Subcontractor.* The Contractor agrees to bind every Subcontractor and every Subcontractor agrees to be bound by the terms of the Agreement, the General Conditions, the Drawings, and Specifications as far as applicable to his work, including the following provisions of this article, unless specifically noted to the contrary in a subcontract approved in writing as adequate by the Owner or Architect.

This does not apply to minor subcontracts.

The Subcontractor agrees—

(*a*) To be bound to the Contractor by the terms of the Agreement, General Conditions, Drawings, and Specifications, and to assume toward him all the obligations and responsibilities that he, by those documents, assumes toward the Owner.

(*b*) To submit to the Contractor applications for payment in such reasonable time as to enable the Contractor to apply for payment under Article 24 of the General Conditions.

(*c*) To make all claims for extras, for extensions of time and for damages for delays or otherwise, to the Contractor in the manner provided in the General Conditions for like claims by the Contractor upon the Owner, except that the time for making claims for extra cost is one week.

The Contractor agrees—

(*d*) To be bound to the Subcontractor by all the obligations that the Owner assumes to the Contractor under the Agreement, General Conditions, Drawings, and Specifications, and by all the provisions thereof affording remedies and redress to the Contractor from the Owner.

(*e*) To pay the Subcontractor, upon the payment of certificates, if issued under the schedule of values described in Article 24 of the General Conditions, the amount allowed to the Contractor on account of the Subcontractor's work to the extent of the Subcontractor's interest therein.

(*f*) To pay the Subcontractor, upon the payment of certificates, if issued otherwise than as in (*e*), so that at all times his total payments shall be as large in proportion to the value of the work done by him as the total amount certified to the Contractor is to the value of the work done by him.

(*g*) To pay the Subcontractor to such extent as may be provided by the Contract Documents or the subcontract, if either of these provides for earlier or larger payments than the above.

(*h*) To pay the Subcontractor on demand for his work or materials as far as executed and fixed in place, less the retained percentage, at the time the certificate should issue, even though the Architect fails to issue it for any cause not the fault of the Subcontractor.

(*j*) To pay the Subcontractor a just share of any fire insurance money received by him, the Contractor, under Article 29 of the General Conditions.

(*k*) To make no demand for liquidated damages or penalty for delay in any sum in excess of such amount as may be specifically named in the subcontract.

(*l*) That no claim for services rendered or materials furnished by the Contractor to the Subcontractor shall be valid unless written notice thereof is given by the Contractor to the Subcontractor during the first ten days of the calendar month following that in which the claim originated.

(*m*) To give the Subcontractor an opportunity to be present and to submit evidence in any arbitration involving his rights.

(*n*) To name as arbitrator under arbitration proceedings as provided in the General Conditions the person nominated by the Subcontractor, if the sole cause of dispute is the work, materials, rights, or responsibilities of the Subcontractor; or, if of the Subcontractor and any other subcontractor jointly, to name as such arbitrator the person upon whom they agree.

The Contractor and the Subcontractor agree that—

(*o*) In the matter of arbitration, their rights and obligations and all procedure shall be analogous to those set forth in this contract.

Nothing in this article shall create any obligation on the part of the Owner to pay to or to see to the payment of any sums to any subcontractor.

ART. 38. *Architect's Status.* The Architect shall have general supervision and direction of the work. He is the agent of the Owner only to the extent provided in the Contract Documents and when in special instances he is authorized by the Owner so to act, and in such instances he shall, upon request, show the Contractor written authority. He has authority to stop the work whenever such stoppage may be necessary to insure the proper execution of the Contract.

As the Architect is, in the first instance, the interpreter of the conditions of the Contract and the judge of its performance, he shall side neither with the Owner nor with the Contractor, but shall use his powers under the contract to enforce its faithful performance by both.

In case of the termination of the employment of the Architect, the Owner shall appoint a capable and reputable Architect, against whom the Contractor makes no reasonable objection, whose status under the contract shall be that of the former Architect; any dispute in connection with such appointment to be subject to arbitration.

ART. 39. *Architect's Decisions.* The Architect shall, within a reasonable time, make decisions on all claims of the Owner or Contractor and on all other matters relating to the execution and progress of the work or the interpretation of the Contract Documents.

The Architect's decisions, in matters relating to artistic effect, shall be final, if within the terms of the Contract Documents.

Except as above or as otherwise expressly provided in the Contract Documents, all the Architect's decisions are subject to arbitration.

If, however, the Architect fails to render a decision within ten days after the parties have presented their evidence, either party may then demand arbitration. If the Architect renders a decision after arbitration proceedings have been initiated, such decision may be entered as evidence but shall not disturb or interrupt such proceedings except where such decision is acceptable to the parties concerned.

ART. 40. *Arbitration.* All disputes, claims, or questions subject to arbitration under this contract shall be submitted to arbitration in accordance with the provisions, then obtaining, of the Standard Form of Arbitration Procedure of The American Institute of Architects, and this agreement shall be specifically enforceable under the prevailing arbitration law, and judgment upon the award rendered may be entered in the court of the forum, state or federal, having jurisdiction. It is mutually agreed that the decision of the arbitrators shall be a condition precedent to any right of legal action that either party may have against the other.

The Contractor shall not cause a delay of the work during any arbitration proceedings, except by agreement with the Owner.

Notice of the demand for arbitration of a dispute shall be filed in writing with the Architect and the other party to the contract. If the arbitration is an appeal from the Architect's decision, the demand therefor shall be made within ten days of its receipt; in any other case the demand for arbitration shall be made within a reasonable time after the dispute has arisen; in no case, however, shall the demand be made later than the time of final payment, except as otherwise expressly stipulated in the contract.

The arbitrators, if they deem that the case requires it, are authorized to award, to the party whose contention is sustained, such sums as they or a majority of them shall deem proper to compensate him for the time and expense incident to the proceeding and, if the arbitration was demanded without reasonable cause, they may also award damages for delay. The arbitrators shall fix their own compensation, unless otherwise provided by agreement, and shall assess the costs and charges of the proceeding upon either or both parties.

ART. 41. *Cash Allowances.* The Contractor shall include in the contract sum all allowances named in the Contract Documents and shall cause the work so covered to be done by such contractors and for such sums as the Architect may direct, the contract sum being adjusted in conformity therewith. The Contractor declares that the contract sum includes such sums for expenses and profit on account of cash allowances as he deems proper. No demand for expenses or profit other than those included in the contract sum shall be allowed. The Contractor shall not be required to employ for any such work persons against whom he has a reasonable objection.

ART. 42. *Use of Premises.* The Contractor shall confine his apparatus, the storage of materials, and the operations of his workmen to limits indicated by law, ordinances, permits, or directions of the Architect and shall not unreasonably encumber the premises with his materials.

The Contractor shall not load or permit any part of the structure to be loaded with a weight that will endanger its safety.

The Contractor shall enforce the Architect's instructions regarding signs, advertisements, fires, and smoking.

ART. 43. *Cutting, Patching, and Digging.* The Contractor shall do all cutting, fitting, or patching of his work that may be required to make its several parts come together properly and fit it to receive or be received by work of other contractors shown upon, or reasonably implied by, the Drawings and Specifications for the completed structure, and he shall make good after them as the Architect may direct.

Any cost caused by defective or ill-timed work shall be borne by the party responsible therefor.

The Contractor shall not endanger any work by cutting, digging, or otherwise, and shall not cut or alter the work of any other contractor save with the consent of the Architect.

ART. 44. *Cleaning up.* The Contractor shall at all times keep the premises free from accumulations of waste material or rubbish caused by his employees or work, and at the completion of the work he shall remove all his rubbish from and about the building and all his tools, scaffolding, and surplus materials and shall leave his work "broom-clean" or its equivalent, unless more exactly specified. In case of dispute the Owner may remove the rubbish and charge the cost to the several contractors as the Architect shall determine to be just.

Short Contract Form.

The foregoing contract forms are more elaborate than is necessary for small construction projects. For work of this classification a condensed form may be sufficient. The "Short Form for Small Construction Contracts" of the American Institute of Architects is a typical example of this type of contract.

EXAMPLE OF SHORT-FORM CONTRACT

THE A.I.A. SHORT FORM
FOR
SMALL CONSTRUCTION CONTRACTS

AGREEMENT AND GENERAL CONDITIONS
BETWEEN CONTRACTOR AND OWNER

ISSUED BY THE AMERICAN INSTITUTE OF ARCHITECTS FOR USE ONLY WHEN THE
PROPOSED WORK IS SIMPLE IN CHARACTER, SMALL IN COST, AND WHEN A STIPU-
LATED SUM FORMS THE BASIS OF PAYMENT. FOR OTHER CONTRACTS THE INSTI-
TUTE ISSUES THE STANDARD FORM OF AGREEMENT BETWEEN CONTRACTOR AND
OWNER FOR CONSTRUCTION OF BUILDINGS AND THE STANDARD GENERAL CONDI-
TIONS IN CONNECTION THEREWITH FOR USE WHEN A STIPULATED SUM FORMS
THE BASIS FOR PAYMENT.

SIXTH EDITION, COPYRIGHT 1936–1951 BY THE AMERICAN INSTITUTE OF
ARCHITECTS, THE OCTAGON, WASHINGTON, D. C.

THIS AGREEMENT made the .... day of ....
in the year Nineteen Hundred and ...., by and between

....

.... hereinafter called the Contractor, and

....

.... hereinafter called the Owner.
WITNESSETH, That the Contractor and the Owner for the considerations
hereinafter named agree as follows:

ART. 1. *Scope of the Work.* The Contractor shall furnish all of the material and perform all of the work for ....

....
(Caption indicating the portion or portions of work covered)
as shown on the drawings and described in the specifications entitled

prepared by .... Architect
all in accordance with the terms of the contract documents.
ART. 2. *Time of Completion.* The work shall be substantially completed

....

ART. 3. *Contract Sum.* The Owner shall pay the Contractor for the performance of the contract subject to the additions and deductions provided
therein in current funds, the sum of ....
.... dollars. ($........................)
ART. 4. *Progress Payments.* The Owner shall make payments on account
of the contract, upon requisition by the Contractor, as follows:

Reprinted with permission of The American Institute of Architects.

ART. 5. *Acceptance and Final Payment.* Final payment shall be due .... days after completion of the work, provided the contract be then fully performed, subject to the provisions of Art. 16 of the General Conditions.

ART. 6. *Contract Documents.* Contract documents are as noted in Art. 1 of the General Conditions. The following is an enumeration of the drawings and specifications:

GENERAL CONDITIONS

ART. 1. *Contract Documents.* The contract includes the AGREEMENT and its GENERAL CONDITIONS, the DRAWINGS, and the SPECIFICATIONS. Two or more copies of each, as required, shall be signed by both parties and one signed copy of each retained by each party.

The intent of these documents is to include all labor, materials, appliances, and services of every kind necessary for the proper execution of the work, and the terms and conditions of payment therefor.

The documents are to be considered as one, and whatever is called for by any one of the documents shall be as binding as if called for by all.

ART. 2. *Samples.* The Contractor shall furnish for approval all samples as directed. The work shall be in accordance with approved samples.

ART. 3. *Materials, Appliances, Employes.* Except as otherwise noted, the Contractor shall provide and pay for all materials, labor, tools, water, power, and other items necessary to complete the work.

Unless otherwise specified, all materials shall be new, and both workmanship and materials shall be of good quality.

All workmen and subcontractors shall be skilled in their trades.

ART. 4. *Royalties and Patents.* The Contractor shall pay all royalties and license fees. He shall defend all suits or claims for infringement of any patent rights and shall save the Owner harmless from loss on account thereof.

ART. 5. *Surveys, Permits, and Regulations.* The Owner shall furnish all surveys unless otherwise specified. Permits and licenses of a temporary nature necessary for the prosecution of the work shall be secured and paid for by the Contractor. Permits, licenses, and easements for permanent structures or permanent changes in existing facilities shall be secured and paid for by the Owner, unless otherwise specified. The Contractor shall comply with all laws and regulations bearing on the conduct of the work and shall notify the Owner if the drawings and specifications are at variance therewith.

ART. 6. *Protection of Work, Property, and Persons.* The Contractor shall adequately protect the work, adjacent property, and the public and shall be responsible for any damage or injury due to his act or neglect.

ART. 7. *Inspection of Work.* The Contractor shall permit and facilitate inspection of the work by the Owner and his agents and public authorities at all times.

ART. 8. *Changes in the Work.* The Owner may order changes in the work, the Contract Sum being adjusted accordingly. All such orders and adjustments shall be in writing. Claims by the Contractor for extra cost must be made in writing before executing the work involved.

ART. 9. *Correction of Work.* The Contractor shall re-execute any work that fails to conform to the requirements of the contract and that appears during the progress of the work, and shall remedy any defects due to faulty materials or workmanship which appear within a period of one year from the date of completion of the contract. The provisions of this article apply to work done by subcontractors as well as to work done by direct employes of the Contractor.

ART. 10. *Owner's Right to Terminate the Contract.* Should the Contractor neglect to prosecute the work properly, or fail to perform any provision of the contract, the Owner, after seven days' written notice to the Contractor, may, without prejudice to any other remedy he may have, make good the deficiencies and may deduct the cost thereof from the payment then or thereafter due the Contractor or, at his option, may terminate the contract and take possession of all materials, tools, and appliances and finish the work by such means as he sees fit, and if the unpaid balance of the contract price exceeds the expense of finishing the work, such excess shall be paid to the Contractor, but if such expense exceeds such unpaid balance, the Contractor shall pay the difference to the Owner.

ART. 11. *Contractor's Right to Terminate Contract.* Should the work be stopped by any public authority for a period of thirty days or more, through no fault of the Contractor, or should the work be stopped through act or neglect of the Owner for a period of seven days, or should the Owner fail to pay the Contractor any payment within seven days after it is due, then the Contractor upon seven days' written notice to the Owner, may stop work or terminate the contract and recover from the Owner payment for all work executed and any loss sustained and reasonable profit and damages.

ART. 12. *Payments.* Payments shall be made as provided in the Agreement. The making and acceptance of the final payment shall constitute a waiver of all claims by the Owner, other than those arising from unsettled liens or from faulty work appearing thereafter, as provided for in Art. 9, and of all claims by the Contractor except any previously made and still unsettled. Payments otherwise due may be withheld on account of defective work not remedied, liens filed, damage by the Contractor to others not adjusted, or failure to make payments properly to subcontractors or for material or labor.

ART. 13. *Contractor's Liability Insurance.* The Contractor shall maintain such insurance as will protect him from claims under workmen's compensation acts and from claims for damages because of bodily injury, including death, which may arise from and during operations under this contract, whether such operations be by himself or by any subcontractor or anyone directly or indirectly employed by either of them. This insurance shall be written for not less than any limits of liability specified as part of this contract. This insurance need not cover any liability imposed by Art. 6 of the General Conditions. Certificates of such insurance shall be filed with the Owner if he so requires.

ART. 14. *Owner's Liability Insurance.* The Owner shall be responsible for and at his option may maintain such insurance as will protect him from his contingent liability to others for damages because of bodily injury, including death, which may arise from operations under this contract, and any other liability for damages which the Contractor is required to insure under any provision of this contract.

ART. 15. *Fire Insurance.* The Owner shall effect and maintain fire insurance upon the entire structure on which the work of this contract is to be done to 100 per cent of the insurable value thereof, including items of labor and materials connected therewith whether in or adjacent to the structure insured, materials in place or to be used as part of the permanent construction including surplus materials, shanties, protective fences, bridges, or temporary structures, miscellaneous materials and supplies incident to the work, and such scaffoldings, stagings, towers, forms, and equipment as are not owned or rented by the contractor, the cost of which is included in the cost of the work. EXCLUSIONS: This insurance does not cover any tools owned by mechanics, any tools, equipment, scaffoldings, stagings, towers, and forms owned or rented by the Contractor, the capital value of which is not included in the cost of the work, or any cook shanties, bunk houses, or other structures erected for housing the workmen. The loss, if any, is to be made adjustable with and payable to the Owner as Trustee for the insureds as their interests may appear, except in such cases as may require payment of all or a proportion of said insurance to be made to a mortgagee as his interests may appear.

The Contractor and all subcontractors shall be named or designated in such capacity as insured jointly with the Owner in all policies, all of which shall be open to the Contractor's inspection. Certificates of such insurance shall be filed with the Contractor if he so requires. If the Owner fails to effect or maintain insurance as above and so notifies the Contractor, the Contractor may insure his own interest and that of the subcontractors and charge the cost thereof to the Owner. If the Contractor is damaged by failure of the Owner to maintain such insurance or to so notify the Contractor, he may recover as stipulated in the contract for recovery of damages. If extended coverage or other special insurance not herein provided for is required by the Contractor, the Owner shall effect such insurance at the Contractor's expense by appropriate riders to his fire insurance policy.

If required in writing by any party in interest, the Owner as Trustee shall, upon the occurrence of loss, give bond for the proper performance of his duties. He shall deposit any money received from insurance in an account separate from all his other funds and he shall distribute it in accordance with such agreement as the parties in interest may reach, or under an award of arbitrators appointed, one by the Owner, another by joint action of the other parties in interest, all other procedure being as provided elsewhere in the contract for Arbitration. If after loss no special agreement is made, replacement of injured work shall be ordered and executed as provided for changes in the work.

The Trustee shall have power to adjust and settle any loss with the insurers unless one of the Contractors interested shall object in writing within three working days of the occurrence of loss, and thereupon arbitrators shall be chosen as above. The Trustee shall in that case make settlement with the insurers in accordance with the directions of such arbitrators, who shall also, if distribution by arbitration is required, direct such distribution.

ART. 16. *Liens.* The final payment shall not be due until the Contractor has delivered to the Owner a complete release of all liens arising out of this contract, or receipts in full covering all labor and materials for which a lien could be filed, or a bond satisfactory to the Owner indemnifying him against any lien.

ART. 17. *Separate Contracts.* The Owner has the right to let other contracts in connection with the work, and the Contractor shall properly cooperate with any such other contractors.

ART. 18. *The Architect's Status.* The Architect shall have general supervision of the work. He has authority to stop the work if necessary to insure its proper execution. He shall certify to the Owner when payments under the contract are due and the amounts to be paid. He shall make decisions on all claims of the Owner or Contractor. All his decisions are subject to arbitration.

ART. 19. *Arbitration.* Any disagreement arising out of this contract or from the breach thereof shall be submitted to arbitration, and judgment upon the award rendered may be entered in the court of the forum, state or federal, having jurisdiction. It is mutually agreed that the decision of the arbitrators shall be a condition precedent to any right of legal action that either party may have against the other. The arbitration shall be held under the Standard Form of Arbitration Procedure of The American Institute of Architects or under the Rules of the American Arbitration Association.

ART. 20. *Cleaning up.* The Contractor shall keep the premises free from accumulation of waste material and rubbish, and at the completion of the work he shall remove from the premises all rubbish, implements, and surplus materials and leave the building broom-clean.

IN WITNESS WHEREOF the parties hereto executed this Agreement, the day and year first above written.

Contractor ....

Owner ....

Subcontracts. As previously noted it is not expected that any one construction contractor will be equally skilled in all the specialized trades and techniques required for a major construction operation, and frequently the services of specialty contractors are utilized under subcontract agreements. In this procedure the general contract known as the prime contract is drawn between the owner and the general contractor. The contractor becomes responsible to the owner for all the work within the scope of the contract. This holds for work actually accomplished by his own forces as well as that performed by subcontractors. Furthermore, the prime contractor assumes responsibility for the management, scheduling, and coordination of the subcontractor's work as well as for his own. The latter element is of great importance in some types of work, especially in the construction of large buildings where many types of specialized trades are required at specific stages of the work. Typical of such trades are electrical work, heating, air conditioning, plumbing, roofing, automatic sprinklers, structural steel erection, and sometimes excavation and foundation construction. The prime contractor usually performs such work as concrete, masonry, carpentry, plastering, and painting.

Frequently the agreement between the prime contractor and the subcontractor is an oral one or an offer and acceptance by letters. For more important work, however, a formal subcontract is recommended in order to define the rights and responsibilities of the parties and thus minimize the likelihood of misunderstandings and disputes during and after the accomplishment of the work. The form of the subcontract should be similar to that between the owner and the prime contractor. It should define the scope of the work to be performed by the subcontractor, the compensation which he is to receive from the prime contractor, and the time during which the work is to be performed. In addition, it should be stipulated that the subcontractor is bound by the terms of the prime contract, the plans, and the specifications in so far as the work included under the subcontract is concerned. The formal subcontract may include other terms required to prescribe the special relationships between the prime contractor and the subcontractor, in which case it should be stated that, if a conflict exists between any of the requirements of the prime contract and those of the subcontract, the terms of the prime contract are to govern, and the parties are so bound. The right to approve or disapprove the selection of any subcontractor should be reserved by the owner in the prime contract.

In connection with the quality and quantity of the work, the subcontractor is not recognized by the owner. In such matters, the owner holds the prime contractor directly responsible for the subcontractor's work. It should be noted, however, that the subcontractor has very definite legal rights which have important implications to the owner. For instance, if the prime contractor should default on payments due the subcontractor, the owner might be held responsible for the indebtedness, and payment could be forced by a mechanic's lien on the structure constructed under the contract. The owner's best protection against such action is to require the prime contractor to furnish a surety bond guaranteeing the payment of all costs for materials, wages, and other expenses, including payments to subcontractors. The owner should also require the prime contractor to furnish releases from all claims and liens in connection with the work, executed by all subcontractors, before the final payment is made to the prime contractor.

EXAMPLE OF LUMP-SUM SUBCONTRACT

THE STANDARD FORM OF SUBCONTRACT *

(Issued by the American Institute of Architects)

FOR USE IN CONNECTION WITH THE SIXTH EDITION OF THE STANDARD FORM OF AGREEMENT AND GENERAL CONDITIONS OF THE CONTRACT.

THE SIXTH EDITION OF THIS FORM HAS RECEIVED THE APPROVAL OF THE ASSOCI-ATED GENERAL CONTRACTORS OF AMERICA ; THE CONTRACTING PLASTERERS' INTER-NATIONAL ASSN. ; THE HEATING, PIPING AND AIR CONDITIONING CONTRACTORS NATIONAL ASSN. ; THE NATIONAL ASSOCIATION OF ORNAMENTAL METAL MANU-FACTURERS ; THE NATIONAL BUILDING GRANITE QUARRIES ASSN., INC. ; THE NATIONAL ELECTRICAL CONTRACTORS ASSN. ; THE PAINTING AND DECORATING CONTRACTORS OF AMERICA, AND THE PRODUCERS' COUNCIL, INC.

THIS AGREEMENT, made this   day of
19   by and between   hereinafter
called the Subcontractor and
hereinafter called the Contractor.

WITNESSETH, That the Subcontractor and Contractor for the considerations hereinafter named agree as follows :

SECTION 1. The Subcontractor agrees to furnish all material and perform all work as described in Section 2 hereof for (Here name the kind of building.)

for   (Here insert the name of the Owner.)

hereinafter called the Owner, at   (Here insert the location of the work.)

in accordance with the General Conditions of the Contract between the Owner and the Contractor and in accordance with the Drawings and the Specifications prepared by
                                        hereinafter called the Architect, all of which General Conditions, Drawings and Specifications signed by the parties thereto or identified by the Architect, form a part of a Contract between the Contractor and the Owner dated,   19
and hereby become a part of this Contract.

* A.I.A. Form C1.

SECTION 2. The Subcontractor and the Contractor agree that the materials to be furnished and work to be done by the Subcontractor are   (Here insert a precise description of the work, preferably by reference to the numbers of the Drawings and the pages of the Specifications.)

SECTION 3. The Subcontractor agrees to complete the several portions and the whole of the work herein sublet by the time or times following:
(Here insert the date or dates, and if there be liquidated damages state them.)

SECTION 4. The Contractor agrees to pay the Subcontractor for the performance of his work the sum of

($) in current funds, subject to additions and deductions for changes as may be agreed upon, and to make payments on account thereof in accordance with Section 5 hereof.

SECTION 5. The Contractor and Subcontractor agree to be bound by the terms of the Agreement, the General Conditions, Drawings and Specifications as far as applicable to this subcontract, and also by the following provisions:

The Subcontractor agrees—

(a) To be bound to the Contractor by the terms of the Agreement, General Conditions, Drawings and Specifications, and to assume toward him all the obligations and responsibilities that he, by those documents, assumes toward the Owner.

(b) To submit to the Contractor applications for payment in such reasonable time as to enable the Contractor to apply for payment under Article 24 of the General Conditions.

(c) To make all claims for extras, for extensions of time and for damages for delays or otherwise, to the Contractor in the manner provided in the General Conditions for like claims by the Contractor upon the Owner, except that the time for making claims for extra cost is one week.

The Contractor agrees—

(d) To be bound to the Subcontractor by all the obligations that the Owner assumes to the Contractor under the Agreement, General Conditions, Drawings and Specifications, and by all the provisions thereof affording remedies and redress to the Contractor from the Owner.

(*e*) To pay the Subcontractor, upon the payment of certificates, if issued under the schedule of values described in Article 24 of the General Conditions, the amount allowed to the Contractor on account of the Subcontractor's work to the extent of the Subcontractor's interest therein.

(*f*) To pay the Subcontractor, upon the payment of certificates, if issued otherwise than as in (*e*), so that at all times his total payments shall be as large in proportion to the value of the work done by him as the total amount certified to the Contractor is to the value of the work done by him.

(*g*) To pay the Subcontractor to such extent as may be provided by the Contract Documents or the subcontract, if either of these provides for earlier or larger payments than the above.

(*h*) To pay the Subcontractor on demand for his work or materials as far as executed and fixed in place, less the retained percentage, at the time the certificate should issue, even though the Architect fails to issue it for any cause not the fault of the Subcontractor.

(*j*) To pay the Subcontractor a just share of any fire insurance money received by him, the Contractor, under Article 29 of the General Conditions.

(*k*) To make no demand for liquidated damages or penalty for delay in any sum in excess of such amount as may be specifically named in the subcontract.

(*l*) That no claim for services rendered or materials furnished by the Contractor to the Subcontractor shall be valid unless written notice thereof is given by the Contractor to the Subcontractor during the first ten days of the calendar month following that in which the claim originated.

(*m*) To give the Subcontractor an opportunity to be present and to submit evidence in any arbitration involving his rights.

(*n*) To name as arbitrator under arbitration proceeding as provided in the General Conditions the person nominated by the Subcontractor, if the sole cause of dispute is the work, materials, rights or responsibilities of the Subcontractor; or, if of the Subcontractor and any other subcontractor jointly, to name as such arbitrator the person upon whom they agree.

The Contractor and the Subcontractor agree that—

(*o*) In the matter of arbitration, their rights and obligations and all procedure shall be analogous to those set forth in this Contract.

Nothing in this article shall create any obligation on the part of the Owner to pay to or to see to the payment of any sums to any Subcontractor.

SECTION 6.

IN WITNESS WHEREOF the parties hereto have executed this agreement the day and year first above written.

Questions

1. Explain the contractual significance of the advertisement, instructions to bidders, and the bid after a formal contract is executed.

2. In planning the general form of a contract, what are the arguments for and against dividing the contract into two parts, the agreement and the general conditions of the contract?

3. Why should the word "bidder" not appear in the contract?

4. What are the factors to be considered in fixing the time of completion?

5. What provisions should be made for changes in the work under a lump-sum contract? Name four methods for arriving at the change in the contract amount to cover the cost of changes in the work, and give the advantages and disadvantages of each.

6. Progress payments to the contractor usually cover 90 per cent of the work completed during the payment period, 10 per cent being retained. What is the owner's justification for this procedure?

7. Progress payments by the owner to the contractor are based on certificates prepared by the engineer and addressed to the owner. Write an engineer's progress payment certificate for the work completed during one month on highway construction under a unit-price contract. The quantities completed during this period and the contract unit prices were as follows.

Earth excavation	1,600 cu. yd. @ $	0.75
Rock excavation	430 cu. yd. @	6.00
Borrow-pit excavation	1,200 cu. yd. @	1.00
Concrete pavement	6,000 sq. yd. @	3.50
Curb and gutter	2,000 lin. ft. @	1.90
12-inch drain pipe	400 lin. ft. @	5.40
Spillway assemblies	4 each @	40.00
Reinforcing steel	60,000 lb. @	0.10

8. Make a comparative analysis of the standard contract forms of the U.S. Government, the American Railway Engineering Association, and the American Institute of Architects.

9. The engineer for a public commission in charge of construction of an elevated railway ruled that a change in the concrete finish required on the columns and some changes in the steel construction required at stations were to be classified and paid for at contract unit prices. The contractor, on the other hand, claimed that these were changes for which he was entitled to extra compensation. The commission insisted that the determination of the engineer was final and binding upon the contractor. The contractor filed suit against the commission and was awarded extra compensation by the court. On what basis could the court make this decision?

10. The contractor on a contract with the U.S. Government for the construction of a dam had set up a conveyor system in gravel pits about 6,000 feet downstream from the dam. The contract made the contractor responsible for all risk of damage to his plant, equipment, and operations at the gravel pit by reason of flood. After the plant had been in operation for several months, a flood occurred. Before the flood, however, the government, by separate contract or by force account, had been clearing the basin above the dam, and much debris including large logs was left in the basin. These were carried away by the flood, lodged against the towers of the conveyor system, and caused the collapse of the towers and a general wrecking of the plant. Was the contractor or the Government liable for the damage? (See *Engineering News-Record*, January 29, 1942, page 63.)

IO

COST-PLUS-FIXED-FEE CONTRACTS

The fundamental concept of the cost-plus-fixed-fee type of negotiated contract is mutual faith and confidence between the owner, the engineer, and the contractor, and the selection of the contractor should be made with this end in view. Likewise, in drawing the contract, these relations should be kept in mind because many of the standard provisions of competitive-bid contracts are not applicable. Though an element of competition may be introduced in connection with the amount of the fixed fee, ordinarily it is advisable to select the contractor on the basis of other considerations, the amount of the fee and the terms of the contract being determined after the contractor is selected. The cost-plus-fixed-fee contract may be adapted to competitive bidding, however, by including a provision whereby the contractor guarantees a ceiling of cost. The proposed amounts of the guaranteed cost and the fixed fee form a suitable basis for the comparison of bids and the award of the contract, assuming all bidders are properly qualified. The cost-plus-fixed-fee contract with an incentive clause in the form of a bonus and penalty provision based upon the bidder's proposed target estimate of cost also may be adapted to competitive bidding.

Selection of the Contractor. In general a contractor's ability can be gauged by his past performances, his reputation, his financial standing, and the personal character of the principals of his firm. Usually a number of contractors will be available for a specific project, all with apparently similar qualifications. In order to make the most satisfactory selections, the following factors should be examined in detail.

a. Previous experience in the particular type of work.

b. Reputation for fairness and excellence in performance.

c. Quality and experience of personnel.

d. Record in the management and coordination of work of subcontractors.

e. Available working capital.

f. Available plant and equipment.

g. Normal volume of work per year.

h. Uncompleted work in progress.

i. Available work capacity (difference between normal volume and uncompleted work in progress, plus capacity to expand).

Information along these lines should be collected and classified systematically, preferably by some form of questionnaire. This procedure is, in effect, similar to prequalification, and prequalification questionnaires are suitable for the purpose. An analysis of these data will usually indicate one or more contractors who are most suitable for the work. Final selection should be made after conferences during which the details of the contractors' qualifications are further developed.

Determination of the Fee. After the contractor has been selected, separate negotiations are conducted to define the terms of the contract and to determine the amount of the fee. This fee should be based on the estimated construction cost, the character and magnitude of the work, the estimated time of completion, and the amount of the work to be accomplished by subcontract. It should be clearly understood that, once the amount of the fee has been agreed upon, it remains unchanged regardless of the accuracy of the estimate of cost since changes in the fee are approved only in the case of material changes in the scope of the work.

Fees for engineering construction work normally vary from about 2 to 10 per cent of the estimated construction cost, with a decrease in the percentage as the magnitude increases. Because of the more specialized skill required, fees will be higher for heavy subaqueous work, for example, than for light construction on dry land. Fees also should be commensurate with the hazards and risks inherent in the work. Projects in remote and primitive locations justify higher compensation to the contractor than those in metropolitan areas because of the greater responsibility assumed by the contractor in the care, maintenance, and security of his personnel and equipment.

In order to determine a fee that will be fair and reasonable to both parties to the contract, it is necessary to have available only rough preliminary plans which are sufficient to indicate the nature and magnitude of the work and to form the basis of a preliminary estimate of cost. The estimate is understood to be approximate, but it should be acceptable to both parties as a consideration in the determination of the fee. In view of these conditions considerable variation between the original estimate and the actual cost is to be expected at the completion of the work.

Terms of the Contract. Because of the elasticity of this type of contract many of the terms, which are necessary in other forms, may be omitted. Specifications may also be condensed or developed as the work progresses. This works no hardship on the contractor since all direct costs are reimbursable. It is not unusual for the specifications to be issued in the form of memorandum instructions as the plans for the work are developed. It is not practicable to prepare a standard contract form which would be suitable for the various types of construction work and for the requirements and ideas of individual owners. A special document must be drawn for each particular project. In general, the essential elements of the agreement are:

a. Definition of the scope of the work, its estimated cost, and the amount of the contractor's fee.

b. Time for completion of the work.

c. Control of the work: responsibilities of the owner, engineer, and contractor.

d. Definition of reimbursable direct costs and overhead.

e. Provisions for plant rental or purchase, maintenance of equipment, and the overhaul of equipment at the end of the job.

f. Subcontracts.

g. Changes.

h. Method of compensation.

i. Termination.

j. Special requirements of the particular project.

Scope of the Work. Since the contractor is to be paid a fixed sum for the accomplishment of a definite amount of work, it is necessary that the scope of the project be defined with great care. It is expected that minor changes will be made in the owner's plans as the work develops, but any material change implies a corresponding change in the contractor's fee and requires special authorization. At the time of the award of the contract, it will be impossible to describe the details of the work, and, generally, detail plans will not be available. In most cases the proper selection of the title of the project, a description of its general features, and a preliminary estimate of its cost will define the owner's intention relative to the scope of the work. Dimensions are desirable when available.

The following may be considered as an illustration of the significance of scope in relation to the contractor's fee. Under a contract for a building, it may be understood that the structure will include concrete floors 4 inches thick. If the thickness is changed later to 6 inches, this is considered to be a detail of design, and no change in scope is involved although the cost may be changed appreciably. It is obvious that the original intent included floors. However, if the building is changed from four stories to five, it is evident that the owner's general intention has changed during the progress of the work, and the contractor is entitled to a change in fee.

Time for Completion of the Work. When the time required to complete the work can be estimated with reasonable accuracy and the completion of the project is of importance to the owner, a date may be established in the contract. If desired, provision for the assessment of liquidated damages may be included. As for competitive-bid contracts, it is important that the amount specified as liquidated damages should be a real measure of the actual damages which the owner would suffer if the contractor should fail to complete the work before the designated date. Otherwise the courts may hold that such liquidated damages are in reality a penalty and therefore void.

Control of the Work. The contract should provide for each party to name a representative in whom is vested responsibility for the carrying out of the requirements of the work and who has authority to act on all matters pertaining to the contract. Provision should also be made for the allocation of responsibility for the preparation of plans, specifications, shop drawings, approval of accounting methods, inspection of materials and workmanship, termination of contract, settlement of disputes, approval of purchases and subcontracts, licenses, permits, patent rights, progress schedules, etc. In general, the owner will be represented by the engineer, and the contractor by his project manager or superintendent.

The contractor's employees and the compensation which they receive are made subject to the approval of the engineer. Upon the signing of the contract the contractor should submit organization charts showing the executive and administration personnel to be assigned to the work with a statement of the duties of each employee and the administrative procedures to be followed for the control and direction of the work. Similar data may be desirable for the field organization. As for other types of contracts, progress charts and schedules are necessary for the coordination of the work.

Reimbursable Direct Costs. Under this form of contract, the contractor is reimbursed in the amount of the actual net cost (out-of-pocket expenditures) of the work paid by him, including overhead or indirect costs chargeable to the work, plus the fixed fee. Direct costs are those incurred on the site of the work and approved for payment by the engineer. Types of direct costs are:

a. The actual net cost for all services and labor, including transportation, hospitalization for occupational injury or illness, vacation pay, payroll assessments, and other similar and usual expenses incident to the employment of labor.

b. The actual net cost of all materials and supplies, including the cost of purchasing, inspection, storage, transportation, salvage, and other similar and usual expenses incident to the procurement and use of materials.

c. The actual net cost for all tools, plant, and equipment purchased or rented by the contractor for use in the construction work, including insurance, transportation, and other collateral costs.

d. The actual net cost of licenses, permits, royalties, patent rights, insurance premiums, bonds, and the like.

e. The actual net cost of keeping and auditing the project records and accounts.

f. The actual net cost of communication services connected with the work, such as telegrams, telephones, radio messages, and postage.

Reimbursable Overhead Costs. The contractor's indirect overhead costs are those not normally charged or allocated to any specific job but which are carried in his accounts as the general expenses of doing business in his central offices. Typical of such costs are salaries of principals of the firm, rent, light, telephones, secretarial services, sales and promotional costs, expediting, and the like. Although they cannot be charged directly against any specific project, overhead costs are none the less real, and the reimbursement of a proportional share of these expenses under a cost-plus contract is justifiable.

A rational breakdown of overhead costs is difficult when a contractor has a large number of contracts in progress concurrently. Frequently on large cost-plus contracts, the contractor is required to set up a special office where all matters related to the contract are handled and all resultant costs are considered to be directly reimbursable. No overhead costs from the contractor's central office are approved for payment under this plan, it being assumed that all such costs not otherwise covered are included in the fee.

If the contractor has only the one project under construction and his entire organization is engaged on the job, all indirect overhead costs can be allocated to the contract and reimbursed. It should be noted, however, that certain types of overhead costs, particularly on U.S. Government contracts, are always considered to be non-reimbursable, and these the contractor must pay out of his fee.

On construction work of a specialized nature, overhead costs are reasonably proportional to direct labor costs. Where this relationship can be determined by acceptable audits of the contractor's books, it offers a method for the reimbursement of overhead costs on an equitable basis without the difficulties and uncertainties which accompany efforts to segregate actual overhead expenses. Under this method the contract provides for the allowance of a fixed percentage of the direct labor costs to cover all of the contractor's indirect overhead. As an alternative a lump sum may be negotiated and included in the contract for this purpose. These methods are particularly suitable on relatively small jobs, and much time and accounting expense may be saved thereby.

Non-Reimbursable Costs. Unless provision is made to the contrary, the expenses of the contractor's central office including the salaries of principals generally are considered non-reimbursable. Also excluded from reimbursement are such items as advertising, donations, entertainment expenses, bad debts, interest paid on loans, commissions and brokerage fees, and costs incurred through failure to take advantage of discounts and credits. Likewise, all costs which arise from unrealized claims and damages and those which arise from the gross negligence of the contractor are non-reimbursable. Any consideration of the excluded items should be made in connection with the determination of the amount of the fixed fee.

Labor and Materials. It is in the owner's interest to prevent labor controversies on the work, and the contract should require the payment of wage rates which are standard or prevalent at the location of the project. All labor costs should be supported by approved time sheets which show how time and costs are determined and allocated to the various parts of the work. All wage rates and salaries of the contractor's employees should be subject to the approval of the engineer, and proof of the payments to workers should be available to the project auditors.

In the procurement of materials for the work, the contractor acts as the purchasing agent for the owner. All materials purchases should be based upon approved purchase orders or storeroom requisitions and classified by major categories of materials. Original invoices, receiving reports, and canceled checks covering payment should be available to substantiate the purchases for the information of the project auditors.

Before approving purchase orders, the engineer should require the contractor to obtain quotations from a number of vendors, usually a minimum of three, to make certain that the lowest practicable price quotations are obtained. All materials bills should be paid in time to take advantage of cash discounts. When through negligence this is not done, the contractor should be held responsible.

Plant and Equipment. When the contract requires the contractor to furnish all plant and equipment required for the work on a rental basis, two methods are commonly used to determine the amounts of the rentals, as elements of the cost of the project. The procedure recommended by the Associated General Contractors of America is based on a flat rental rate for each type of equipment. This is simple to apply, but it has the disadvantage that insufficient distinction is made in the payment for similar pieces of equipment of different ages and production capacities. The Bureau of Yards and Docks determines rental rates on the basis of actual annual cost of ownership to the contractor. Payment under this method includes insurance premiums, depreciation, property taxes, interest on investment, and general administration and operating expenses. Neither method should allow a rental profit to the contractor, all profit being covered by the fixed fee. The contract should provide for operating repairs during the course of the work and a general overhaul at the end of the work to replace the equipment in its original condition less normal wear and tear. In view of the uncertainties that are likely to arise at the end of the work relative to the original condition, it is advisable that a survey and appraisal of the equipment be made both at the beginning and at the end of the work.

When it becomes necessary to use a piece of equipment not owned by the contractor the owner may either have to purchase and furnish it himself, or the contractor must rent it from a third party. Such a rental will be on a different basis than that of contractor-owned equipment in that the rental rate will include a profit for the owner of the equipment. Third-party rentals require close control since they work to the advantage of the contractor and the disadvantage of the owner. Where possible during the preliminary negotiations a definite understanding should be reached relative to specific items of equipment to be purchased by the owner or rented from third parties.

The contract should be explicit where maximum total rental and recapture provisions are desired. On large contracts provision is sometimes made for rental payments to stop when the total rentals paid equal some specified percentage of the purchase price of the equipment. Recapture clauses permit the owner to purchase the equipment from the contractor at any time during the life of the contract, previously paid rental being applied to the purchase price.

For large projects, when conditions in the equipment market permit, it usually will be economical for all plant and equipment to be purchased for the owner by the contractor as items of reimbursable cost. This will result in increased efficiency and less total cost through the use of new equipment on the job, as compared with the rental of contractor-owned, used equipment. Inasmuch as the efficient life of most types of construction equipment will be exhausted at the end of a major project, the problems of resale or salvage will be minor and all questions of repair and overhaul will be avoided.

In all jobs of appreciable magnitude, it will be desirable to have all small tools purchased by the contractor as items of reimbursable cost. Small tools normally are used up during the work, and the problems of accounting are such that the rental method is impracticable. As an alternative to the purchase of small tools for the account of the owner, the contractor may be paid a negotiated lump sum to furnish all small tools for the work. For minor jobs it is customary for the contractor to be required to furnish all small tools on a non-reimbursable basis.

Subcontracts. Under a cost-plus-fixed-fee contract, it is usually anticipated that the general contractor will accomplish about 50 to 80 per cent of the work with his own forces and that the remainder will be done by specialty subcontractors. However, as the trend towards specialization in the construction industry continues, a change in these proportions can be expected. The amount of the work to be performed by specialty subcontractors will also depend upon the understanding with the general contractor when the prime contract is negotiated. It is not uncommon for the general contractor to be retained primarily as a construction manager and administrator, with the understanding that practically all of the work will be performed by subcontractors.

Ordinarily all subcontracts are drawn on a lump-sum or unit-price basis and are awarded by competitive bidding with a minimum of three bidders. Occasionally it may be desirable to use cost-plus-fixed-fee subcontracts to meet special conditions. When this is contemplated, the prime contract should specify the methods to be followed in the selection of subcontractors and the determination of subcontractors' fees. Like purchases, all subcontracts should be subject to approval by the engineer.

Method of Compensation. At predetermined intervals of from seven to thirty days the contractor should submit payment vouchers accompanied by certified payrolls, paid invoices, bills, and other expenses, all previously approved by the engineer. Payment to the contractor should be made as soon as possible after receipt of the requisition. The payment should also include a proportional payment on account of the fixed fee.

Changes. Changes may be made at any time in the drawings and specifications within the general scope of the work without altering the conditions of the contract. However, if such changes cause a material change in the amount or character of the work, a corresponding adjustment should be made in the contractor's fee. All changes should be made formally by written order. If the change is minor in character and related directly to the contract work, the engineer may handle the matter by the issuance of a written change order, accepted by the owner and by the contractor, in which a proportional adjustment is made in the fee. When the change involves a large alteration in plans or when new projects of considerable magnitude are introduced into the contract, it is recommended that the work be covered by a supplemental agreement prepared for formal acceptance by both parties.

Termination. In all U.S. Government contracts and in many classifications of private work, provision is made for the termination of the contract at the owner's initiative. Generally, when the work is being performed under a unit-price or lump-sum agreement, termination before completion of the work is enforced only in extreme circumstances. On a cost-plus-fixed-fee basis, however, termination may be more readily effected, and it is usually agreed that the owner may terminate at will without fault on the part of the contractor. When this is done the contractor must be compensated equitably for the expense of mobilizing, organizing, and demobilizing the work as well as for all costs arising directly out of the work. If the contractor can show that he has suffered expense through the disorganization of his business resulting from the cancellation of the contract, the owner may be liable for damages.

Accounting Methods and Control of Costs. On minor projects the contractor's vouchers are usually accepted for reimbursement by the owner, and no check is made of the contractor's expenditures for labor and materials at the site of the work. For important work, however, the owner should employ a competent auditor, experienced in construction accounting practice, to keep a constant check and control on the contractor's accounting methods.

The contractor should be required to keep accurate records and books of accounts in accordance with standard accounting procedures. Reports should be required periodically with all essential details of the receipt and disbursement of all funds, the outstanding obligations and commitments, the cost and use of all items of labor, materials, equipment, supplies, services, and all other charges under the contract. Concurrently, estimates of cost should be made for the completion of all classifications of work under the contract. These reports serve to keep the owner advised of the financial status of the project as compared with his budget for the work. They also serve as a constant check on the accuracy and efficiency of the contractor's performance.

A detailed discussion of accounting procedures and cost control methods is outside the scope of this book. For an excellent treatment of these subjects, the reader is referred to the manual *Con-*

struction Cost Control prepared and published by the American Society of Civil Engineers, 33 West 39th Street, New York, N. Y.

Some Precautions in the Use of Cost-Plus-Fixed-Fee Contracts. As in other types of construction contracts, uncertainties and disputes will arise unless the administrative control of the work is properly planned and executed. Experience in the use of the cost-plus-fixed-fee form indicates that controversies are most likely to occur in connection with the following.

a. Changes in scope. All matters related to changes in the scope of the work and the corresponding changes in the contractor's fee should be settled definitely at the time the change is ordered and before the work is completed.

b. Equipment rental and overhaul. The methods for determining the amounts of equipment rental rates and the basis for repair and overhaul allowances should be definitely established at the beginning of the work in order to avoid controversies when the contract is being closed out.

c. Subcontracts and third-party rentals. If it is intended that any subcontracts are to be approved on a fixed-fee basis, the conditions relative to the duplication of fees paid to the prime contractor and the subcontractor for the same services should be agreed upon at the beginning of the work. This also applies to equipment rented from third parties on rates different from those established for the equipment owned by the prime contractor.

d. Cost accounting system. Uncertainties in cost finding and delays due to controversies over reimbursable costs may be expected unless the details of the accounting methods are worked out and agreed upon in advance.

e. On large projects the work may be broken down into several separate contracts, some of which may be on a cost-plus basis, whereas others are let by competitive bidding. On such work no contractor should be allowed to have contracts of both types concurrently because of the difficulty in separating the non-reimbursable costs under the competitive-bid contract from the reimbursable costs under the cost-plus contract.

f. Closing out of the contract. The contractor, either voluntarily or involuntarily, tends to slow down production and drag out the work toward the completion of the contract. This tendency may be more pronounced if the contractor has no other work in view in order to maintain his organization on a reimbursable basis as long as possible. Therefore, increasing overhead and production costs are to be expected near the end of the work. If this tendency should develop, its objectionable results may be avoided by terminating the cost-plus-fixed-fee contract and arranging for the completion of the unfinished portion of the work on a lump-sum or unit-price basis under a new contract with the same contractor.

U.S. Government Cost-Plus-Fixed-Fee Contracts. It should be noted that all government contracts are subject to review by the General Accounting Office. This agency has the authority to reverse previous findings by contracting officers. In the case of cost-plus-fixed-fee contracts, reimbursable items of cost previously approved and paid under the contract may be held unjustifiable by the General Accounting Office. Such action results ultimately in a claim against the contractor for the return of the disallowed funds. If good faith on the part of the contractor and the Government Contracting Officer is assumed, this situation results in a unique hazard to the contractor which must be anticipated when he enters into the contract. A similar hazard exists with regard to renegotiation proceedings which may be instituted at the completion of the work under war conditions, with the view of ascertaining the extent to which the contractor's profit under the contract may be considered excessive. The contractor is obliged to return to the government the full amount of the excess profit.

For the use of the Government in the review and examination of the contract, the contractor is required to retain and preserve at his own expense all records and accounts pertaining to the project for a minimum period of six years after the completion of the work. After the lapse of this period, if the contractor desires to dispose of the records he may notify the contracting officer who will either authorize their destruction or arrange for their transfer to the Government for disposition.

Example of Cost-Plus-Fixed-Fee Contract. To illustrate the wording and format of a cost-plus-fixed-fee contract, the standard agreement of the Bureau of Yards and Docks, U.S. Navy, is reproduced herein. Excepting the clause covering the contractor's fee, this document also serves to illustrate the variations of the cost-plus type of contract such as the sliding-scale-of-fees, the guaranteed ceiling of cost, and the fixed fee with bonus and penalty provisions. As in all Government contracts this form contains many unilateral provisions designed to protect the Government. In general such provisions are not encountered in private contracts.

EXAMPLE OF COST-PLUS-FIXED-FEE CONTRACT

CONTRACT FOR CONSTRUCTION *

THIS NEGOTIATED CONTRACT, entered into as of the   day of   , 1951, by THE UNITED STATES OF AMERICA, hereinafter called the GOVERNMENT, represented by the CHIEF OF THE BUREAU OF YARDS AND DOCKS, DEPARTMENT OF THE NAVY, hereinafter called the CONTRACTING OFFICER, and

of the city of

hereinafter called the CONTRACTOR,
WITNESSETH:

the parties hereto do mutually agree as follows:

WORK TO BE DONE AND FIXED-FEE

ARTICLE 1. (a) The Contractor shall construct or otherwise accomplish the completion of the following public works project or projects at the locations indicated, said project(s) being designated by general titles and the estimated cost being stated to indicate generally the degree of magnitude and not a limit of cost, viz.:

Project No.   *Description*

Net Amount ....
Plus Fixed-Fee ....

Estimated Construction Cost, Including Fee......

(b) The amount of the fixed-fee is

(c) The above-mentioned projects designated by general titles may be further indicated by lists of individual projects to be prepared and approved by the Contracting Officer to define with more particularity the scope of the work contemplated by the contract, and such lists may be furnished to the Contractor and be considered to be a part of the contract.

* Revised March, 1952.

(*d*) The Contractor shall proceed immediately with the organization of office and field forces to be engaged upon the work under this contract and shall direct his efforts toward early purchases and transportation of materials and the initiation of actual construction work on the site and shall concentrate upon rapid progress and the completion of the entire work at the earliest possible date. It is the policy of the Government as declared by the Congress to bring about the greatest utilization of small business concerns which is consistent with efficient production. The Contractor agrees to accomplish the maximum amount of subcontracting to small business concerns that the Contractor finds to be consistent with the efficient performance of this contract.

PLANS AND SPECIFICATIONS

ARTICLE 2. (*a*) The projects indicated in Article 1, as and when accomplished, shall conform to plans and specifications which may be furnished by the Government to the Contractor from time to time during the progress of the work covered by the contract.

(*b*) The Contractor shall furnish all plans and specifications to supplement those furnished by the Government, which may be determined by the OFFICER IN CHARGE to be necessary for the accomplishment of the projects. All plans and specifications so furnished by the Contractor shall be subject to approval by the OFFICER IN CHARGE and become the property of the Government.

CHANGES

ARTICLE 3. The Contracting Officer may at any time, by a written order, make changes in approved drawings and/or specifications within the general scope of the work required under this contract or suspend, omit, or add projects or parts thereof. If such changes or the addition of any project or omission or suspension of any part or portion of the original project cause a material increase or decrease in the amount or character of the work to be done under this contract, an equitable adjustment of the amount of the fixed-fee to be paid to the Contractor shall be made by the Contracting Officer and upon written notice thereof to the Contractor the contract shall be deemed to be modified in writing accordingly. Any claim by the Contractor for modification of such adjustment must be asserted within 30 days from the date he receives such written notice of adjustment; PROVIDED, HOWEVER, that the Contracting Officer, if he determines that the facts justify such action, may receive and consider, and adjust any such claim asserted at any time prior to the date of final settlement of the contract. If the parties fail to agree upon the adjustment to be made the dispute shall be determined as provided in Article 21 hereof. Nothing provided in this article shall excuse the Contractor from proceeding with the prosecution of the work so changed.

GOVERNMENT OFFICERS

ARTICLE 4. The Government will designate an officer of the Civil Engineer Corps, U.S. Navy as "Officer in Charge of Construction," herein referred to as "Officer in Charge," who, under the direction of the Contracting Officer, shall have complete charge, on behalf of the Government, of the work under this contract in the field.

CONTRACTOR'S ORGANIZATION AND METHODS

ARTICLE 5. (*a*) The Contractor shall designate a Project Manager or Superintendent who, on behalf of the Contractor, shall have complete charge of all work under this contract.

(*b*) The Contractor shall also designate such qualified and experienced engineers as may be required, each of whom, under the direction of the Project Manager, shall have charge in the field, on behalf of the Contractor, of such projects or sections or divisions of projects as may be designated by the Contractor.

(*c*) The Contractor agrees that no representative provided for in this article, nor any other employee occupying a key position in the field organization will be withdrawn or separated from his assignment during the progress of the work except for cogent reasons and after full consultation with the Officer in Charge.

(*d*) Upon the execution of this contract the Contractor shall submit to the Contracting Officer a chart showing the Executive and Administrative personnel to be regularly assigned for full or part time service in connection with the work under the contract, together with a written statement of the duties of each person and the administrative procedure to be followed by the Contractor for the control and direction of the work; and the data so furnished shall be supplemented as additional pertinent data become available. There shall also be submitted to the Contracting Officer by the Contractor charts of the various field organizations showing all personnel, other than artisans, mechanics, helpers, and laborers, to be assigned for full or part time service outside the central office organization, together with a written statement of the duties and rates of pay of each person and the procedure proposed to be followed by the Contractor for the accomplishment of all field work, including temporary requirements; and the data so furnished shall be supplemented as additional pertinent data become available.

(*e*) The Government may award other contracts for work at the site of any work under this contract or may perform work with Government labor, and the Contractor shall fully cooperate with any other Contractor and/or Government labor forces and carefully fit his own work to that provided under other contracts or performed by Government labor as may be directed by the Officer in Charge. The Contractor shall not commit or permit any act which will interfere with the performance of work by any other Contractor or by Government employees.

SERVICES AND LABOR

ARTICLE 6. (*a*) All services and labor, including personal services of every character, except such as may be furnished by the Government, required outside the central office organization of the Contractor for the accomplishment of the work under this contract shall be furnished by the Contractor. The salaries or compensation paid for services and labor shall be subject to the approval of the Contracting Officer.

(*b*) No person shall be assigned to service by the Contractor as Superintendent of Construction, Chief Engineer, Chief Purchasing Agent, Chief Accountant, or similar position in the field organizations, or as principal assistant to any such person, until his employment has been approved by the

Officer in Charge.   There shall be submitted to the Officer in Charge such information as he may request as to the experience, qualifications, and former compensation of such persons.   Contracts for such services, so far as they shall constitute items of cost to the Government, shall include provisions as to salary or compensation and as to travelling, lodging, subsistence, leaves of absence, and special (if any) allowances, and shall be subject to approval by the Contracting Officer.

(c) The Officer in Charge may require the Contractor to dismiss from the work any employee that the Officer in Charge deems incompetent, careless, insubordinate, trouble making, or otherwise objectionable.

(d) No persons other than those engaged upon the work under this contract shall be lodged or quartered upon any Naval reservation within which work under the contract is being carried on or be allowed to visit the site of such work except when so authorized by a pass issued by the Officer in Charge or superior authorities.

(e) The Contractor shall provide and maintain at the site an efficient and reliable police and guard force, if and as directed by the Officer in Charge, for the protection of personnel and property.   No person shall be assigned to such force except after approval by the Officer in Charge.   In the event the site of work to be performed hereunder, as indicated in Article 1, is within the confines of a naval or military activity, all rules and regulations established by the Commanding Officer of the activity relating to the performance of guard and police duties shall be strictly observed.

PLANT AND EQUIPMENT

ARTICLE 7. (a) The rental compensation for items of plant and equipment owned, or controlled for use hereunder as if owned, by the Contractor which are furnished for performance of the work hereunder shall be calculated on the basis of cost to the Contractor with no allowance for profit.   Rental compensation under this paragraph shall be calculated in accordance with "Uniform Rental Rates for Contractor-Owned Construction Plant under Cost-Plus-a-Fixed-Fee Contracts," as prescribed by the Contracting Officer.

(b) All equipment shall be delivered to the work in first class working condition.   During the progress of the work repairs shall be made as required by the exigencies of the work.   At the end of the work, the equipment shall be placed in as good condition as when delivered to the work, minus ordinary wear and tear.   The cost of these repairs shall constitute items of cost under the contract.

(c) No item of equipment costing in excess of $1,000 shall be purchased without the prior written approval of the Contracting Officer; similarly, written approval of the Contracting Officer shall be obtained prior to the purchase or rental of any passenger automotive vehicle, typewriter, office machine, or other office type labor-saving device regardless of cost.

(d) No item of equipment shall be rented by the Contractor from a third party without prior approval of the Officer in Charge.   Prior to granting such approval, the Officer in Charge will first estimate the total rentals to be paid for the use of the item of equipment and if such estimate exceeds 20% of the valuation of said item of equipment the rental shall not be authorized except with the written approval of the Contracting Officer.   If the estimated total rentals to be paid for the item of equipment does not exceed 20% of its valuation, the Officer in Charge may approve rental thereof by the Contractor.

(*e*) The Contracting Officer may in his discretion and on behalf of the Government, take possession at any place he may elect of any item of plant or equipment for the purpose of transporting it to the site where it is to be used or held for further disposition and may subsequently return any such item to the possession of the Contractor for use on the work. Final disposition of all items of Government plant and equipment shall be made as directed by the Contracting Officer.

(*f*) The Contractor agrees to use such items of plant and equipment and such shop, storage, transportation, communication, and other facilities owned by the Government as may be available to him and as directed by the Officer in Charge.

MATERIALS—PURCHASES

ARTICLE 8. (*a*) Except as provided in Article 7 or where provision is otherwise made by the Officer in Charge, all materials, articles, supplies, and equipment required for the accomplishment of the work under this contract shall be procured by the Contractor for the Government. The Contractor shall act as the purchasing agent of the Government in effecting such procurement, and the Government shall be directly liable to the vendors for the purchase price. The exercise of this agency is subject to the obtaining of approval in the instances and in the manner required by subparagraph (*c*) of this Article. The Contractor shall negotiate and administer all such purchases and shall advance all payments therefor unless the Officer in Charge shall otherwise direct.

(*b*) Title to all such materials, articles, supplies, and equipment, the cost of which is reimbursable to the Contractor hereunder, shall pass directly from the vendor to the Government without vesting in the Contractor, and such title (except as to property to which the Government has obtained title at an earlier date) shall vest in the Government at the time payment is made therefor by the Government or by the Contractor or upon delivery thereof to the Government or the Contractor, whichever of said events shall first occur. This provision for passage of title shall not relieve the Contractor of any of its duties or obligations under this contract or constitute any waiver of the Government's right to absolute fulfillment of all of the terms hereof.

(*c*) No purchase of materials, articles, or supplies in excess of $500 shall be made hereunder without the prior written approval of the Officer in Charge, except that the Officer in Charge may, in his discretion, either reduce the limitation on the amount of any purchase which may be made without such prior approval or authorize the Contractor to make purchases in amounts not in excess of $2,500 for any one purchase without obtaining such prior approval.

(*d*) No subcontract shall be entered into by the Contractor without the prior written approval of the Officer in Charge.

(*e*) Each cost-plus-a-fixed-fee subcontract shall, unless otherwise directed by the Contracting Officer, provide that the subcontractor shall act as the purchasing agent of the Government to the extent and with the same authority as that specified in subparagraph (*a*) of this article in regard to such action by the Contractor, and shall likewise provide that title to purchases for which the subcontractor is entitled to reimbursement shall pass to the Government in the same manner as that specified in subparagraph (*b*) of this article in regard to purchases for the cost of which the Contractor is entitled to re-imbursement.

(*f*) The Officer in Charge may take charge of any materials, articles, sup-plies, or equipment, procured under this contract for the purpose of transport-ing them to the site where they are to be used or held for further disposition.

(*g*) Final disposition of excess items furnished hereunder shall be made as directed by the Officer in Charge.

LICENSES AND PERMITS

ARTICLE 9. The Contractor shall determine what permits or licenses are required in connection with the accomplishment of the work under this con-tract and, with the approval of the Officer in Charge, shall take the necessary action to secure the same as required: PROVIDED, That all sites of the work covered by this contract and essential working spaces adjacent thereto which are owned by or under the control of the Government will be made available to the Contractor by the Government. Any such permits or licenses issued by any Federal Government Department or agency will, upon request made by the Contractor, be procured by the Contracting Officer.

COMPENSATION

ARTICLE 10. (*a*) As compensation for the performance of the work here-under, the Government shall pay to the Contractor the allowable net cost thereof as determined in accordance with Part 4 of Section XV of the Armed Services Procurement Regulation provided such cost has been incurred in strict conformity with the provisions of this contract, plus the fixed-fee stated in Article 1 (*b*). Notwithstanding the foregoing, the Contractor shall be reimbursed for payments made to cover pension, retirement, group health, accident and life insurance plans for employees, (except the Contractor's central office personnel), as approved by the Officer in Charge; provided, however, that the Contractor shall not be reimbursed for payment of pre-miums for insurance of lives of directors, officers, proprietors, or other persons where the Contractor is the beneficiary directly or indirectly.

(*b*) In determining the allowable cost of articles and materials of every kind required for the purposes of this contract there shall be deducted from the gross cost thereof all cash and trade discounts, rebates, allowances, credits, salvage values, and commissions which have accrued to the benefit of the Contractor, or would have so accrued except for the fault or neglect of the Contractor. Such benefits lost through no fault of the Contractor shall not be deducted from gross cost. The Contractor shall, to the extent of his ability, take advantage of all such benefits and if unable to do so in any instance; shall promptly notify the Officer in Charge in writing to that effect and the reason therefor.

PAYMENTS

ARTICLE 11. (*a*) The Contractor may submit to the Officer in Charge at intervals of not less than seven calendar days, unless otherwise authorized by the Officer in Charge, payment requisitions accompanied by duly certified and approved paid payrolls, paid invoices, paid bills, and other substantiating documents, including the Contractor's bill on account of the prescribed fixed-fee, all equaling the amount of the requisition. Payment requisitions covering all items shall be submitted to the Officer in Charge. As soon as practicable after the receipt and audit by the Officer in Charge of such payment requisition there will be prepared a Public Voucher, in the amount of such requisition as is approved by the Officer in Charge, which voucher shall be signed by a duly authorized representative of the Contractor, and thereafter by the Officer in Charge or Resident Officer in Charge, and thereupon transmitted to such Naval Disbursing Officer as may be designated by the Contracting Officer for issuance and delivery of a Government check drawn payable to the order of the Contractor. The Officer in Charge shall have the right to defer approval of payments at any time in an amount not to exceed ten percentum of all payments previously made on account of the Contractor's fees if in his judgment such action is necessary to protect properly the interests of the Government.

(b) Subject to the withholding provisions of the preceding paragraph, partial payments on account of the fixed-fee shall be in the same proportion to such fee as payments of reimbursable costs to the Contractor, plus the value of Government-furnished materials in place, are to the total estimated cost of the work at the time the partial payment is to be made, including the total value of materials to be furnished by the Government, but exclusive of fee.

(c) After payment of eighty per cent (80%) of the total estimated cost, further payment shall be withheld until a reserve of not less than shall have been set aside, such reserve to be paid to the Contractor at the time of final payment except when otherwise authorized by the Contracting Officer. The Contractor and each assignee under an assignment in effect at the time of final payment shall execute and deliver, at the time of and as a condition precedent to final payment, a release in form and substance satisfactory to and containing such exceptions as may be found appropriate by the Contracting Officer, discharging the Government, its officers, agents, and employees of and from liabilities, obligations, and claims arising under this contract. The estimated cost referred to in the first sentence of this paragraph is the estimated cost exclusive of the estimated cost of the property to be furnished by the Government (if any) as such estimated costs may be determined or modified from time to time by the Contracting Officer.

PROPERTY LOSS OR DAMAGE

ARTICLE 12. (a) The Contractor is requested not to carry or incur the expense of any insurance against any form of loss of or damage to equipment or materials furnished by the Government hereunder, or any property to which the Government has taken and continues to hold title hereunder, and no reimbursement will be allowed for such insurance premium expenses.

(b) The Government assumes the risk of loss of or damage to such property, whether or not caused by the negligence of the Contractor, his agents, servants, or employees, including expenses incidental to such loss or damage. Notwithstanding the foregoing assumption of risk, the Contractor shall be responsible for any loss or damage for which he is expressly made responsible under any other provision of this contract, or which results from willful misconduct or lack of good faith on the part of any of the Contractor's directors, officers, or any of his other representatives having supervision or direction individually or collectively of all or substantially all of the Contractor's operations under this contract.

LIABILITY TO THIRD PERSONS

ARTICLE 13.   (a) The Contractor shall procure and thereafter maintain workmen's compensation, employer's liability, and bodily injury liability insurance, with respect to work done hereunder, and such other liability insurance with respect to such work as the Contracting Officer may from time to time require or approve. All such insurance shall be in such form, in such amounts, for such periods of time, and with such insurers, as the Contracting Officer may from time to time require or approve.

(b) The Contractor shall procure any special insurance or bonds as may be specifically required by the Contracting Officer.

(c) The Contractor shall be reimbursed (1) for the cost of such insurance of the character described in paragraph (a) of this article as may be required or approved by the Contracting Officer, (2) for the cost of any bonds or special insurance procured under paragraph (b), and (3) for liabilities to third persons for loss of or damage to property, death, or bodily injury not compensated by insurance or otherwise, arising out of and incurred during the performance of this contract, whether or not caused by the negligence of the Contractor, its agents, servants, or employees, provided such liabilities are represented by final judgments or by settlements approved in writing by the Contracting Officer, and expenses (as authorized by the Contracting Officer), incidental to such liabilities, except liabilities (i) for which the Contractor is otherwise responsible under the express terms of this contract, or (ii) with respect to which the Contractor has failed to insure as required or approved by the Contracting Officer, or (iii) which result from willful misconduct or lack of good faith on the part of any of the Contractor's directors or officers or of any of its other representatives having supervision or direction of all or substantially all of the Contractor's operations under this contract.

FEDERAL, STATE, AND LOCAL TAXES

ARTICLE 14.   (a) Any tax or similar charge (other than a tax upon income of the Contractor or any subcontractor and other than a tax imposed upon, by reason of, or measured by the Contractor's or subcontractor's fee) paid by the Contractor or any cost or cost-plus-a-fixed-fee subcontractor hereunder (provided that there is no fixed-price subcontractor intervening between such subcontractor and the prime Contractor) which was imposed legally by the Federal Government or any state or local Government and is directly applicable to the completed articles, or work to be furnished under this contract, or to articles incorporated in such completed articles or work, or to the materials or services required or used in the production of such supplies or work, or to the importation, transportation, production, processing, manufacture, construction, sale, or use of such articles, services, or materials shall constitute an item of Allowable Cost under this contract or such cost or cost-plus-a-fixed-fee subcontract, PROVIDED, however, that the Contractor shall request, and, at its option, the Government may issue appropriate tax exemption certificates or furnish other proof of exemption for use by the Contractor and any such cost or cost-plus-a-fixed-fee subcontractor hereunder in obtaining exemption from any such state or local tax incurred in the performance of this contract, or such subcontract hereunder, and upon issuance of such tax exemption certificate or furnishing of such proof of exemption with respect to any such tax or charge, such tax or charge shall not be considered as an item of Allowable Cost under this contract.

(b) Notwithstanding the foregoing, amounts paid by the Contractor in respect of any such state or local tax for which a tax exemption certificate or other proof of exemption has been furnished hereunder will be considered items of Allowable Cost under this contract if, after the state or local taxing authorities refuse to recognize such tax exemption certificate or other proof of exemption, the Contractor shall notify the Contracting Officer of the state or local taxing authorities' refusal to recognize such tax exemption certificate or other such proof of exemption and shall take such steps as may be requested by the Government to cause such tax to be paid under protest, to preserve and to cause to be assigned to the Government any and all rights to any refund of such tax, to permit the Government to prosecute any claim, litigation, or proceeding in the name of the Contractor and to furnish to the Government all reasonable assistance and cooperation requested by the Government in the prosecution of any such claim, litigation, or proceeding for the recovery of such tax.

Losses

ARTICLE 15. Specific items of loss or expense not compensated by insurance or otherwise, actually sustained by the Contractor in connection with the work and found and certified by the Contracting Officer to be just and reasonable and to be reasonably incident to work of the nature and magnitude of that contemplated by this contract, shall constitute items of Allowable Cost under this contract, unless reimbursement therefor is especially prohibited by the terms of this contract or is expressly unallowable under the provisions of Part 4, Section XV, of the Armed Services Procurement Regulation: PROVIDED, That such reimbursement shall not include any amount for which the Contractor would have been indemnified or compensated by insurance except for failure of the Contractor to procure or maintain bonds or insurance in accordance with the requirements of the Contracting Officer pursuant to the provision of Article 13 of this contract.

Notice of Claims

ARTICLE 16. The Contractor shall give the Contracting Officer immediate notice of any suit or action filed, or any claim made, against the Contractor arising out of the performance of this contract, the cost and expense of which is reimbursable to the Contractor under the provisions of this contract, and the risk of which is then uninsured or in which the amount claimed exceeds the amount of insurance coverage. The Contractor shall furnish immediately to the Contracting Officer copies of all pertinent papers received by the Contractor. If the amount of the liability claimed exceeds the amount of insurance coverage, the Contractor shall authorize representatives of the Government to collaborate with counsel for the insurance carrier, if any, in settling or defending such claim. If the liability is not insured, the Contractor shall, if required by the Contracting Officer, authorize representatives of the

Government to settle or defend any such claim and to represent the Contractor in or take charge of any litigation in connection therewith.

GRADE OF MATERIALS AND WORKMANSHIP

ARTICLE 17.   Unless otherwise authorized by the Officer in Charge, all workmanship, equipment, materials, and articles incorporated in the work covered by this contract or provided for temporary use are to be of the most suitable grade of their respective kinds for the purpose. Where equipment, materials, or articles are referred to in specifications as "equal to" any particular standard, the Contracting Officer shall decide the question of equality. When required by specifications, or when called for by the Officer in Charge, the Contractor shall furnish to the Contracting Officer for approval full information concerning the materials or articles which he contemplates incorporating in the work.   Samples of materials shall be submitted for approval when so directed by the Officer in Charge.

INSPECTION

ARTICLE 18.   (a) All materials and workmanship, except as may be otherwisewise provided herein or otherwise authorized by the Officer in Charge, shall be subject to inspection, examination, and test by Government inspectors, or inspectors employed by the Contractor with the approval of the Officer in Charge, at any and all times during manufacture and/or construction and at any and all places where such manufacture and/or construction are carried on.   The Government shall have the right to reject defective material and workmanship or require its correction.   Rejected workmanship shall be satisfactorily corrected and rejected material shall be satisfactorily replaced with the proper material, and the Contractor shall promptly segregate and remove the rejected items from the premises.

(b) The Contractor shall furnish promptly all reasonable facilities, labor, and materials necessary for the safe and convenient inspection and tests that may be required by the inspectors.   All inspection and tests by the Government shall be performed in such manner as not to delay the work unnecessarily.   Special, full size, and performance tests shall be as specified.

(c) Should it be considered necessary or advisable by the Government at any time before final acceptance of the entire work to make an examination of work already completed by removing or tearing out same, the Contractor shall on request of the Officer in Charge promptly furnish all necessary facilities, labor, and material for the purpose.

(d) Inspection of material and finished articles to be incorporated in the work at the site shall be made at the place of production, manufacture, or shipment, whenever the quantity justifies it, unless otherwise stated in specifications; and such inspection and acceptance in writing unless otherwise stated in specifications shall be final, except as regards latent defects, departures from specific requirements of the contract and the specifications and drawings, damage or loss in transit, fraud, or such gross mistakes as amount to fraud.   Subject to the requirements contained in the preceding sentence, the inspection of material and workmanship for final acceptance as a whole or in part shall be made at the site.

(*e*) The actual net cost to the Contractor of labor, facilities, and material necessarily incident to the inspections provided in this article including the correction or replacement of defective material and workmanship and of reconstruction, shall be allowed the Contractor as a reimbursable cost, provided, however, that in the event of disclosure of defects resulting from gross negligence or fraud on the part of any of the Contractor's directors or officers or of any of its other representatives having supervision or direction of all or substantially all of the Contractor's operations under this contract, the cost of the examination and of satisfactory correction, replacement or reconstruction in respect of such defects shall be borne by the Contractor, and he is expressly made responsible therefor without right of reimbursement hereunder.

Records and Accounts

Article 19. (*a*) The Contractor shall keep books, records, documents, and other evidence (herein collectively called "the records") bearing on its costs and expenses under this contract and in respect of any termination of work hereunder. The Contractor's method of accounting shall be in accordance with the Bureau of Yards and Docks' *Manual of Accounting, Auditing, and Control for Negotiated Cost-Plus-a-Fixed-Fee Contracts,* subject to such modifications therein as may be authorized or directed by the Contracting Officer.

(*b*) The Contractor shall furnish to the Department of the Navy upon request such of the records as may be required by the General Accounting Office.

(*c*) Except for such of the records furnished by the Contractor to the Department of the Navy pursuant to paragraph (*b*) of this Section and retained by the Department of the Navy, or the General Accounting Office, the Contractor shall preserve the records pertaining to this contract and keep the Contracting Officer advised of the location of such records; provided, however, that if the Contractor, at any time after the lapse of five years following the date upon which final payment under this contract becomes due, desires to dispose of the records, he shall so notify the Contracting Officer, who shall either authorize their destruction to the extent permitted by law or notify the Contractor to turn them over to the Government for disposition, and the Contractor shall promptly comply with such notice. The cost of storage and all other charges incidental to the preservation of the records incurred after the completion of this contract as determined by the Officer in Charge shall not be reimbursable under the terms of this contract.

(*d*) The Officer in Charge shall at all reasonable times have access to the records pertaining to this contract.

(*e*) The provisions of this article shall be applicable to and included in each fixed-price adjusted cost or cost-plus-a-fixed-fee contract entered into by the Contractor incident to the performance of this contract unless otherwise directed by the Contracting Officer.

Termination

Article 20. (*a*) The performance of work under the contract may be terminated by the Government in accordance with this clause in whole, or from time to time in part,

(1) whenever the Contractor shall default in performance of this contract in accordance with its terms (including in the term "default" any such failure by the Contractor to make progress in the prosecution of the work hereunder as endangers such performance), and shall fail to cure such default within a period of ten days (or such longer period as the Contracting Officer may allow) after receipt from the Contracting Officer of a notice specifying the default, or

(2) whenever for any reason the Contracting Officer shall determine that such termination is in the best interests of the Government. Any such termination shall be effected by delivery to the Contractor of a Notice of Termination specifying whether termination is for the default of the Contractor or for the convenience of the Government, the extent to which performance of work under the contract is terminated, and the date upon which such termination becomes effective. If, after notice of termination of this contract for default under (1) above, it is determined that the Contractor's failure to perform or to make progress in performance is due to causes beyond the control and without the fault or negligence of the Contractor pursuant to the provisions of the clause of this contract relating to excusable delays, the Notice of Termination shall be deemed to have been issued under (2) above, and the rights and obligations of the parties hereto shall in such event be governed accordingly.

(b) After receipt of a Notice of Termination and except as otherwise directed by the Contracting Officer, the Contractor shall (1) stop work under the contract on the date and to the extent specified in the Notice of Termination; (2) place no further orders or subcontracts for materials, services, or facilities except as may be necessary for completion of such portion of the work under the contract as is not terminated; (3) terminate all orders and subcontracts to the extent that they relate to the performance of work terminated by the Notice of Termination; (4) assign to the Government, in the manner and to the extent directed by the Contracting Officer, all of the right, title, and interest of the Contractor under the orders or subcontracts so terminated; (5) with the approval or ratification of the Contracting Officer, to the extent he may require, which approval or ratification shall be final and conclusive for all purposes of this clause, settle all outstanding liabilities and all claims arising out of such termination of orders and subcontracts, the cost of which would be reimbursable, in whole or in part, in accordance with the provisions of this contract; (6) transfer title (to the extent that title has not already been transferred) and, in the manner, to the extent, and at the times directed by the Contracting Officer, deliver to the Government (i) the fabricated or unfabricated parts, work in process, completed work, supplies, and other material produced as a part of, or acquired in respect of the performance of, the work terminated by the Notice of Termination, (ii) the completed or partially completed plans, drawings, information, and other property which, if the contract had been completed, would be required to be furnished to the Government, and (iii) the jigs, dies, and fixtures, and other special tools and tooling acquired or manufactured for the performance of this contract for the cost of which the Contractor has been or will be reimbursed under this contract; (7) use its best efforts to sell in the manner, at the times, to the extent, and at the price or prices directed or authorized by the Contracting Officer, any property of the types referred to in provision (6) of this paragraph, PROVIDED, HOWEVER, that the Contractor (i) shall not be required to extend credit to any purchaser, and (ii) may acquire any such property under the conditions prescribed by and at a price or prices approved by the Contracting Officer; and PROVIDED further that the proceeds of any such transfer or disposition shall be applied in reduction of any payments to be made by the Government to the Contractor under this contract or shall otherwise be credited to the price or cost of the work covered by this contract or paid in such other manner as the Contracting Officer may direct; (8) complete performance of such part of the work as shall not have been terminated by the Notice of Termination; and (9) take such action as may be necessary, or as the Contracting Officer may direct, for the protection and preservation of the property related to this contract which is in the possession of the Contractor and in which the Government has or may acquire an interest. The Contractor shall proceed immediately with the performance of the above obligations notwithstanding any delay in determining or adjusting the amount of the fixed-fee, or any item of reimbursable cost, under this clause. At any time after expiration of the plant clearance period, as defined in Section VIII, Armed Services Procurement Regulation, as it may be amended from time to time, the Contractor may submit to the Contracting Officer a list, certified as to quantity and quality, of any or all items of termination inventory not previously disposed of, exclusive of items the disposition of which has been directed or authorized by the Contracting Officer, and may request the Government to remove such items or enter into a storage agreement covering them. Not later than fifteen (15) days thereafter, the Government will accept title to such items and remove them or enter into a storage agreement covering the same, provided that the list submitted shall be subject to verification by the Contracting Officer upon removal of the items, or if the items are stored, within forty-five (45) days from the date of submission of the list, and any necessary adjustment to correct the list as submitted shall be made prior to final settlement.

(c) After receipt of a Notice of Termination, the Contractor shall submit to the Contracting Officer its termination claim in the form and with the certification prescribed by the Contracting Officer. Such claim shall be submitted promptly but in no event later than two years from the effective date of termination, unless one or more extensions in writing are granted by the Contracting Officer upon request of the Contractor made in writing within such two-year period or authorized extension thereof. However, if the Contracting Officer determines that the facts justify such action, he may receive and act upon any such termination claim at any time after such two-year period or any extension thereof. Upon failure of the Contractor to submit its termination claim within the time allowed, the Contracting Officer may determine, on the basis of information available to him, the amount, if any, due to the Contractor by reason of the termination and shall thereupon pay to the Contractor the amount so determined.

(d) Subject to the provisions of paragraph (c), the Contractor and the Contracting Officer may agree upon the whole or any part of the amount or amounts to be paid (including an allowance for the fixed-fee) to the Contractor by reason of the total or partial termination of work pursuant to this clause. The contract shall be amended accordingly, and the Contractor shall be paid the agreed amount.

(e) In the event of the failure of the Contractor and the Contracting Officer to agree in whole or in part, as provided in paragraph (d) above, as to the amounts with respect to costs and fixed-fee, or as to the amount of the fixed-fee, to be paid to the Contractor in connection with the termination of work pursuant to this clause, the Contracting Officer shall determine, on the basis of information available to him, the amount, if any, due to the Contractor by reason of the termination and shall pay to the Contractor the amount determined as follows: (1) If the settlement includes cost and fixed-fee (i) There shall be included therein all costs and expenses reimbursable in accordance with this contract, not previously paid to the Contractor for the performance of this contract prior to the effective date of the Notice of Termination, and such of these costs as may continue for a reasonable time thereafter with the approval of or as directed by the Contracting Officer, provided, however, that the Contractor shall proceed as rapidly as practicable to discontinue such costs. (ii) There shall be included therein, so far as not included under (i) above, the cost of settling and paying claims arising out of the termination of work under subcontracts or orders, as provided in paragraph (b) (5) above, which are properly chargeable to the terminated portion of the contract. (iii) There shall be included therein the reasonable costs of settlement, including accounting, legal, clerical, and other expenses reasonably necessary for the preparation of settlement claims and supporting data with respect to the terminated portion of the contract and for the termination and settlement of subcontracts thereunder, together with reasonable storage, transportation, and other costs incurred in connection with the protection or disposition of termination inventory; provided, however, that if the termination is for default of the Contractor there shall not be included any amounts for the preparation of the Contractor's settlement proposal. (iv) There shall be included therein a portion of the fixed-fee payable under the contract determined as follows: (A) In the event of the termination of this contract for the convenience of the Government and not for the default of the Contractor, there shall be paid a percentage of the fee equivalent to the percentage of the completion of work contemplated by the contract, less fixed-fee payments previously made hereunder. (B) In the event of the termination of this contract for the default of the Contractor, the total fixed-fee payable shall be such proportionate part of the fee as the actual work in place bears to the total work contemplated by the contract.

If the amount determined under this paragraph is less than the total payment of fixed-fee theretofore made to the Contractor, the Contractor shall repay to the Government the excess amount. (2) If the settlement includes only the fixed fee, the amount thereof will be determined in accordance with subparagraph (e) (1) (iv) above.

(f) The Contractor shall have the right of appeal, under the clause of this contract entitled "Disputes," from any determination made by the Contracting Officer under paragraphs (c) or (e) above, except that if the Contractor has failed to submit its claim within the time provided in paragraph (c) above and has failed to request extension of such time he shall have no such right of appeal. In any case where the Contracting Officer has made a determination of the amount due under paragraph (c) or (e) above, the Government shall pay to the Contractor the following: (i) if there is no right of appeal hereunder or if no timely appeal has been taken, the amount so determined by the Contracting Officer, or (ii) if an appeal has been taken, the amount finally determined on such appeal.

(g) In arriving at the amount due the Contractor under this clause there shall be deducted (1) all unliquidated advance or other unliquidated payments theretofore made to the Contractor, (2) any claim which the Government may have against the Contractor in connection with this contract, and (3) the agreed price for, or the proceeds of sale of, any materials, supplies, or other things acquired by the Contractor or sold pursuant to the provisions of this clause and not otherwise recovered by or credited to the Government.

(*h*) In the event of a partial termination, the portion of the fixed-fee which is payable with respect to the work under the continued portion of the contract shall be equitably adjusted by agreement between the Contractor and the Contracting Officer, and such adjustment shall be evidenced by an amendment to this contract.

(*i*) The Government may from time to time, under such terms and conditions as it may prescribe, make partial payments and payments on account against costs incurred by the Contractor in connection with the terminated portion of the contract whenever in the opinion of the Contracting Officer the aggregate of such payments shall be within the amount to which the Contractor will be entitled hereunder. If the total of such payments is in excess of the amount finally determined to be due under this clause, such excess shall be payable by the Contractor to the Government upon demand, together with interest computed at the rate of 6% per annum, for the period from the date such excess payment is received by the Contractor to the date on which such excess is repaid to the Government; provided, however, that no interest shall be charged with respect to any such excess payment attributable to a reduction in the Contractor's claim by reason of retention or other disposition of termination inventory until ten days after the date of such retention or disposition.

(*j*) The provisions of this clause relating to the fixed-fee shall be inapplicable if this contract does not provide for payment of a fixed-fee.

(*k*) Unless otherwise provided for in this contract or by applicable statute, the Contractor, from the effective date of termination and for a period of six years after final settlement under this contract, shall preserve and make available to the Government at all reasonable times at the office of the Contractor, but without direct charge to the Government, all its books, records, documents, and other evidence bearing on the cost and expenses of the Contractor under this contract and relating to the work terminated hereunder, or, to the extent approved by the Contracting Officer, photographs, microphotographs, or other authentic reproductions thereof.

DISPUTES

ARTICLE 21.   Except as otherwise provided in this contract, any dispute concerning a question of fact arising under this contract which is not disposed of by agreement shall be decided by the Contracting Officer, who shall reduce his decision to writing and mail or otherwise furnish a copy thereof to the Contractor. Within 30 days from the date of receipt of such copy, the Contractor may appeal by mailing or otherwise furnishing to the Contracting Officer a written appeal addressed to the Secretary, and the decision of the Secretary or his duly authorized representative for the hearing of such appeals shall be final and conclusive; provided, that, if no such appeal is taken, the decision of the Contracting Officer shall be final and conclusive. In connection with any appeal proceeding under this clause, the Contractor shall be afforded an opportunity to be heard and to offer evidence in support of its appeal. Pending final decision of a dispute hereunder, the Contractor shall proceed diligently with the performance of the contract and in accordance with the Contracting Officer's decision.

Military Security Requirements

ARTICLE 22. (a) The provisions of the following paragraphs of this clause shall apply only if and to the extent that this contract involves access to classified matter, which as used in this clause shall mean information or material classified "Top Secret," "Secret," "Confidential," or "Restricted."

(b) The Contractor (i) shall be responsible for safeguarding all classified matter and shall not supply or disclose classified matter to any unauthorized person, (ii) shall not make or permit to be made any reproductions of matter classified "Top Secret" except with the prior written authorization of the Contracting Officer, (iii) shall not make or permit to be made any reproductions of matter classified "Secret," "Confidential," or "Restricted" except as may be essential to performance of the contract, (iv) shall submit to the Contracting Officer, at such times as the Contracting Officer may direct, an accounting of all reproductions of matter classified "Top Secret," "Secret," or "Confidential," and (v) shall not incorporate in any other project any special features of design or construction which will disclose classified matter, except with the prior written authorization of the Contracting Officer.

(c) Except with the prior written consent of the Secretary or his duly authorized representative, the Contractor (i) shall not permit any alien to have access to classified matter and (ii) shall not permit any individual to have access to matter classified "Top Secret" or "Secret."

(d) The Contractor agrees (i) to submit immediately to the Contracting Officer a complete confidential report of any information which the Contractor may have concerning existing or threatened espionage, sabotage, or subversive activity, (ii) to submit to the Contracting Officer, upon written request, any and all information which the Contractor may have concerning any of its employees engaged in any work at any plant, factory, or site at which work under this contract is being performed, and (iii) to exclude from its plant, factory, site, or part thereof, at which work under this contract is being performed, any person or persons whom the Secretary or his duly authorized representative, in the interest of security, may designate in writing.

(e) The Contractor is authorized to rely on any letter or other written instrument signed by the Contracting Officer, changing or entirely removing the classification of this contract or of any classified matter.

(f) The obligations of the Contractor under this clause shall be in addition to any obligations of the Contractor to comply with all the terms and provisions of any applicable security or secrecy agreement theretofore or hereafter entered into between the Contractor and the Government.

(g) The Contractor agrees to insert, in all subcontracts hereunder which involve access to classified matter, provisions which shall conform substantially to the language of this clause, including this paragraph (g): PROVIDED, that such provisions need not be included in any subcontract as to which the Contracting Officer shall consent to the omission of such provisions.

Transfer of Contract and Assignment of Contractor's Claims

ARTICLE 23. (a) Neither this contract nor any interest herein, except as otherwise provided in this article, shall be transferred by the Contractor to any other party or parties.

(*b*) Pursuant to the provisions of the Assignment of Claims Act of 1940 as amended (31 U.S. Code 203, 41 U.S. Code 15), if this contract provides for payments aggregating $1,000 or more, claims for moneys due or to become due the Contractor from the Government under this contract may be assigned to a bank, trust company, or other financing institution, including any Federal lending agency, and may thereafter be further assigned and reassigned to any such institution. Any such assignment or reassignment shall cover all amounts payable under this contract and not already paid, and shall not be made to more than one party, except that any such assignment or reassignment may be made to one party as agent or trustee for two or more parties participating in such financing. Notwithstanding any other provision of this contract, payments to an assignee of any moneys due or to become due under this contract shall not, to the extent provided in said Act as amended, be subject to reduction or setoff.

(*c*) In no event shall copies of this contract or of any plans, specifications, or other similar documents relating to work under this contract, if marked "Top Secret," "Secret," "Confidential," or "Restricted," be furnished to any assignee of any claim arising under this contract or to any other person not entitled to receive the same; PROVIDED, that a copy of any part or all of this contract so marked may be furnished, or any information contained therein may be disclosed, to such assignee upon the prior written authorization of the Contracting Officer.

DAVIS-BACON ACT

ARTICLE 24. This contract, to the extent that it is of a character specified in the Davis-Bacon Act as amended (40 U.S. Code 276a), is subject to all provisions and exceptions of said Davis-Bacon Act, including in particular the following:

(*a*) The Contractor and his subcontractor shall pay all mechanics and laborers employed directly upon the site of the work, unconditionally and not less often than once a week, and without subsequent deduction or rebate on any account (except such payroll deductions as are permitted by applicable regulations prescribed by the Secretary of Labor), the full amounts accrued at time of payment computed at wage rates not less than those stated in the appendix hereto, regardless of any contractual relationship which may be alleged to exist between the Contractor or subcontractor and such laborers and mechanics; and the scale or wages to be paid shall be posted by the Contractor in a prominent and easily accessible place at the site of the work.

(*b*) The Contracting Officer shall have the right to withhold from the Contractor so much of accrued payments as may be considered necessary by the Contracting Officer to pay to laborers and mechanics employed on the work the difference between (i) the rates of wages required by the contract to be paid laborers and mechanics on the work and (ii) the rates of wages received by such laborers and mechanics and not refunded to the Contractor, subcontractors, or their agents.

(*c*) In the event it is found by the Contracting Officer that any laborer or mechanic employed by the Contractor or any subcontractor directly on the site of the work covered by the contract has been or is being paid a rate of wages less than the rate of wages required by the contract to be paid as aforesaid, the Contracting Officer may (i) by written notice to the Contractor, terminate his right to proceed with the work, or such part of the work as to which there has been a failure to pay said required wages, and (ii) prosecute the work to completion by contract or otherwise, whereupon the Contractor and his sureties shall be liable to the Government for any excess costs occasioned the Government thereby.

EIGHT-HOUR LAW

ARTICLE 25.   This contract, to the extent that it is of a character specified in the Eight-Hour Law of 1912 as amended (40 U.S. Code 324–326) and is not covered by the Walsh-Healey Public Contracts Act (41 U.S. Code 35–45), is subject to the following provisions and exceptions of said Eight-Hour Law of 1912 as amended, and to all other provisions and exceptions of said Law:

No laborer or mechanic doing any part of the work contemplated by this contract, in the employ of the Contractor or any subcontractor contracting for any part of the said work, shall be required or permitted to work more than eight hours in any one calendar day upon such work, except upon the condition that compensation is paid to such laborer or mechanic in accordance with the provisions of this clause.   The wages of every such laborer and mechanic employed by the Contractor or any subcontractor engaged in the performance of this contract shall be computed on a basic day rate of eight hours per day; and work in excess of eight hours per day is permitted only upon the condition that every such laborer and mechanic shall be compensated for all hours worked in excess of eight hours per day at not less than one and one-half times the basic rate of pay.   For each violation of the requirements of this clause a penalty of five dollars shall be imposed upon the Contractor for each such laborer or mechanic for every calendar day in which such employee is required or permitted to labor more than eight hours upon said work without receiving compensation computed in accordance with this clause; and all penalties thus imposed shall be withheld for the use and benefit of the Government.

CONVICT LABOR

ARTICLE 26.   In connection with the performance of work under this contract, the Contractor agrees not to employ any person undergoing sentence of imprisonment at hard labor.

NON-DISCRIMINATION IN EMPLOYMENT

ARTICLE 27.   In connection with the performance of work under this contract, the Contractor agrees not to discriminate against any employee or applicant for employment because of race, creed, color, or national origin; and further agrees to insert the foregoing provision in all subcontracts hereunder except subcontracts for standard commercial supplies or for raw materials.

COPELAND ACT

ARTICLE 28.   To the extent that this contract is of a character specified in the Copeland ("Anti-Kickback") Act as amended (18 U.S. Code 874 and 40 U.S. Code 276c), the Contractor agrees to comply with the regulations, rulings, and interpretations of the Secretary of Labor pursuant to said

Copeland Act, which Act makes it unlawful to induce any person employed in the construction or repair of public buildings or public works to give up any part of the compensation to which he is entitled under his contract of employment; and the Contractor agrees to insert a like provision in all subcontracts hereunder.

Labor Statistics

Article 29. The Contractor shall report, and shall by agreement require his subcontractors to report, at such times, in such manner, and covering such matters as the Officer in Charge may direct, such labor statistics, applicable only to work performed under this contract at the site of the work, as may be required for transmittal to the Department of Labor.

Officials Not to Benefit

Article 30. No member of or delegate to Congress, or resident commissioner, shall be admitted to any share or part of this contract, or to any benefit that may arise therefrom; but this provision shall not be construed to extend to this contract if made with a corporation for its general benefit.

Covenant Against Contingent Fees

Article 31. The Contractor warrants that no person or selling agency has been employed or retained to solicit or secure this contract upon an agreement or understanding for a commission, percentage, brokerage, or contingent fee, excepting bona fide employees or bona fide established commercial or selling agencies maintained by the Contractor for the purpose of securing business. For breach or violation of this warranty the Government shall have the right to annul this contract without liability or in its discretion to deduct from the contract price or consideration the full amount of such commission, percentage, brokerage, or contingent fee.

Buy-American Act

Article 32. The Contractor agrees that in the performance of the work under this contract the Contractor, subcontractors, material men, and suppliers shall use only such unmanufactured articles, materials, and supplies (which term "articles, materials, and supplies" is hereinafter referred to in this clause as "supplies") as have been mined or produced in the United States and only such manufactured supplies as have been manufactured in the United States substantially all from supplies mined, produced, or manufactured, as the case may be, in the United States. Pursuant to the Buy-American Act (41 U.S. Code 10a–d), the foregoing provisions shall not apply (i) with respect to supplies excepted by the Secretary from the application of that Act, (ii) with respect to supplies for use outside the United States, or (iii) with respect to the supplies to be used in the performance of work under this contract which are of a class or kind determined by the Secretary or his duly authorized representative not to be mined, produced, or manufactured, as the case may be, in the United States in sufficient and reasonably available commercial quantities and of a satisfactory quality, or (iv) with respect to such supplies, from which the supplies to be used in the performance of work under this contract are manufactured, as are of a class or kind determined by the Secretary or his duly authorized representative not to be mined, produced, or manufactured, as the case may be, in the United States in sufficient and reasonably available commercial quantities and of a satisfactory quality, provided that this exception (iv) shall not permit the use in the performance of work under this contract of supplies manufactured outside the United States if such supplies are manufactured in the United States in sufficient and reasonably available commercial quantities and of a satisfactory quality.

RENEGOTIATION

ARTICLE 33. (a) This contract is subject to the Renegotiation Act of 1951 (P.L. 9, 82nd Congress) and shall be deemed to contain all the provisions required by Section 104 of said Act.

(b) The Contractor (which term as used in this clause means the party contracting to furnish the materials or perform the work required by this contract) agrees to insert the provisions of this clause, including this paragraph (b), in all subcontracts as required by Section 104 of the Renegotiation Act of 1951; PROVIDED, that the Contractor shall not be required to insert the provisions of this clause in any subcontract of a class or type described in Section 106(a) of the Renegotiation Act of 1951.

PATENT INDEMNITY

ARTICLE 34. (a) The Contractor agrees to indemnify the Government and its officers, agents, and employees against liability, including costs and expenses, for infringement upon any Letters Patent of the United States (except Letters Patent issued upon an application which is now or may hereafter be, for reasons of national security, ordered by the Government to be kept secret or otherwise withheld from issue) arising out of the performance of this contract or out of the use or disposal by or for the account of the Government of supplies furnished or construction work performed hereunder. The foregoing indemnity shall not apply unless the Contractor shall have been informed as soon as practicable by the Government of the suit or action alleging such infringement, and shall have been given an opportunity to present recommendations as to the defense thereof; and further, such indemnity shall not apply in any one of the following situations: (i) Any infringement resulting from the addition to any such supplies or other supplies not furnished by the Contractor for the purpose of such addition; (ii) any settlement of a claim of infringement made without the consent of the Contractor, unless required by final decree of a court of competent jurisdiction; (iii) any claim of infringement arising from use or disposal outside the scope of any license limitation under which the Contractor is bound, provided that the Contractor has notified the Government of the limitation prior to first delivery under this contract; (iv) any infringement necessarily resulting from changes (other than the substitution of another standard commercial part or component manufactured or supplied by the Contractor) ordered pursuant to this contract, or from specific written instructions given by the Contracting Officer directing a manner of performing the contract not normally utilized by the Contractor.

Authorization and Consent for Use of Patents

Article 35. The Government hereby gives its authorization and consent (without prejudice to its rights of indemnification, if such rights are provided for in this contract) for all use and manufacture, in the performance of this contract or any part hereof or any amendment hereto or any subcontract hereunder (including any lower-tier subcontract), of any patented invention (1) embodied in the structure or composition of any article the delivery of which is accepted by the Government under this contract, or (ii) utilized in the machinery, tools, or methods the use of which necessarily results from compliance by the Contractor or the using subcontractor with (a) specifications or written provisions now or hereafter forming a part of this contract, or (b) specific written instructions given by the Contracting Officer directing the manner of performance.

Notice and Assistance Regarding Patent Infringement

Article 36. (a) The Contractor agrees to report to the Contracting Officer, promptly and in reasonable written detail, each claim of patent infringement based on the performance of this contract and asserted against it, or against any of its subcontractors if it has notice thereof.

(b) In the event of litigation against the Government on account of any claim of infringement arising out of the performance of this contract or out of the use of any supplies furnished or construction work performed hereunder, the Contractor agrees that it will furnish to the Government upon request, all evidence and information in its possession pertaining to the defense of such litigation. Such information shall be furnished at the expense of the Government except in those cases in which the Contractor has agreed to indemnify the Government against the claim being asserted.

Compliance

Article 37. The failure of the Government, in any one or more instances, to insist upon strict performance of any of the terms of this contract or to exercise any option herein conferred, shall not be construed as a waiver or relinquishment for the future of any such terms or options.

Composition of Contractor

Article 38. If the term "Contractor" as used herein includes more than one legal entity, then each entity so included shall be jointly and severally liable for the undertakings of the Contractor hereunder. Plant or equipment owned directly or indirectly by any one or combination of such entities shall be considered as owned by the Contractor for the purposes of the provisions of this contract.

Supersedure

Article 39. If this contract has been preceded by a Letter or Dispatch of Intent or a Notice of Award, anticipating the execution of this contract, then such Letter or Dispatch or Notice and all rights and obligations of the parties thereunder are superseded and merged into this contract. All Acts of the Contractor and the Government and all payments, if any, made by the Government under said Letter or Dispatch or Notice shall be deemed to have been under this contract.

DEFINITIONS

ARTICLE 40. (*a*) The term "Secretary" means the Secretary, the Under Secretary, or any Assistant Secretary of the Department of the Navy and the head or any assistant head of the executive agency; and the term "his duly authorized representative" means any person or persons or board (other than the Contracting Officer) authorized to act for the Secretary.

(*b*) The term "Contracting Officer" means the person executing this contract on behalf of the Government, and any other officer or civilian employee who is a properly designated Contracting Officer; and the term includes, except as otherwise provided in this contract, the authorized representative of a Contracting Officer acting within the limits of his authority.

(*c*) The term "Officer in Charge" as used in this contract shall include the duly appointed successor of such designated officer and likewise any person or persons authorized by notice in writing to the Contractor to act for him or his successor.

EXAMINATION OF RECORDS

ARTICLE 41. (*a*) The Contractor agrees that the Comptroller General of the United States or any of his duly authorized representatives shall, until the expiration of three years after final payment under this contract, have access to and the right to examine any directly pertinent books, documents, papers, and records of the Contractor involving transactions related to this contract.

(*b*) The Contractor further agrees to include the following provision, with appropriate insertions, in all his subcontracts hereunder:

(Name of Subcontractor) agrees that the Comptroller General of the United States or any of his duly authorized representatives shall, until the expiration of three years after final payment under prime contract (Contract Symbol and Number) between the United States of America and (Name of Contractor), have access to and the right to examine any directly pertinent books, documents, papers, and records of (Name of Subcontractor) involving transactions related to this contract.

GRATUITIES

ARTICLE 42. (*a*) The Government may, by written notice to the Contractor, terminate the right of the Contractor to proceed under this contract if it is found, after notice and hearing, by the Secretary or his duly authorized representative, that gratuities (in the form of entertainment, gifts, or otherwise) were offered or given by the Contractor, or any agent or representative of the Contractor, to any officer or employee of the Government with a view toward securing a contract or securing favorable treatment with respect to the awarding or amending, or the making of any determinations with respect to the performing, of such contract; PROVIDED, that the existence of the facts upon which the Secretary or his duly authorized representative makes such findings shall be in issue and may be reviewed in any competent court.

(*b*) In the event this contract is terminated as provided in paragraph (1) hereof, the Government shall be entitled (i) to pursue the same remedies against the Contractor as it could pursue in the event of a breach of the contract by the Contractor, and (ii) as a penalty in addition to any other damages to which it may be entitled by law, to exemplary damages in an amount (as determined by the Secretary or his duly authorized representative) which shall be not less than three nor more than ten times the costs incurred by the Contractor in providing any such gratuities to any such officer or employee.

(*c*) The rights and remedies of the Government provided in this clause shall not be exclusive and are in addition to any other rights and remedies provided by law or under this contract.

ALTERATIONS

ARTICLE 43. The following additional provisions are hereby incorporated in and made a part of this contract.

Add to Article 6:

"(*e*) The Contractor will not establish a work week in excess of 40 hours or a multiple shift or other accelerated form of operation without prior authorization of the Contracting Officer.

(*f*) No overtime work shall be performed without the prior authorization of the Officer in Charge.

(*g*) Travel and transportation of contract personnel will not be permitted without prior authorization and approval of the Officer in Charge."

AUTHORITY

This contract is entered into under authority of Section 2(c)(1) of Public Law 413, 80th Congress, (Armed Services Procurement Act of 1947), and any required determination and findings with respect thereto have been made.

IN WITNESS WHEREOF, the parties hereto have executed this contract as of the day and year first above written.

UNITED STATES OF AMERICA

By..

Civil Engineer Corps, U.S.N., for Chief, Bureau of Yards and Docks, Department of the Navy.

CONTRACTOR..

.... By ....

(Witness)

....

(Witness)   (Official Title)

....

(Business Address)

Corporate Certificate

I,   , certify that I am the Secretary of the corporation named as Contractor in the foregoing contract; that
             who signed said contract on behalf of the Contractor was then   of said corporation; that said contract was duly signed for and in behalf of said corporation by authority of its governing body and is within the scope of its corporate powers.

.... (CORPORATE SEAL)

Partnership Certificate

State of ⎫
County of ⎰ ss.

On this   day of   , 19........., before me personally appeared   , known to me and known by me to be the person who executed the above instrument, who, being by me first duly sworn, did depose and say that he is a general partner in the firm of
                ; that said firm consists of himself and   ; and that he executed the foregoing instrument on behalf of said firm for the uses and purposes stated herein.

....

Notary Public in and for the County of

...., State of ....

My Commission Expires:

.... (Notarial Seal)

Questions

1. Given a list of contractors considered to be qualified to construct a large office building under a negotiated cost-plus-fixed-fee contract, how should a selection be made as to the award of the contract?

2. Under normal conditions, why would the amount of the fixed fee not be a major consideration in the award of the contract?

3. What factors should govern the amount of the fixed fee?

4. Under what conditions would a sliding scale of fees be advisable in lieu of one fixed fee, and what changes would be necessary in the wording of the cost-plus-fixed-fee contract to convert it to a sliding-scale-of-fees basis?

5. Explain the nature of changes in the work on a large dam which would justify changes in the amount of the contractor's fixed fee. What types of changes would not warrant changes in the fee?

6. Which of the following items of cost should be considered reimbursable and which are non-reimbursable under a cost-plus-fixed-fee contract?

a. The salary of the construction superintendent.

b. The salary of the concrete foreman.

c. Cement, sand, and gravel.

d. The salary of the president of the contracting company.

e. The cost of a new bulldozer.

f. The rental of a steam shovel.

g. The complete overhaul of a dragline at the end of the job.

h. The salary of the contractor's chief cost accountant in his home office.

i. Interest on funds borrowed by the contractor for use as working capital.

7. Write an article for a cost-plus-fixed-fee contract to cover the rental to be paid to the contractor for the use of his equipment. Base the amounts of rentals on the standard schedule of the Associated General Contractors reduced to eliminate the contractor's profit. Why should the contractor's profit be eliminated from equipment rental? How would the terms of this article differ from those of an agreement for the rental of equipment from a third party?

8. Write an article for a cost-plus type contract covering reimbursement to the contractor of the costs of the work. Consider both direct and indirect costs.

9. Write an article for a cost-plus type contract providing for the contractor to receive a fixed fee based on the estimated cost of the work and containing the following additional provisions.

a. A ceiling price beyond which additional costs will not be reimbursed.

b. Payment to the contractor, in addition to the fixed fee, of 25 per cent of any saving in the total cost of the work as compared with the estimated cost.

10. Is it implied that the contractor under a cost-plus type contract is an agent of the owner for the purchase of materials and that the owner is therefore liable to the material vendor for the price?

11. A municipal construction contract gave the contractor a fixed fee equal to 15 per cent of the guaranteed estimate of construction cost plus additional compensation for the rental of equipment furnished by him. Was the contract void and therefore subject to attack by taxpayers because it gave the contractor a dual interest in the contract and made him his own employer?

I I

CONSTRUCTION INSURANCE

The many types of risk and liability connected with construction work are matters of importance to the engineer. Most of these are the responsibility of the contractor, but the owner's interests are concerned, and the engineer is usually called upon for advice relative to insurance coverage. Moreover, the contract and the specifications should be explicit as to types of insurance required of the contractor, and they should be equally definite in stating the risks to be assumed by the owner and to be covered by insurance obtained by him.

Under the law of torts, an independent contractor is liable for damages caused by his acts during construction operations, but in most instances it is advisable for the contract to require insurance against well known hazards as additional financial protection to the owner. The satisfactory completion of the construction work and the payment of all of the contractor's expenses for labor and materials in connection with the work normally are guaranteed by performance and payment bonds, but protection against the risks of loss due to fire, storms, explosions, water leakage, and the like is provided separately by insurance policies. Such risks are covered under the general classification of casualty insurance. Risks of loss due to the operation of public liability and workmen's compensation laws are likewise insurable under the various forms of liability and workmen's compensation insurance. Burglary, robbery, and theft insurance provides protection against losses from these sources, and fidelity insurance covers the risks of loss through the dishonesty of employees.

Insurance Contracts. An insurance policy is a conditional contract whereby the insurance company agrees to pay the insured or his designated beneficiary for a specified loss or liability, contingent on the occurrence of some event. Like all contracts the policy must meet the essential requirements as to agreement, lawful subject matter, consideration, competency of the contracting parties, and

245

form in order to be valid. Its primary purpose is the payment of indemnity for loss, and, although it is enforceable in the event of negligence on the part of the insured, a loss resulting from an intentional act of the insured will usually invalidate the policy.

The insured must have an insurable interest as a condition to the contract, and the amount of insurance protection obtainable is limited to the valuation of the insurable interest. That is to say, one may not recover more than the loss suffered.

Under the insurance policy the company promises to protect the insured against a specified loss. The consideration of the contract is the payment by the insured of a stipulated premium. In the case of mutual insurance companies, the consideration may also require the payment of assessments as well as the premium. Failure to pay the premium when due will usually mean that the insured automatically forfeits his rights under the policy. Some state statutes require the company to notify the insured when premium payments are due, and under this condition failure to pay the premium on time is not held to be cause for invalidating the policy unless such notification has been made.

In transactions relative to insurance policies the insurance company is usually represented by an agent, and the laws of agency are applicable. The agent has the authority to enter into contract for the company, and the company is bound by the acts and promises of its agent, within the scope of the authorized agency. It is the responsibility of the insured to inform himself of any limitations or restrictions placed on the agent by the company. However, any fraudulent act, promise, or representation by the agent will bind the company even though not authorized by the company, provided that the act was within the apparent scope of the agency and that the insured was not a party to the act.

Insurance Coverage for Construction Work. The types and amounts of insurance policies required to protect the parties to a construction contract must be determined for each specific project after full consideration of the perils inherent in the work. Many of the liabilities which may arise out of a construction project may devolve upon the owner unless he is adequately protected contractually, and for this reason it is common practice for the construction contract to require the contractor to obtain appropriate insurance coverage against known risks. Other questions of insurance protection may be matters for the contractor to decide upon his own option when the owner's interests are not concerned.

The types of insurance normally required in connection with construction work may be outlined broadly as follows.

1. Insurance coverage needed by the contractor to protect himself.
   a. Property damage caused by fire, lightning, tornadoes, cyclones, riots, explosions, machinery hazards, water leakage, and the like.
   b. Loss or damage of equipment while in transit or in use.
   c. Burglary, robbery, and theft.
   d. Dishonesty of employees through forgery, embezzlement, and the like.
2. Insurance coverage needed by the contractor to protect third parties.
   a. Contractor's public liability.
   b. Contractor's protective liability.
   c. Contractual liability.
   d. Automobile and truck liability.
   e. Completed operation liability.
   f. Workmen's compensation liability.
3. Insurance coverage needed to protect the owner.
   a. Owner's protective liability.
   b. Damage to the owner's property, not otherwise protected.

For both property damage and public liability, it is possible to obtain comprehensive insurance policies in which all contemplated risks of each classification are covered in one contract. These have many advantages, particularly in that the possibility of gaps between coverages is reduced. It should be noted, however, that there are no all-risk policies, and it is necessary to make certain that each individual risk is adequately covered.

Fire Insurance. The standard fire insurance policy obligates the insurance company to assume liability for fire damage under three classifications, namely, the building, household furniture, and stock and fixtures. For construction-contract purposes the first form is of primary interest inasmuch as it provides coverage for the building, additions, extensions, and appurtenances, such as mechanical equipment, sidewalks, fences, and other permanent equipment. It should be noted that liability for loss of plans, specifications, and similar papers contained in the building is usually excluded from this form, as is damage to foundations, unless special provision is made for such coverage. The standard policy does not cover loss due to fire caused by storms, war damage, riots, strikes, theft in connection with fire, explosions, lightning, negligence of the insured, or change in hazard, as, for instance, alterations to a building making it less fire resistant. Special conditions and requirements may be written into the standard policy, however, by negotiation with the insurance company. The policy may cover liability for the full value of the property or any percentage of its full value. Partial loss in the latter case would be paid in full up to the fractional limit of the policy. Before fire insurance is purchased all exceptions and waivers in the company's standard agreement should be checked to make certain that the proper coverage is being obtained.

Advice relative to fire insurance problems and measures to be taken for fire prevention is available from the National Board of Fire Underwriters, an organization sponsored by the insurance companies in the interests of fire prevention and the reduction of losses due to fire.

Whenever acquiring fire insurance, it is necessary to consult representatives of insurance companies to determine the forms of insurance and the carriers most suitable, the rates for specific protection, and the terms under which the policy is to be written. Fire insurance rates vary with the locality, the fire fighting facilities available, the types of construction, the preventative measures in operation, and, in the case of new construction, with the construction methods contemplated and fire precautions observed.

Miscellaneous Forms of Property Damage Insurance. In addition to loss by fire property damage may occur due to numerous other causes, such as water leakage, explosions, riots, wind, cyclones, and earthquakes. Insurance protection against such risks may be obtained under individual policies or as endorsements on fire insurance policies. In addition, comprehensive all-purpose policies are written by some insurance companies to cover a combination of a number of different hazards. Like all insurance, these policies and endorsements are predicated on the probabilities of loss. Rates and premiums, therefore, are based on local conditions. In general, these policies provide for the insurance company to assume liability only for damages due directly to the hazard for which the policy is written, excluding possible damage incidental thereto, except by special agreement for its coverage.

Extended Form of Fire Insurance. Standard fire insurance policies do not include liability for loss or damage due to demolition, required by law or safety ordinances, of a portion of a building remaining after a fire. An ordinary policy may be extended to cover this possibility, however. In a similar manner such hazards as tornadoes, cyclones, hailstorms, lightning, riots, strikes, water leakage, operation and maintenance of machinery, and explosions may be included. In standard agreements coverage of each of these risks in turn is accompanied by waivers, exceptions, and conditions and is subject to negotiation with the insurance company when the policy is written. Thus, all types of property damage may be protected under one comprehensive policy. This is generally desirable and is particularly effective in the case of associated risks such as explosion and fire where questions may arise as to which casualty occurred first and thereby caused the others and which casualty actually caused the damage. Some fire insurance policies, for example, do not cover fire damage if the fire results from an explosion.

To avoid the possibility of conflicts and gaps in property damage insurance, it is recommended that all associated risks such as fire and explosion be covered in the same amounts, either in a comprehensive form of policy or in separate policies. If they are to be covered by separate policies, the policies should be in the same amounts and carried with the same insurance company. Methods for determining and justifying the insurable value of the property also should be specific.

Marine and Inland Marine Insurance. Marine insurance is designed to meet the risks of loss incurred through water transportation on the high seas. Policies are written in connection with loss of, or damage to, ships, ships' cargoes, and freight charges. Policies may also be classified according to transportation on the sea, along the coast, or on inland waterways. With reference to construction work, this form of insurance is chiefly of interest to contractors transporting floating equipment, and materials, supplies, or machinery by ship. By a modification of marine insurance, designed as inland marine insurance, protection may be extended to property under other forms of transportation such as by railroad, motor truck, parcel post, and railway express. Under this type of policy, property and equipment of contractors may be insured against loss due to fire, storms, wrecks, and other risks while in transit on land carriers. This form of insurance is sometimes called the Contractor's Equipment Floater Policy.

Inasmuch as bridges are related to transportation, bridge insurance is classified as the inland marine type. It is written in three forms, namely, bridge builder's risk, bridge property damage, and bridge use and occupancy insurance. The bridge builder's risk form covers damage to the structure, during construction, due to fire, lightning, floods, ice, collision, explosions, riots, vandalism, wind, tornadoes, and earthquakes. Other risks may be included by special agreement. Usually excluded are losses due to suspension of construction through operation of the law or governmental order, suspension, termination, or lapse of any license, permit, or lease, or by injunction. The amount of the insurance company's liability is proportional to the actual value of the work completed at the time of the loss and is based on the most recent report of the progress of the work. Coverage may be for the entire value of the completed work or any desired percentage thereof.

Bridge property damage insurance is written for a specific bridge at a specific location and is similar to the Bridge Builder's Risk Policy as to the liability assumed by the insurance company. In addition, loss due to collapse is included.

Bridge use and occupancy insurance protects the owner against loss due to inability to operate a bridge as a result of damage of the type covered by the Bridge Builder's Risk Policy. This form is of particular interest to owners of toll bridges and provides insurance for the actual loss suffered from the suspension of tolls less normal operating charges. Usually losses due to suspensions of less than seven days are deductible.

Comprehensive Builder's Risk Insurance. For new construction the comprehensive Builder's Risk Policy has been devised to provide insurance protection for the increasing value of the property as construction progresses. This type of insurance is designed to be as comprehensive in scope as possible, and when it is used no other form of property damage insurance will be needed. It is necessary in each instance, however, to make certain that all contemplated risks are covered without conflicts and gaps.

Two types of builder's risk insurance are available, namely, the Builder's Risk Reporting Form and the Builder's Risk Completed Value Policy. Under the former the insured forwards monthly progress reports to the insurance company, and the premiums and the liability of the company are computed on the basis of the last progress report received. As in the standard policy, coverage may be obtained for the full value of the completed work or any fraction thereof. Under this form the insurance company assumes liability for fire damage to the work and to appurtenances, excepting foundations. All materials, supplies, equipment, construction plant, and temporary structures to be used in the construction are included. Tools, construction equipment, and machinery are also included when in the structure, in cars on side tracks on the premises, or in adjacent streets or alleys. The policy is operative only while the structure is under construction and not occupied.

To avoid the inconvenience of detailed monthly progress reports to the insurance company, the Builder's Risk Completed Value Policy may be used. Under this form the insurance is based on the entire completed value of the property with the provision of liability at any specific time for only the amount of the actual value the property bears at that date. The rates and premiums are based on a schedule, drawn up before construction begins, of the average amount of liability at each stage during the construction period.

Burglary, Robbery, and Theft Insurance. As distinguished from real property and equipment, the contractor's negotiable money and securities may be subject to loss through burglary, robbery, theft, disappearance, destruction, or wrongful abstraction; and insurance protection against such risks may be obtained under separate policies or by extension of other forms of insurance. The Broad Form Money and Security Policy is designed to cover losses of this character.

Fidelity Insurance. Insurance against loss through the dishonesty of employees may be obtained by one of three types of fidelity bonds, namely,

1. A Schedule Bond which covers those persons whose names or positions are listed in the bond and for the stated amount for each.

2. A Blanket Position Bond which affords protection on all employees to the full amount of the bond.

3. A Primary Commercial Bond which covers all employees although the insurer's liability is the limit of the bond for any loss caused by one or more employees.

The blanket form is usually preferred since it is the most efficient and frequently is less expensive.

Insurance against loss through forged, raised, or altered checks may be obtained under the Depositor's Forgery Policy. Forgery may be committed by employees or others, but the most frequent forgery losses are those caused by employees. Therefore, this coverage may be used to supplement fidelity blanket protection.

Public Liability Insurance. Extraordinary risks to persons and property are inherent in construction work, even though such persons and property may not be directly associated with the work in progress. Under the laws of torts, any damages considered due to negligence become the responsibility of those causing them. In construction work, the amounts of resultant damages may be large, and protection through public liability insurance is practically essential. Public liability may be direct so that the insured's own legal liability is involved; it may be contractual so that the insured assumes the legal liability of another by the terms of a contract; or it may be contingent and protective, and involve the indirect liability of the insured for the acts of others for which the law may hold him liable.

A construction contractor acting as an independent contractor and not as an agent of the owner is responsible for any damages incidental to the work being performed under the contract. This responsibility is applicable to adjacent property and to persons in the vicinity of the construction operations including persons on the site whether entitled to be there or not. Likewise, the owner is liable for damages to others by reason of any defective condition, such as faulty construction, repair, maintenance, or operation of his own property. Furthermore, by reason of contingent liability, he may be held jointly liable for the acts of an independent contractor with whom he has entered into contract. That is to say, the owner may delegate by contract the responsibility for risks existing during construction operations, but he may not completely absolve himself of responsibility thereby. In a similar manner, an independent contractor may be held contingently liable for the acts of one of his subcontractors who is acting as an independent contractor and not as an agent of the prime contractor. A principal, of course, is fully liable for all damages committed by his agent while the agent is carrying out the obligations of the agency agreement. The various forms of public liability insurance may be covered under separate policies, or the Comprehensive Liability Policy can be written to cover all of the liability risks usually required in connection with construction work.

Contractor's Liability Insurance. Under the Contractor's Liability Policy, the insured is protected against liability for injury to persons by reason of his construction operations. This policy may be extended to cover damage to the property of others as well. Under this type of policy, the insurance company assumes responsibility for any claims arising from accidental injury to or death of any person except employees of the insured and others for whom the insured is liable under workmen's compensation laws. Property damage liability may arise from accidental damage to property including the loss of its use. Property owned, leased, rented, or used by the insured or any of his employees is excluded, however. That is to say, contractor's liability insurance applies only to persons and property not connected with the insured.

Contractor's Protective Liability Insurance. A contractor employing a subcontractor is subject to risks of contingent liability with respect to acts of the subcontractor in the same manner and to the same extent as the owner. The Contractor's Protective Liability Insurance covers the contractor for legal liability because of bodily injury, including death resulting therefrom, and property damage caused by accidents arising out of operations performed for the contractor by independent subcontractors. The insurance also protects the contractor against liability which may be imposed upon him by law for damages caused by accident and arising out of omissions or supervisory acts of the contractor in connection with work performed for him by independent subcontractors. Protection is also afforded for accidents (except accidents due to misdelivery) which occur after completion or abandonment of such operations and which arise out of pickup or delivery operations or the existence of tools, uninstalled equipment, and abandoned or unused materials. A claimant may sue not only the subcontractor and owner but also the general contractor for injuries sustained or for death resulting therefrom. If a subcontractor in turn sublets work, he would also need this type of protection.

Contractual Liability Insurance. It is not uncommon for the contractor to be required by a construction contract to save the owner harmless and to assume all the owner's liability including contingent liability for damage to persons and property arising from construction operations. This is an assumption of liability on the part of the contractor which is outside the scope of liability imposed by law. In view of the fact that the legal liability types of insurance previously discussed specifically exclude coverage for assumed liability, a special policy providing contractual liability insurance is required. Under these conditions, owner's protective liability insurance may not be necessary.

Automobile and Truck Liability Insurance. The use of automobiles and trucks in connection with construction work involves risks of two types, namely, the risks of loss due to collision, fire, theft, and the like and the risks due to bodily injury and property damage to other persons. The former are classified as property damage risks, and the latter are liability risks. Each may be covered under separate insurance policies, or they both may be combined in one policy.

In addition to property damage and liability insurance on his automobiles and trucks, the contractor needs two additional types of coverage, as follows.

1. Non-Ownership Liability Insurance which protects the contractor while an employee is using his automobile while on a business trip for him.

2. Hired Vehicle Liability Insurance which covers the risks which arise out of the use of hired vehicles.

Workmen's Compensation Insurance. With the enactment of workmen's compensation laws which placed responsibility on employers for damages due to industrial accidents, it became evident that losses so heavy that they jeopardized an employer's financial stability were possible. To meet this hazard, the standard Workmen's Compensation and Employers' Liability Policy was devised. By this type of policy, the insurance company assumes the risk of an employer in consideration of the payment of a premium. The amount of the premium is determined on the basis of accident probability statistics which consider the location and type of work being performed and the number of workmen employed. Under most state laws, such insurance is made compulsory.* Protection is furnished against losses resulting from personal injury or death of employees, as imposed by the Work-

---

* An exception is made when an employer can qualify as a self-insurer, however.

men's Compensation Act and also losses due to legal liability for such losses as imposed by the Employer's Liability Act.

The details of the coverage of workmen's compensation insurance vary somewhat with the law of the state involved. In addition, special endorsements may be added to meet individual requirements. Typical of these endorsements is the coverage of occupational diseases which are not usually included. Also, the protection of subcontractors' employees as well as those of the prime contractor is a matter for special consideration.

Owner's Protective Liability Insurance. An owner's protective liability policy is a form of public liability insurance in which the insurance company covers the owner's contingent liability for damages due to bodily injury, including death resulting therefrom, and property damage caused by accident arising from construction operations performed for the owner by independent contractors and subcontractors. This kind of insurance also protects the interest of the owner against liability which may be imposed upon him by law because of his supervisory acts or omission thereof in connection with the work performed by the general contractor and subcontractors.

Although it is usual to hold the contractor primarily liable for his own operations, the owner also may be subjected to lawsuits, either singly or jointly with the contractor. Furthermore, consideration must be given to the possibility of default by the contractor in which case the owner may become fully liable. The protection provided by this form of insurance is similar to that of the Contractor's Protective Liability Insurance.

Where work is performed for an owner on a cost-plus-percentage or cost-plus-fixed-fee basis, the protection of the owner may be added by endorsement to the contractor's insurance at no additional premium charge. This eliminates the need for owner's protective liability insurance if the owner is satisfied that the contractor carries adequate limits of liability to protect both interests.

Insurance Underwriting. The issuance of an insurance policy whereby the insurer assumes the risk of the insured and becomes answerable for a designated loss or damage is called underwriting. The insurance company, in order to properly consider the contractor as an insurance risk, must know a great deal about him and his operations. The information necessary for such consideration differs with the type of insurance, but the following may be considered representative of the information which must be made available to the insurance company.

a. Contractor's character, moral habits, and personal integrity.

b. Experience of the contractor, and his standing both in his home community and in the industry.

c. Loss experience in accidents, bodily injury, property damage, and the like.

d. Safety program for the prevention of accidents.

e. Financial standing of the contractor.

f. Contractor's bid price in relation to the engineer's estimates and the bids of other contractors.

g. Site of operations in relation to the location of the service facilities of the insurance company.

h. Maintenance of records of payrolls, materials costs, and work sublet for auditing purposes to determine the premium.

i. Conditions of operation care and maintenance of equipment.

State Supervision of Insurance Companies. The laws of all states provide for the examination and supervision of insurance companies with regard to business methods, financial security, regulation of rates, licensing of brokers and agents, terms of policies, and like matters. State supervision is considered desirable as a means of protecting purchasers of insurance and investors in insurance companies. Administrative control of state supervision is usually vested in a commissioner or superintendent who is authorized to administer the law in the best interests of the people. Thus he serves as a referee or arbiter whose general rulings and decisions have weight about equal to those of a judge in a court of law. Some states give the insurance commissioner the authority to intervene in disputes between an insurance company and an insured in connection with claims under insurance policies.

Questions

1. In drawing the general contract for the construction of an office building, what insurance policies should the prime contractor be required to carry for the owner's protection? What policies should he carry of his own volition to protect his own interests? What insurance protection should the owner obtain?

2. The owners of a toll bridge insured the structure against collapse. The agent of the insurance company, believing that the occurrence of such

an event was unlikely, did not report to his company that he had written the policy, and he retained the premiums paid by the owners. Later the bridge did collapse, and when the owners presented their claim the insurance company had no record of the policy. Who was liable for the loss?

3. What are the advantages of comprehensive insurance policies, such as builder's risk, in construction work?

4. When a number of risks such as fire, tornado and explosion are to be covered by insurance, what are the advantages to the insured of having all policies written by the same company?

5. A truck trailer was unloading gravel. As the trailer was hoisted in front, all of the gravel on one side ran out. The resulting weight on the other side caused the trailer to tip over, and it was damaged as it came in contact with the ground. Was the owner entitled to collect on an insurance policy covering "direct and accidental damage to the automobile caused by upset"?

6. A prime contractor agreed to carry liability insurance for the benefit of the owner and his employees, and to require subcontractors to carry liability insurance. An employee of the owner was injured through negligent driving of a subcontractor's truck. The subcontractor carried no insurance. The injured employee was partly compensated for his injury by workmen's compensation insurance. Did he and the owner's compensation insurer have a right to sue the prime contractor for their respective losses resulting from the contractor's failure to require the subcontractor to carry liability insurance?

12

SURETY BONDS, PREQUALIFICATION OF CONTRACTORS

This chapter deals with some of the non-technical features of contracting practice which are intended to insure the satisfactory completion of the work. If the contractor were always competent, honest, and financially responsible and if the contract documents were always perfect, questions would never arise. Unfortunately, this combination of conditions is improbable. Therefore, it is necessary to provide some means for the prevention or completion of defaulted contracts which are costly and cause delays in the progress of the work. Defaulted contracts are best avoided by means of surety bonds which guarantee the integrity of the contract under the penalty on the surety of performance and payment of all contract obligations defaulted by the contractor.

SURETY BONDS

Contractors are generally required to furnish surety bonds as a guarantee of the faithful performance of the contract and the payment of bills for labor and materials. Surety bonds are usually issued by a bonding company, although private individuals may serve as sureties, or the contractor may furnish his own surety by depositing acceptable property or collateral with the owner. Corporate surety bonds are the general rule, however, inasmuch as their cost is relatively low. There are three parties to the execution of a bond.

a. The principal (contractor) in whose behalf the bond is written and whose performance is guaranteed.

b. The obligee (owner) in whose favor the bond is written.

c. The surety (bonding company) who acts as guarantor for the principal and who is obligated to make good to the obligee any default on the part of the principal.

Three common types of surety bonds are normally used in connection with competitive-bid contracts, namely, bid bonds, performance bonds, and payment bonds. Surety bonds are not often required with negotiated contracts since these are usually on a cost-plus basis in which the owner selects the contractor and assumes partial responsibility for the control of the work. For contracts negotiated on a lump-sum or unit-price basis, however, it may be advisable to require payment bonds in order to protect the owner against loss or mechanic's liens arising from the contractor's possible failure to pay his bills in connection with the work.

Bid Bond. The bid bond is submitted with the proposal. It guarantees that the bid has been made in good faith and that the bidder will enter into the contract if his proposal is accepted. It also guarantees that he will furnish performance and payment bonds. In the event that the bidder should refuse to sign the contract in accordance with his bid, the surety is obligated to pay to the owner the damages caused thereby subject to the penalty of the bond. Under current practice the penal sum to be required in the bid bond varies from 5 to 10 per cent of the bid amount. The standard form of bid bond used by the U.S. Government stipulates that, if the approved bid is withdrawn and the contract refused, the Government shall be paid the difference between the amount specified in the bid and the amount for which the government may procure the required work under the second-lowest acceptable bid.

A certificate of surety is similar in some respects to the bid bond. It accompanies the bid and guarantees that if the contract is awarded to the bidder the surety will furnish performance and payment bonds. The certificate of surety is rarely used.

Performance Bond. This bond guarantees that the contractor will perform the contract in accordance with its terms. Usually it covers whatever obligations the contractor assumes in the contract, including protection to the owner against defective work by the contractor. It is to be distinguished from the payment bond which guarantees that the contractor will pay for his labor and materials although sometimes the obligation to pay for labor and materials is included in the performance bond itself. This type of bond is called a combined performance and payment bond, and such a form was used by the U.S. Government until 1938. On account of certain technical and procedural difficulties arising under a single bond, the Miller Act provided for the superior dual system of performance and payment bonds which is in use today. It is advisable to have the performance bond executed in the full amount of the contract.

Payment Bond. This bond guarantees that the contractor will pay all bills and obligations incurred under the contract, thus rendering the owner harmless from claims and liens which might be filed after the completion of the work and after all payments had been made to the contractor. It is always advisable that the payment bond be a separate instrument from the performance bond, and it should be executed in the amount of at least 50 per cent of the contract price.

Miscellaneous Types of Bonds. Maintenance bonds guarantee the quality or maintenance cost of the construction work for a definite period of time. Labor and material bonds are designed to protect workmen, material dealers, and subcontractors. They are in effect a form of payment bond. Supply bonds guarantee that the contractor will furnish materials and equipment required under supply contracts. Fidelity bonds guarantee the honesty of individuals, companies, or corporations. They may also guarantee the faithful performance of duty.

The Surety Bond as a Contract. A surety bond is explicitly a contract between the surety and the owner in which the surety promises to make good a deficiency on the part of the contractor, the obligations of the contractor being defined in the construction contract to which the surety is not a party. Construction contracts generally provide for changes and modifications as desired by the owner. Thus, in the case of changes, the surety may be placed in the position of being obligated by the contractual agreements of persons other than himself. This is contrary to the common law inasmuch as two contracting parties may not bind a third without his consent. Therefore, a change in the contract may discharge the bond and should never be undertaken without first obtaining the approval of the surety, unless the terms of the bond stipulate some other procedure. In this connection it should be noted that the standard performance bond for U.S. Government contracts provides for changes in the contract without advance permission by the surety. The surety thereby waives his rights under the common law in this respect. The ultimate test of a surety bond arises when a claim is filed. Until then it is a contract document containing written promises redeemable when the contract for which the bond was furnished is breached by the contractor. When a loss or default occurs and a proper claim under the bond is made, the promises explicit in the bond must be fulfilled.

Warranty. A warranty is frequently required in construction contracts in connection with the furnishing of certain types of materials and equipment. By means of a warranty the contractor certifies that the material or equipment will perform as required. If the material or equipment so covered should fail to perform as required, the contractor is liable to a suit for damages on the grounds of breach of warranty. A warranty promises results and implies that the contractor may have control over the sufficiency of the plans. If the owner retains control over the details of the plans, the warranty may not be enforceable. Warranties are usually required with the purchase of machinery, such as motors and pumps, and in connection with the use of certain processed materials where the designs are furnished by the vendor and the products are manufactured by him to meet the performance specifications. A warranty unlike a guarantee does not have to be in writing. Warranties required under construction contracts constitute separate agreements and do not conflict with or duplicate the functions of surety bonds. If desired, however, surety bonds may be made to guarantee the compliance of a warranty. This is usually required under U.S. Government contracts.

Completion of Defaulted Contracts. If a contractor fails to complete a contract the surety is called upon to finish the work. This he may elect to accomplish with his own forces, those of the defaulted contractor, or by letting a new contract for the uncompleted portion of the work. Furthermore, it is not unusual to include a provision for the owner or the surety to take over and use the contractor's equipment to complete the work in case of default by the contractor. If he so elects the surety may simply pay the penal amount and leave the responsibility for the completion of the work to the owner. Regardless of the method chosen, delays and extra expenses to the owner are to be expected. Therefore, it is desirable that the engineer exert every possible effort to avoid default. This may sometimes be accomplished by compromise on issues that arise during the course of the work or even by yielding on questions of interpretation of the contract when such action can be justified in the light of the implications of default.

Advantages of Surety Bonds. The basic purpose of a bond is the elimination of financial risk to the owner. In addition to providing this indemnity for loss, however, there are many valuable by-products of a bonded competitive system. The loss-prevention activities of surety companies often avert a pending default. The technical skill of surety underwriters is made available to bonded contractors to assist them in difficult situations. By eliminating unqualified contractors, the bond system has a stabilizing influence on the entire construction industry. Also it gives subcontractors and furnishers of labor and materials assurance that they will be paid, and this results in an economical and preferential flow of labor and materials to the project.

Bond Underwriting. Before a corporate surety will furnish a construction bond, the character, ability, and financial status of the contractor are carefully investigated. The education, training, experience, and the ability of the contractor himself, his superintendent, and his estimator are reviewed since they are the ones upon whom the success of the contract chiefly depends. All of the contractor's key personnel should be men of good moral habits and personal integrity.

The contractor's financial condition is of prime importance in the performance of the contract. His financial statement should be sufficiently complete for the surety to understand and evaluate his worth and credit. Any items not self-explanatory should be supplemented by notes. The contractor's bank credit and his relationships with his sources of credit and supply are important factors. It is not uncommon for the officers of corporate contractors to guarantee the successful completion of the contract by giving their personal indemnity to the surety.

The contractor must have sufficient equipment to handle the work he has in progress and any additional work contemplated. His equipment should be insured and in good condition.

The surety will examine the contract to determine the method of payment, amount of retained percentages, the use of patented articles, the time of completion, the penalty for delay, and the like. With respect to the contract price, the contractor's recent bids will be examined to ascertain if they are generally in line with those of other reputable contractors and with the engineer's estimates. The portions of the work to be sublet to specialty subcontractors and the contractor's relationships with such subcontractors will be determined.

General conditions affecting the contract are studied to determine the accessibility of the site and unusual hazards of the project. Underground conditions should have been investigated by the contractor by means of borings. Flood hazards are always present in certain types of work near, on, or under, water and should be evaluated. The availability of labor at the site and the contractor's relations with labor unions are investigated. The surety will make certain that the contractor is carrying the proper types and amounts of insurance to protect himself adequately against all reasonable liabilities.

Bond Forms. The standard forms for bid, performance, and payment bonds for both private and U.S. Government contracts are reproduced herein as examples of current bonding practice. The forms for private works are those recommended by the Surety Association of America and approved by the American Institute of Architects. The standard U.S. Government forms are used with all Government contracts and must be interpreted in the light of the statutes under which they were adopted. Surety bonds for all U.S. Government contracts for construction must be executed in accordance with the following instructions.

1. This [standard] form shall be used for construction work or the furnishing of supplies or services, whenever a [surety] bond is required. There shall be no deviation from this form except as authorized by the General Services Administration.

2. The surety on the bond may be any corporation authorized by the Secretary of the Treasury to act as surety, or two responsible individual sureties. Where individual sureties are used, this bond must be accompanied by a completed Affidavit of Individual Surety for each individual surety (Standard Form 28).

3. The name, including full Christian name, and business or residence address of each individual party to the bond shall be inserted in the space provided therefor, and each such party shall sign the bond with his usual signature on the line opposite the scroll seal, and if signed in Maine or New Hampshire, an adhesive seal shall be affixed opposite the signature.

4. If the principals are partners, their individual names shall appear in the space provided therefor, with the recital that they

are partners composing a firm, naming it, and all the members of the firm shall execute the bond as individuals.

5. If the principal or surety is a corporation, the name of the State in which incorporated shall be inserted in the space provided therefor, and said instrument shall be executed and attested under the corporate seal as indicated in the form. If the corporation has no corporate seal the fact shall be stated, in which case a scroll or adhesive seal shall appear following the corporate name.

6. The official character and authority of the person or persons executing the bond for the principal, if a corporation, shall be certified by the secretary or assistant secretary, according to the form herein provided. In lieu of such certificate there may be attached to the bond copies of so much of the records of the corporation as will show the official character and authority of the officer signing, duly certified by the secretary or assistant secretary, under the corporate seal, to be true copies.

7. The date of this bond must not be prior to the date of the instrument in connection with which it is given.

EXAMPLES OF BOND FORMS

Standard Form 24 Revised November 1950 Prescribed by General Services Administration General Regulation No. 5 **BID BOND** **(For U.S. Government Contracts)**	Date Bond Executed

Principal

Surety

Penal Sum of Bond (*express in words and figures*)

Date of Bid

KNOW ALL MEN BY THESE PRESENTS, That we, the PRINCIPAL and SURETY above named, are held and firmly bound unto the United States of America, hereinafter called the Government, in the penal sum of the amount stated above, for the payment of which sum well and truly to be made, we bind ourselves, our heirs, executors, administrators, and successors, jointly and severally, firmly by these presents.

THE CONDITION OF THIS OBLIGATION IS SUCH, that whereas the principal has submitted the accompanying bid, dated as shown above, for

Now, THEREFORE, if the principal shall not withdraw said bid within the period specified therein after the opening of the same, or, if no period be specified, within sixty (60) days after said opening, and shall within the period specified therefor, or, if no period be specified, within ten (10) days

after the prescribed forms are presented to him for signature, execute such
further contractual documents, if any, as may be required by the terms of
the bid as accepted, and give bonds with good and sufficient surety or sureties,
as may be required, for the faithful performance and proper fulfillment of the
resulting contract, and for the protection of all persons supplying labor and
material in the prosecution of the work provided for in such contract, or in
the event of the withdrawal of said bid within the period specified, or the
failure to enter into such contract and give such bonds within the time
specified, if the principal shall pay the Government the difference between
the amount specified in said bid and the amount for which the Government
may procure the required work, supplies, and services, if the latter amount
be in excess of the former, then the above obligation shall be void and of no
effect, otherwise to remain in full force and virtue.

IN WITNESS WHEREOF, the above-bounden parties have executed this instrument under their several seals on the date indicated above, the name and
corporate seal of each corporate party being hereto affixed and these presents
duly signed by its undersigned representative, pursuant to authority of its
governing body.

In Presence of:

	WITNESS	INDIVIDUAL PRINCIPAL
1.	.... as to	....[SEAL]
2.	.... as to	....[SEAL]
3.	.... as to	....[SEAL]
4.	.... as to	....[SEAL]
	WITNESS	INDIVIDUAL SURETY
1.	.... as to	....[SEAL]
2.	.... as to	....[SEAL]

Attest:

Corporate Principal

Business Address

By

Title

AFFIX CORPORATE SEAL

Attest:

Corporate Surety

Business Address

By

Title

AFFIX CORPORATE SEAL

The rate of premium on this bond is .... per thousand.

Total amount of premium charged, $.....................

(The above must be filled in by corporate surety)

CERTIFICATE AS TO CORPORATE PRINCIPAL

I, ..., certify that I am the

.... secretary of the corporation named as principal in

the within bond; that ...., who signed the said

bond on behalf of the principal, was then .... of

said corporation; that I know his signature, and his signature thereto is genuine; and that said bond was duly signed, sealed, and attested for and in behalf of said corporation by authority of its governing body.

.... [CORPORATE SEAL]

BID BOND

(Private Construction)

KNOW ALL MEN BY THESE PRESENTS: THAT..

....

(hereinafter called the Principal) and the..

(hereinafter called the Surety) are held and firmly bound unto.......................

....

....(hereinafter called the Obligee) in the full

and just sum of ....Dollars good and lawful

money of the United States of America, to the payment of which said sum of money, well and truly to be made and done, the said Principal binds himself, his heirs, executors, administrators, successors and assigns, and the said Surety binds itself, its successors and assigns, jointly and severally, firmly by these presents.

Signed, sealed and dated this....................day of.............................., A.D., 19.........

THE CONDITIONS OF THIS OBLIGATION ARE SUCH, that if any awards made, within sixty (60) days from the date of this instrument, by said Obligee, to the above-bounden Principal under a public invitation for..............................

....

....

....

shall be accepted by said Principal and said Principal shall enter into a contract for the completion of said work, and give Bond with the

....

as surety, or with other surety or sureties to be approved by the
Obligee for the faithful performance thereof, then this obligation shall
be null and void; otherwise to remain in full force and effect.

PROVIDED: *First:*—That the liability of the Surety shall in no event exceed
the penalty of this bond.

       Second:—That any suits at law or proceedings in equity brought
or to be brought against said Surety to recover any claim hereunder, must be
instituted within six (6) months from the date of this instrument.

    ....(Seal)
      Principal

    ....(Seal)
      Surety Company

Standard Form 25
Revised November 1950
Prescribed by General **PERFORMANCE BOND**
Services Administration
General Regulation No. 5
    (For U.S. Government Contracts)

Date Bond Executed

Principal

Surety

Penal Sum of Bond (*express in words and figures*) | Contract No. | Date of Contract

KNOW ALL MEN BY THESE PRESENTS, That we, the PRINCIPAL and SURETY
above named, are held and firmly bound unto the United States of America,
hereinafter called the Government, in the penal sum of the amount stated
above, for the payment of which sum well and truly to be made, we bind
ourselves, our heirs, executors, administrators, and successors, jointly and
severally, firmly by these presents.

THE CONDITION OF THIS OBLIGATION IS SUCH, that whereas the principal
entered into a certain contract with the Government, numbered and dated
as shown above and hereto attached;

NOW, THEREFORE, if the principal shall well and truly perform and fulfill all
the undertakings, covenants, terms, conditions, and agreements of said contract during the original term of said contract and any extensions thereof
that may be granted by the Government, with or without notice to the surety,
and during the life of any guaranty required under the contract, and shall
also well and truly perform and fulfill all the undertakings, covenants, terms,
conditions, and agreements of any and all duly authorized modifications of
said contract that may hereafter be made, notice of which modifications to
the surety being hereby waived, then, this obligation to be void; otherwise
to remain in full force and virtue.

IN WITNESS WHEREOF, the above-bounden parties have executed this instrument under their several seals on the date indicated above, the name and
corporate seal of each corporate party being hereto affixed and these presents
duly signed by its undersigned representative, pursuant to authority of its
governing body.

In Presence of:

	WITNESS		INDIVIDUAL PRINCIPAL	
1.	....	as to	....	[SEAL]
2.	....	as to	....	[SEAL]
3.	....	as to	....	[SEAL]
4.	....	as to	....	[SEAL]

	WITNESS		INDIVIDUAL SURETY	
1.	....	as to	....	[SEAL]
2.	....	as to	....	[SEAL]

Attest:	Corporate Principal	
	Business Address	
	By	AFFIX CORPORATE SEAL
	Title	

Attest:	Corporate Surety	
	Business Address	
	By	AFFIX CORPORATE SEAL
	Title	

The rate of premium on this bond is .... per thousand.

Total amount of premium charged, $............................

(The above must be filled in by corporate surety)

CERTIFICATE AS TO CORPORATE PRINCIPAL

I, ..., certify that I am the

.... secretary of the corporation named as principal in

the within bond; that ...., who signed the said

bond on behalf of the principal, was then .... of
said corporation; that I know his signature, and his signature thereto is
genuine; and that said bond was duly signed, sealed, and attested for and
in behalf of said corporation by authority of its governing body.

.... [CORPORATE SEAL]

PERFORMANCE BOND *

(Private Construction)

KNOW ALL MEN BY THESE PRESENTS, THAT...
.... *(Here insert name and address,*
...., as Principal, hereinafter called
or legal title, of the Contractor)
Contractor, and..., as Surety,
....*(Here insert the legal title of Surety)*
hereinafter called Surety, are held and firmly bound unto.................................
....*(Here insert the name*
...., as Obligee, hereinafter called
and address, or legal title, of the Owner)
Owner, in the amount of...Dollars
($) for the payment whereof Contractor and Surety bind
themselves, their heirs, executors, administrators, successors and assigns,
jointly and severally, firmly by these presents.

WHEREAS, Contractor has by written agreement dated.................................
....entered into a contract with Owner
for..
....
in accordance with drawings and specifications prepared by.................................
....*(Here insert full name*
...., which contract is by reference
and title)
made a part hereof, and is hereinafter referred to as the CONTRACT.

NOW, THEREFORE, THE CONDITION OF THIS OBLIGATION is such that, **if
Contractor shall promptly and faithfully perform said CONTRACT,**
then this obligation shall be null and void; otherwise it shall remain in full
force and effect.

Whenever Contractor shall be, and declared by Owner to be in default
under the CONTRACT, the Owner having performed Owner's obligations thereunder, the Surety may promptly remedy the default, or shall promptly

(1) Complete the CONTRACT in accordance with its terms and conditions, or
(2) Obtain a bid or bids for submission to Owner for completing the CON-
TRACT in accordance with its terms and conditions, and upon determination by Owner and Surety of the lowest responsible bidder, arrange for
a contract between such bidder and Owner and make available as work
progresses (even though there should be a default or a succession of
defaults under the contract or contracts of completion arranged under
this paragraph) sufficient funds to pay the cost of completion less the
balance of the contract price; but not exceeding, including other costs
and damages for which the Surety may be liable hereunder, the amount
set forth in the first paragraph hereof. The term "balance of the contract price," as used in this paragraph, shall mean the total amount payable by Owner to Contractor under the CONTRACT and any amendments
thereto, less the amount properly paid by Owner to Contractor.

Any suit under this bond must be instituted before the expiration of two
(2) years from the date on which final payment under the CONTRACT falls due.

(* *Approved by the Executive Committee of The American Institute of
Architects—August, 1946.*)
Reprinted with permission of The American Institute of Architects.

No right of action shall accrue on this bond to or for the use of any person or corporation other than the owner named herein or the heirs, executors, administrators or successors of Owner.

Signed and sealed this...................... day of ....A.D. 19........

In presence of:

....(Seal)
　　　　　　　　　　Principal
....(Seal)
　　　　　　　　　　Surety Company

Standard Form 25A Revised November 1950 Prescribed by General Services Administration General Regulation No. 5	**PAYMENT BOND** (For U.S. Government Contracts)	Date Bond Executed

Principal

Surety

Penal Sum of Bond (*express in words and figures*)	Contract No.	Date of Contract

KNOW ALL MEN BY THESE PRESENTS, That we, the PRINCIPAL and SURETY above named, are held and firmly bound unto the United States of America, hereinafter called the Government, in the penal sum of the amount stated above, for the payment of which sum well and truly to be made, we bind ourselves, our heirs, executors, administrators, and successors, jointly and severally, firmly by these presents.

THE CONDITION OF THIS OBLIGATION IS SUCH, that whereas the principal entered into a certain contract with the Government, numbered and dated as shown above and hereto attached;

NOW, THEREFORE, if the principal shall promptly make payment to all persons supplying labor and material in the prosecution of the work provided for in said contract, and any and all duly authorized modifications of said contract that may hereafter be made, notice of which modifications to the surety being hereby waived, then this obligation to be void; otherwise to remain in full force and virtue.

IN WITNESS WHEREOF, the above-bounden parties have executed this instrument under their several seals on the date indicated above, the name and corporate seal of each corporate party being hereto affixed and these presents duly signed by its undersigned representative, pursuant to authority of its governing body.

In Presence of:

	WITNESS	INDIVIDUAL PRINCIPAL	
1.	.... as to	....	[SEAL]
2.	.... as to	....	[SEAL]
3.	.... as to	....	[SEAL]
4.	.... as to	....	[SEAL]

WITNESS	INDIVIDUAL SURETY
1. .... as to	.... [SEAL]
2. .... as to	.... [SEAL]

Attest:	Corporate Principal	
	Business Address	
	By	AFFIX
	Title	CORPORATE SEAL

Attest:	Corporate Surety	
	Business Address	
	By	AFFIX
	Title	CORPORATE SEAL

Total amount of premium charged, $............................

(The above must be filled in by corporate surety)

CERTIFICATE AS TO CORPORATE PRINCIPAL

I, ..., certify that I am the

.... secretary of the corporation named as principal in

the within bond; that ...., who signed the said

bond on behalf of the principal, was then .... of said corporation; that I know his signature, and his signature thereto is genuine; and that said bond was duly signed, sealed, and attested for and in behalf of said corporation by authority of its governing body.

.... [CORPORATE SEAL]

LABOR AND MATERIAL PAYMENT BOND *

(Private Construction)

NOTE: *This bond is issued simultaneously with another bond in favor of the owner conditioned for the full and faithful performance of the contract.*

KNOW ALL MEN BY THESE PRESENTS, THAT..
(Here insert name and address,

(*Approved by the Executive Committee of The American Institute of Architects—August, 1946.)*
Reprinted with permission of The American Institute of Architects.

...., as Principal, hereinafter called
or legal title, of the Contractor)
Principal, and..as Surety, hereinafter
               (*Here insert the legal title of Surety*)
called Surety, are held and firmly bound unto..
                (*Here insert the name and address,*
...., as Obligee, hereinafter called Owner, for
or legal title, of the Owner)
the use and benefit of claimants as hereinbelow defined, in the amount of
....Dollars ($), for
(*Here insert a sum equal to one-half of the contract price*)
the payment whereof Principal and Surety bind themselves, their heirs, executors, administrators, successors and assigns, jointly and severally, firmly by these presents.

WHEREAS, Principal has by written agreement dated...
..........................entered into a contract with Owner for...
....

in accordance with drawings and specifications prepared by.................................
                (*Here insert full*
...., which contract is by reference made a
name and title)
part hereof, and is hereafter referred to as the CONTRACT.

NOW, THEREFORE, THE CONDITION OF THIS OBLIGATION is such that if the Principal **shall promptly make payment to all claimants as hereinafter defined, for all labor and material used or reasonably required for use in the performance of the Contract,** then this obligation shall be void; otherwise it shall remain in full force and effect, subject, however, to the following conditions:

1. A claimant is defined as one having a direct contract with the Principal or with a sub-contractor of the Principal for labor, material, or both, used or reasonably required for use in the performance of the contract, labor and material being construed to include that part of water, gas, power, light, heat, oil, gasoline, telephone service or rental of equipment directly applicable to the contract.

2. The above named Principal, and Surety hereby jointly and severally agree with the Owner that every claimant as herein defined, who has not been paid in full before the expiration of a period of ninety (90) days after the date on which the last of such claimant's work or labor was done or performed, or materials were furnished by such claimant may sue on this bond for the use of such claimant in the name of the Owner, prosecute the suit to final judgment for such sum or sums as may be justly due claimant, and have execution thereon, provided, however, that the Owner shall not be liable for the payment of any costs or expenses of any such suit.

3. No suit or action shall be commenced hereunder by any claimant,

(*a*) Unless claimant shall have given written notice to any two of the following: The Principal, the Owner, or the Surety above named, within ninety (90) days after such claimant did or performed the last of the work or labor, or furnished the last of the materials for which said claim is made, stating with substantial accuracy the amount claimed and the name of the party to whom the materials were furnished, or for whom the work or labor was done or performed. Such notice shall be served by mailing the same by registered mail, postage prepaid, in an envelope addressed to the Principal, Owner or Surety, at any place where an office is regularly maintained for the transaction of business, or served in any manner in which legal process may be served in the state in which the aforesaid project is located, save that such service need not be made by a public officer.

(b) After the expiration of one (1) year following the date on which Principal ceased work on said CONTRACT.

(c) Other than in a state court of competent jurisdiction in and for the county or other political subdivision of the state in which the project, or any part thereof, is situated, or in the United States District Court for the district in which the project, or any part thereof, is situated and not elsewhere.

4. The amount of this bond shall be reduced by and to the extent of any payment or payments made in good faith hereunder, inclusive of the payment by Surety of mechanics' liens which may be filed of record against said improvement, whether or not claim for the amount of such lien be presented under and against this bond.

Signed and sealed this.....................day of...A.D. 19..........

In presence of:

....(Seal)
Principal

....(Seal)
Surety Company

PREQUALIFICATION OF CONTRACTORS

The necessity to advertise for bidders, to accept bids from all who are inclined to compete, and to award contracts to the lowest responsible bidder introduces problems in the awarding of contracts for public works which are not encountered in private construction where the list of bidders may be selected without restriction. For a bidder to establish responsibility under open competition usually means the furnishing of the required surety and a record free from defaults or proved dishonesty. Thus incompetent and overextended contractors and those with inadequate financial resources may be placed more or less on an equal basis with responsible bidders in the competition for the award of a contract. Although bonding companies ordinarily exercise care in providing bonds for contractors, it should be noted that the ability of a contractor to obtain bond does not necessarily indicate competency or financial responsibility. This is true because surety underwriting depends upon judgment factors and underwriters like engineers and contractors will never be infallible.

The employment of an unqualified contractor usually leads to difficulties during the operation of the contract. Also, slow progress, unsatisfactory quality of work, and excessive costs may result. Moreover, incompetency is one very important factor in contract defaults which always cause inconvenience, delays, and extra cost to the owner. The Bureau of Public Roads estimated that over 60 per cent of the cost of delays and unsatisfactory progress on federal aid road projects was due to poor management and inadequate equipment. Likewise, a large surety company estimated that 75 per cent of the cost of losses sustained by it through defaulted contracts was due to inadequate financial responsibility, overextension, and incompetency. To avoid or reduce these difficulties the prequalification of bidders has been recommended.

The purpose of prequalification is to determine before a contractor is allowed to bid whether he is responsible and competent to satisfactorily complete a given construction contract. In some instances the prequalification procedure is sufficiently comprehensive so that it is considered safe to reduce the amount of surety bonds, or to omit them altogether, in the award of contracts, it being assumed that all unqualified contractors are eliminated from the approved list of bidders. When carried to this extent, a direct reduction in construction costs results inasmuch as the cost of surety is reflected in the contractor's bid.

Prequalification Procedure. Prequalification procedure requires the contractor to submit a formal application to bid. The application contains sufficient information, usually in the form of answers to a questionnaire, to determine the contractor's competency and fitness to perform the contemplated work. The application should contain detailed data on the contractor's past experience, current work, financial status, quality of organization, and available plant and equipment. A consideration of these factors will usually eliminate unfit contractors from the list of those permitted. For instance, the value of contracts completed in the preceding five years is a measure of a contractor's average annual capacity. The difference between this amount and the value of incomplete contracts in progress will indicate his available capacity for new work. If his available capacity is not at least equal to the estimated value of the new contract, his organization should be closely examined to determine his ability to expand to meet the additional load. Likewise, his financial condition should be carefully examined in the light of the estimated amount of working capital required for the proposed work, and his past experience in the type of work in question should be of appropriate quality.

Example of a Prequalification Questionnaire.

1. Name of firm; state whether individual, partnership, corporation, or joint venture. Give date of organization, the state in which incorporated, and names and addresses of officers or partners.

2. Official address.

3. How many years has your firm been in business as a general contractor under your present business name?

4. List by years the contracts completed during the past five or ten years, giving the character of work and contract amounts.

5. Amount of uncompleted work on contracts now in progress. List the total value of each contract, percentage uncompleted, estimated date of completion, and value of uncompleted portion.

6. Financial resources available as working capital for new work. State both cash on hand and sources of credit. Attach recent financial statement and letters from banks regarding credit.

7. List items of plant and equipment suitable and in good condition which can be made available for new work; state whether owned by you, where located, and, if not owned by you, what assurance you have that the equipment will be available when needed. Also give the total present value of available plant.

8. List of officials and key personnel who would devote full or part time to a new contract. Give name, title, age, experience, and salary of each.

9. Knowledge of local conditions; transportation facilities, material and labor markets, living conditions for workmen and families, etc.

10. To what extent would you expect to employ subcontractors?

11. Does your organization include men experienced in securing cooperation of materials dealers in prompt filling of orders, and in securing cooperation of labor? Give particulars.

12. Have you ever failed to complete any work awarded to you? If so, give particulars.

13. Has an officer or partner of your firm ever been an officer or partner of some other organization that failed to complete a construction contract? If so, give particulars.

14. Has your firm ever engaged in litigation for the settlement of claims or disputes arising out of a construction contract? If so, give particulars.

Advantages and Disadvantages of Prequalification. The prequalification of contractors is still a somewhat controversial matter although the general consensus of opinion among both engineers and contractors appears to be that it is very desirable. It is required by law for public works projects in several states and has been adopted by several federal agencies. The legality of prequalification seems to be definitely established by rulings of various courts. It should be noted, however, that there is no unanimity of opinion with regard to prequalification even among officials who are operating under this system.

Some of the obvious advantages of prequalification are that lists of competent bidders may be established in advance when there is sufficient time to investigate contractor's qualifications. When all bidders are qualified, the contract is simply awarded to the lowest bidder, and the public official who awards the contract is saved the embarrassment of rejecting a low bid from an unqualified bidder. Contractors are saved the time and expense of preparing bids for work in which they are unqualified by inexperience and lack of financial or other resources. Failures and defaults of contracts during construction are minimized by the elimination of unfit contractors which results in a saving of time and cost of construction work.

Opponents of prequalification object that the questionnaires are too complicated for small contractors, many of whom have incomplete records; that the procedure requires an undue expenditure of time; that it places unlimited authority on public officials and offers an opportunity for favoritism; that it restricts competition to the large and rich contractors; and that it tends to eliminate young contractors who are starting in business. It is sometimes contended that financial statements and other details of a contractor's business or organization are personal and private matters and that it is unconstitutional to require that they be made public in a prequalification statement. Opponents of prequalification also contend that prequalification standards may be arbitrary and judgment may be unreasonable. Both may result in restriction of construction to local contractors, thereby limiting competition and increasing costs.

The disqualification of a contractor is a serious action and must be taken only for a well established reason. There is a possibility that such action will be construed as a reflection upon the integrity and competency of the disqualified contractor even though the usual causes are inadequate capacity to undertake new work, insufficient working capital to finance the job, lack of the specialized experience required, and the like. The value of prequalification frequently is lost when well-meaning public officials reduce the procedure to an ineffective routine in order to avoid adverse publicity.

The owner's interests probably are best protected when prequalification and the bonding of construction contracts are supplementary but not duplicate means of determining the qualification of bidders. Prequalification at best is a judgment of a contractor's qualifications as of the day submitted. Subsequent contracts undertaken might and often do change a contractor's financial condition drastically. Whereas, if the contract is bonded the owner continues to obtain the collateral advantages of guaranteed credit to suppliers of labor and material and the guarantee against loss which the bond penalty affords.

Bureau of Contract Information. The Associated General Contractors of America has approved and sponsored the adoption of prequalification of contractors because of its benefits to all parties in construction work. In 1929 this organization established a Bureau of Contract Information with headquarters at 1420 New York Ave., N.W., Washington, D. C. The objective of the Bureau of Contract Information is to study the performance records of every general contractor in the United States with the view to eliminating irresponsibility in the field of competition in the construction industry through the discouragement of the award of contracts to irresponsible bidders. This bureau is a non-profit fact-finding organization which cooperates with those responsible for the award of construction contracts.

Questions

1. What is the purpose of a bid bond? a performance bond? a payment bond?

2. Explain the contract relationships between the principal, the obligee, and the surety in a surety bond.

3. Under what conditions may a change order under a construction contract invalidate a surety bond and relieve the surety of responsibility?

4. Explain the difference between a surety bond and a warranty.

5. How may the information obtained from a prequalification questionnaire, such as that on page 275, be used to determine a contractor's fitness to perform construction work?

6. What objectionable types of contractors may be eliminated by the prequalification procedure?

7. A contractor had furnished a payment bond in connection with his undertaking to construct certain dwelling units for the U.S. Government. He placed an order for some necessary building materials with Material-man A. Materialman A filled the order after purchasing the materials from Materialman B. A was paid in full by the prime contractor, but A failed to pay B in full. B thereupon brought suit on the bond. Was the surety liable for the debt? Was B entitled to a mechanic's lien on the dwelling units? (See *Engineering News-Record,* August 21, 1947, page 92.)

8. A highway contractor did not complete the road base on a project within reasonable time. He therefore lost the right to require performance of a subcontract for surfacing. The contractor's surety took over the work as provided by its bond and took an assignment of the contractor's rights. (1) Could the surety hold the subcontractor liable for refusal to perform after the base was provided? (2) Was the subcontractor justified in treating the subcontract as terminated when the surety took over the work?

9. A municipal contractor gave a bond that was so worded as to make the contractor liable for damage to a third party's property in blasting. The city did not require that so broad a bond be given, and there was no law requiring it. Was the third party's suit to collect damages under the bond properly dismissed?

10. Under statutory requirement, Philadelphia was bound to award an incinerator and garbage-plant construction contract to the lowest responsible bidder, under regulations provided by ordinance. An ordinance empowered the director of public works to secure information from bidders as to their ability to perform. Specifications called for bids for construction embodying alternative types of incinerators, and each bidder was required to show that he or a named subcontractor had designed and constructed a plant of each type in operation for at least two years. Did imposition of that requirement invalidate a contract awarded to the successful bidder as an arbitrary and unreasonable suppression of competitive bidding?

13

CONTRACTS FOR ENGINEERING AND ARCHITECTURAL SERVICES

As the term is used in the United States, consulting engineering includes not only consultation, advice, and expert testimony by consulting engineers but also the furnishing by engineers in private practice of the more extensive engineering services required from the conception to the completion of major engineering projects. The practice of architecture follows similar lines. The business relations between the client and the engineer or architect are of a contractual nature, and the obligations of each should be set forth clearly in a written agreement.

Services of the Engineer. The services to be furnished by the engineer will depend on the size and the complexity of the work, the client's technical staff and the basis on which the fee is determined. Complete engineering services for important projects normally are accomplished in three phases as follows.

1. Preliminary report.
   a. Preliminary surveys and supervision of explorations and borings.
   b. Preliminary designs of alternative types of construction.
   c. Comparative estimates of cost for alternative types of construction.
   d. Recommendations as to the most suitable solution of the problem.*
   e. Economic justification for the project.
2. Preparation of contract plans and specifications.
   a. Preparation of final design drawings and detailed specifications.

---

* When construction work is to be accomplished under a cost-plus type contract, the preliminary design drawings may be made sufficiently complete to serve as the basis for the award of the construction contract, and the detail drawings can be prepared after the beginning of construction operations.

      b. Estimates of quantities and costs.

      c. Preparation of instructions to bidders, bid forms, and assistance in preparation of contract forms.

      d. Assistance in obtaining and analyzing bids and the award of contracts.

  3. Supervision of construction.

      a. Field layout, inpection and supervision of the work during construction.

      b. Checking of shop and working drawings furnished by contractors.

      c. Supervision of mill and shop inspections and tests.

      d. Progress and final estimates and reports.

      e. Issuance of certificates for progress and final payments to contractors.

      f. Revision of contract drawings to show changes introduced during the progress of the work.

Complete engineering services may be furnished by one firm, or different firms may be retained for the various parts of the work. Where complete engineering services are required, the advantages derived from continuity by retaining one firm throughout are apparent. Frequently, however, the client may elect to accomplish parts of the work with his own staff, retaining an engineering firm for the remainder. Several agencies of the U.S. Government retain private engineering firms for design and the preparation of contract plans and specifications but use Government staff engineers in the supervision of construction. In similar circumstances private clients frequently desire general consulting services and the checking of shop drawings during construction, but use their own staffs for inspection and supervision of the work.

It is common practice for one engineering firm or the client's own staff to perform the engineering services, with one or more consulting engineers retained as advisors on various aspects of the work. On large projects a board of consultants may be convened as advisors and general supervisors. Such consulting engineers are selected upon the basis of their professional experience and reputation as authorities in the engineering specialties involved.

Investment bankers generally have the view that preliminary reports, which include economic studies leading to the economic justification of the project, should be prepared by disinterested engineers who are not associated otherwise in the engineering of the work.

Selection of the Engineer. The selection of the engineer or architect by the client should be based upon sound technical qualifications, broad professional experience, and personal and professional integrity. These qualities are intangible and cannot be defined exactly in a precise specification. Therefore, bidding on contracts for engineering and architectural services is never engaged in by ethical engineers and architects because bid prices do not reflect the competence and efficiency of the bidder. Similar conditions prevail in all other professions such as law and medicine. For this reason all contracts for professional services should be of the negotiated type. The American Society of Civil Engineers recommends the following procedure in obtaining the services of an engineer.

1. From a list of engineers recommended by qualified sources, such as other employers or engineering societies, select one or more engineers to be interviewed.

2. Determine which one of the engineers interviewed is best qualified for the particular engagement under consideration.

3. Negotiate with the engineer so selected for services of the nature and extent required.

4. The reasonableness of fees to be charged may be checked with the sources contacted during step 1 above.

5. Engagements involving preliminary investigation and reports should commit the engineer to limiting fees in case additional engineering services are required at a later date on the same project.

Types of Contracts. Excepting the omission of the element of competitive bidding, contracts for engineering and architectural services are closely analogous to those for construction. Like construction contracts, they are generally designated by the method used in arriving at the amount of the compensation. The types of engineering contracts in common usage are as follows.

a. Fee Based on a Percentage of the Actual Net Construction Cost. For many years this type of contract has been the most convenient and the most widely used for both engineering and architectural services. Reasonable percentages for fees for the various classes of work have been established by professional societies and are generally accepted as the basis for negotiating contracts. The cost of the work is defined as the actual total construction cost in place, including labor, materials, and both installed and collateral equipment. Excluded as items of net cost are the cost of the engineer's fee and the cost of financing, commissions, real estate, legal, and other similar expenses.

If the services cover only design and preparation of plans and specifications or if the construction is postponed or canceled, the engineer's fee is based on the estimated construction cost as determined by the engineer. The estimated construction cost is also used as the basis of progress payments to the engineer during the design period pending the determination of the actual cost when the construction contract is awarded.

This type of contract has the advantages that the client pays and the engineer receives compensation in direct proportion to the work performed and that adjustments in the engineer's fee to cover changes in construction are provided for automatically without additional alteration in the engineering contract. The principal disadvantage is that the client does not know the exact cost of the engineering work when he signs the contract, and thus he does not know the extent of his obligations. Another objection occasionally raised is that the engineer may deliberately inflate the cost of the work in order to increase his own fee proportionately. Needless to say, such practice would be extremely unprofessional if not actually fraudulent and would not be tolerated by ethical engineers.

b. Lump-Sum Fee Based on a Percentage of the Estimated Construction Cost. The engineer's fee under this method may be arrived at on the basis of a percentage of the estimated construction cost which is then considered to be an unchangeable lump sum irrespective of variations between the actual and estimated construction costs. An alternate method is based upon the estimated engineering costs, plus an allowance to cover the engineer's net fee.

This form of contract has the advantage that the cost of engineering work is known definitely in advance. It has the disadvantage, however, that errors in the estimated cost are reflected in the engineer's fee. Furthermore, each change order on the construction contract indicates the necessity for a corresponding change in the engineering contract.

c. Cost Plus Fixed Fee. Under the preceding types of contracts the fee covers the engineer's normal costs and operating expenses connected with the work and his profit. The cost-plus-fixed-fee type provides a fee for profit only, since all engineering costs of the project, including overhead expenses, are reimbursed by the client. The fee is based on the estimated construction cost and remains fixed regardless of any variation between the estimated and actual costs. This type of contract has many advantages when the scope of the engineer's services cannot be accurately determined in advance, as in the case of alteration work and projects for which the client's requirements are not definitely established. Its disadvantages lie in the multiplicity of accounting records necessary to determine the engineer's true costs and in the difficulty in segregating costs when more than one project is being handled in the engineer's office concurrently.

d. Cost Plus Fixed Fee with a Guaranteed Ceiling. Under this form of contract the engineer is reimbursed for the actual costs of the engineering work provided that the total amount does not exceed the maximum limit established in the contract. If this should occur the engineer is held responsible for the excess and receives no compensation over the guaranteed ceiling cost. As compared with the ordinary cost-plus types of contracts, this form removes some of the uncertainties concerning the total cost to the client. As compared to the lump-sum type, it has the advantage that the client receives the benefit of the saving if the actual cost of the work should be less than that estimated when the contract was executed. For these reasons it is attractive to owners. On the other hand, the character of the work must be such that a reasonable estimate of engineering costs can be made in advance, or a suitable margin of safety should be allowed in arriving at the guaranteed ceiling. Otherwise the engineer would be accepting undue speculative risks. With these limitations this type of contract has similar advantages to those of the cost-plus-fixed-fee type.

e. Fee Based on a Time Rate. Contracts drawn on this basis may be for full- or part-time work or may provide for complete engineering services inclusive of assistants and overhead. More frequently, however, this type of contract is used for personal services only as, for instance, consultation and advisory services or expert testimony. The fee ordinarily is determined on a per diem basis although hourly, monthly, and annual rates are sometimes used. When the extent of such services cannot be determined accurately in advance, a minimum retainer is usually provided to cover the availability of the consulting engineer whether he is called upon for services or not. Such contracts sometimes base the fee on a retainer plus a per diem charge.

Essentials of the Contract. In most instances the contract is drawn as a formal legal document similar to those used for construction purposes. Many engineers, however, prefer to set forth the desired services and obligations in the form of a letter proposal which becomes a contract when accepted and signed by the client. Some corporations use their standard purchase order form as the contractual means of employing engineering services. The selection of form is a matter of personal preference since the various forms are equally valid. For important projects, however, it is advisable to use the formal legal type, and the advice of a professional lawyer should be obtained in the drafting of the agreement.

As for construction contracts the document should be written in simple language for clarity. As a minimum it should contain an introduction or preamble, definition of the parties to the agreement, the conditions precedent to the agreement if any, the services to be furnished and the scope of the work, the time of completion, the compensation to be paid to the engineer, and the signatures of the parties and the witnesses. In addition to the basic essentials for a valid contract, supplementary articles are usually required to define in more particularity the objectives of the contract.

The American Society of Civil Engineers considers that any contract or letter agreement for engineering services should contain the following essential provisions.*

a. Date of execution of the agreement.

b. Names and descriptions of the two parties to the agreement with their addresses, and in case of a corporate body, the legal description of the incorporation. If the client is a commission or public body, state the authority under which it acts.

c. Brief but precise description of the scope of the engineer's services and the obligations of the engineer to the client.

* *Manual of Engineering Practice* No. 8 and No. 29, American Society of Civil Engineers.

d. The re-use of plans, ideas, or copyrights. Design drawings are instruments of service and should remain the property of the engineer.

e. Provisions for termination of service before final completion, and proper compensation for such termination to be paid by the client to the engineer.

f. Money consideration for the work to be executed, including times and methods of payments on account, retaining fees, interim payments, both for professional service and expenses; and final payment in full settlement.

g. Statement as to when work is to be commenced and when concluded, in case such provisions and obligations are required.

h. Additional compensation for re-design after receiving approval of preliminary plans.

Scope of the Work. Of utmost importance in the drawing of contracts for professional services is the definition of the scope of the services to be furnished. The statement should be definite and precise and should establish the limits of the engineer's responsibilities beyond question. To be avoided are such general statements as "The engineer shall do all engineering work and perform all engineering services required in connection with the construction of the project." Such a requirement may obligate the engineer, at his own expense, to represent the client in lengthy and costly investigations and court procedure if litigation or arbitration proceedings should develop from the work. This may be true even though the dispute might have no direct connection with the plans and specifications. The contract should be specific in providing for additional compensation for any services required other than those contemplated by the agreement and on which the original fee was based.

Ordinarily in engineering practice the costs of field surveys, foundation explorations, materials testing, shop and mill inspection of materials, and similar costs are borne by the client and are not included in the fee. Such costs are sometimes covered by arranging for them to be included within the scope of the construction contract and paid by the contractor. Travel and subsistence expenses of the engineer and his representatives when trips away from the engineer's central office are demanded by the work should be made reimbursable. The contract should be specific in placing the responsibility for such costs and similar expenses on the client, or the engineer may find himself liable for the full amount. The standard methods for the determination of fees do not contemplate such expenses, so that a large loss to the engineer may be incurred. If it is desired that the engineer furnish services involving travel or personal expenses, a precise description of them should be included in the scope of the work and appropriate allowances should be made for their cost when the fee is determined.

The engineer should pay special attention to the wording of the terms of the contract concerning the estimates of construction costs to avoid any possibility of allowing himself to become liable for their accuracy. If care is not taken, it may be construed that the estimate is a guarantee of a limit of cost, in which case the engineer could be held responsible for any difference between the estimate and the actual cost of the work. Such a condition is, of course, never intended in view of the constantly fluctuating wage rates and materials costs and considering the fact that engineer's estimates are usually made well in advance of actual construction.

Time of Completion. The contract should state the date on which the work is to be started and the estimated time in which it is to be completed. This is primarily for the benefit of the client in planning the financing and construction of the project and its use after completion. It is also important to the engineer in budgeting his time and engineering costs. Its greatest significance to the engineer, however, occurs in the case of cost-plus-fixed-fee types of contracts. Unless a definite time of completion is established in the contract, the work may be drawn out over an unreasonable length of time by additions and changes with no corresponding adjustment in the engineer's fixed fee. Such contracts should always provide for an increase in the fee if the time of completion exceeds a predetermined limit because of factors beyond the control of the engineer.

Provisions for bonds and the assessment of penalties or liquidated damages are never acceptable in contracts for engineering and architectural services on the grounds that such services are professional in nature and require a high degree of technical training and experience. The engineer himself is the best judge of the time required to perform the required services although he may thus find that the professional requirements of the work may cause him to exceed the time specified in the contract. The client should realize that this is in his own best interests inasmuch as engineering work hurried in order to meet a rigid contract date might result in inferior work. Nevertheless, the contract should contain a reasonable estimate of the time required, and the engineer should make a conscientious effort to complete the work within that period.

Determination of the Fee. The amount of the engineer's fee will depend upon the cost and the complexity of the work and the scope of the services desired. The previous experience and performance record of the engineer also may be factors. No standard schedule of fees can anticipate and provide for all the differences in detail that arise in connection with the wide variety of projects in engineering practice. Conditions and practices also change from time to time, and it is necessary to adjust methods of compensation to meet these new situations. Any schedule of fees, therefore, can be expressed only in general terms and can be applied to a particular engagement only after full consideration of all the relevant facts. Nevertheless, such schedules are of value as a guide to engineers and their clients in arriving at fair and reasonable fees for particular projects. Standard schedules of fees have been established by the various engineering and architectural societies. These have been derived from the past experience of many engineers and architects in many classes of professional work and are generally accepted as the basis for the negotiation of contracts for engineering and architectural services.

Occasionally an engineer will undertake work on which he will be obliged to accept stock in the project as partial compensation for his services. Such a situation implies many dangers to the engineer, and he should enter into an agreement of this type only after careful consideration and after obtaining legal advice. For instance, the ownership of some kinds of stock may make the owner liable for financial obligations of the company or corporation which issued the stock. There is also the matter of the par value of the stock as compared with the value at which the engineer accepted it. This may have important implications for the assessment of taxes.

Fees Based on a Percentage of Construction Cost. The chart on page 289 shows graphically the schedules of fees recommended by the American Society of Civil Engineers.* Curves *A* and *B* apply to projects of average complexity in each classification. In general the percentage fee taken from the curves is intended to cover the following items.

a. Preliminary investigations and report.

b. Assistance in application for public funds.

c. Preparation of specifications for and office advice concerning test borings or other subsurface investigations.

d. Preparation of detail plans and specifications for construction.

e. Estimate of quantities and cost.

f. Assistance in securing bids.

g. Analysis of bids.

h. Assistance in letting contracts.

i. Checking shop and working drawings furnished by contractors.

j. Consultation and advice during construction.

k. Reviewing estimates for progress and final payments to contractors.

l. Assistance in tune-up and test of equipment.

m. Preparation of record drawings (if required by the specific engineering contract).

n. Final inspection and report.

When the fee is based on a percentage of the construction cost, the contract should describe the construction work to be included. Unless this description appears, controversies are likely to occur in connection with the right of the engineer to collect a fee on the purchase of equipment specified by him and on other work incidental to the project but not specifically included in the construction contract. Typical of the latter class is landscaping, the furnishing and decorating of buildings, the purchase of turbines for power plants, and motors for movable bridges. Usually such items are closely related to the planning of the project, and the engineer's advisory services, if not actual drawings, are required. Therefore, it is generally considered that he is entitled to a fee based on the total cost of all the work, whether it is all designed in detail or not, it being understood that professional advice in connection with the procurement of appurtenances is not to be distinguished

* *Manual of Engineering Practice* No. 29, American Society of Civil Engineers.

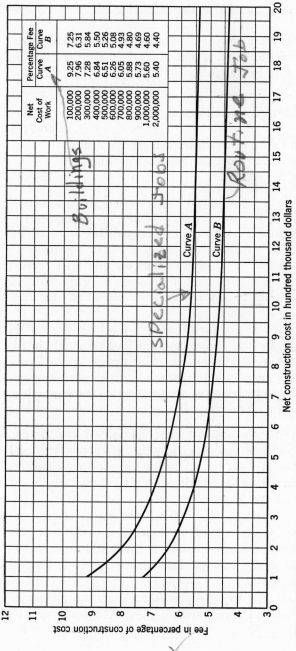

Net Cost of Work	Percentage Fee Curve A	Curve B
100,000	9.25	7.25
200,000	7.96	6.31
300,000	7.28	5.84
400,000	6.84	5.50
500,000	6.51	5.26
600,000	6.26	5.08
700,000	6.05	4.93
800,000	5.88	4.80
900,000	5.73	4.69
1,000,000	5.60	4.60
2,000,000	5.40	4.40

Fig. 2. Median fees for professional engineering services reported for work within the continental United States. Curve A applies to plants and engineering structures, such as power plants, pumping stations; incinerators; foundations; water, sewage and industrial wastes treatment plants; complicated bridges and dams; grade crossings eliminations; large airports with extensive facilities; complicated waterfront and marine terminal improvements; intercepting and relief sewers in large cities; air-pressure tunnels; through highways in congested districts; and additions to or reconstruction of curve B projects, etc. Curve B applies to general engineering projects, such as sewer and storm drains; water distribution systems; simple bridges and dams; small and medium-size airports; irrigation works; railways; highways; conventional levees and flood walls and sewer and water tunnels (free air), etc. Both curves are to be adjusted to suit special conditions.

from detailed design as far as the calculation of the fee is concerned.

For a project not involving extraordinary investigations, construction supervision, or other special services, the total fee can be broken down roughly as follows.

1. Preliminary investigation and report.   5–20%
2. Preparation of contract plans and specifications, etc.   60–75%
3. Advice on award of construction contracts, consultation during construction, checking shop drawings, etc.   15–30%

These allocations depend upon the nature of the project and should be used with caution. Preliminary investigations frequently include studies of many projects, and this phase of the work may represent a much larger part of the total. On the other hand, the construction work may be drawn out over a long period of time with a large number of shop drawings to be checked. In this case the engineer's cost for the later phases of the work would be relatively greater.

The American Society of Civil Engineers concludes that certain items of engineering cost cannot be determined accurately in advance and are not within the sole control of the engineer. Therefore, they are not included in the percentage fees from the standard curves and ordinarily are paid separately by the client. The following items are of this classification.

1. Field surveys, property, boundary, and right-of-way surveys, flow gauging, specialized subsurface investigations and borings, and similar instrumental work for preliminary investigations and report; instrument surveys and borings for design; and services of resident engineers and inspectors.

A fee based on salary cost plus 100 per cent, plus reimbursement for actual traveling and subsistence expense, long-distance telephone and telegraph charges, and similar direct field expense.

2. Services of locally employed field staff additional to resident engineer and inspectors.

A charge based on salary cost plus 50 per cent, plus reimbursement for necessarily incurred traveling and subsistence expense and supplies.

3. Furnishing reproduction of drawings or of detailed plans and specifications.

At cost plus 50 per cent service charge, or at price agreed upon with the client.

4. Services during readvertisement for bids for construction.

A charge based on salary cost plus 50 per cent, plus reimbursement for necessarily incurred traveling and subsistence expense, long-distance telephone and telegraph charges, and payment at recommended rate (see 3 above) for needed additional copies of plans and specifications.

Limiting lump sums for the above services are often included in engineering contracts.

Progress or interim payments on account of the engineer's fee are made at stated intervals. The amounts of such payments may be determined on the basis of the net cost expended, the percentage of the engineering work physically completed, or stipulated percentages of the total fee at various stages of the work.

Lump-Sum Fees Based on a Percentage of the Estimated Construction Cost. The standard schedules prepared for contracts in which the fee is determined as a percentage of the actual net cost of construction may be used to determine lump-sum fees by using the estimated cost of construction in lieu of the actual cost. The amounts so determined, however, should be increased by 10 to 25 per cent to allow for contingencies which are unforeseen at the time the contract is negotiated. Such unforeseen contingencies always occur and cannot be anticipated specifically. Therefore, this allowance represents actual engineering costs which are as real as any predictable cost. If the contingency allowance is not made the engineer is obliged to assume undue business risks.

Lump-Sum Fees Based on Estimated Engineering Costs. Engineering costs, as a basis for determining lump-sum fees, can be estimated conveniently under four main headings:

a. Direct payroll costs.
b. Indirect overhead costs.
c. Net fee.
d. Travel, subsistence, and other out-of-pocket expenses.

To estimate direct payroll costs, a detailed analysis of the job is required. The time and salary rates must be estimated for all the personnel to be assigned to the work. This will include the time of partners or principals of the firm, project engineer, designers, detailers, draftsmen, resident engineers, inspectors, and the like. Where the partners or principals of the firm are not paid actual salaries, the cost of their time on the work may be determined on the basis of their drawing account or imputed salaries. There should be included a contingency item to cover unforeseen delays, expenses, and possible salary changes. Data on personnel requirements for engineering services can be obtained only from the engineer's past experience and cost records.

Engineering overhead classifications vary somewhat with the accounting procedure used. Normally such indirect costs include the following.

Social security taxes.
Workmen's compensation insurance.
Pension, retirement, group health, accident, and life insurance.
Annual bonus and incentive payments.
Vacation.
Holiday and severance pay.
Sick leave and military leave pay to the extent that they are not included in direct salary costs.
Salaries of supervisory personnel to the extent that they are not included as direct costs.
Legal and corporate expense.
New business activities.
Clerical and stenographic work.

Filing and mailing.
General accounting.
Office and drafting room supplies.
Depreciation of office and drafting room furniture and equipment.
Local telephone service.
Rent of office space.
Utilities.
Technical organization and publication dues and subscriptions.
Non-allocable time of engineers, draftsmen, and other technical employees for standby time, research, etc.
Blue prints, photostats, mimeographing, etc.

Accounting practice varies in the allocation of some of the foregoing costs between the overhead account and direct payroll. Likewise, the time of partners or principals spent on the work is carried by some firms as a direct payroll charge against the job at drawing account rates whereas others carry such costs under the overhead account.

The past experience of many engineering firms confirms that overhead expenses have a fairly uniform relationship to direct costs in an efficiently managed office. However, they do vary somewhat with the volume of work performed. Therefore, it is customary to determine the overhead allowance from an average of such expenses audited over a three- to five-year period to reflect normal variations in the volume of the engineer's practice.

The average is expressed as a percentage of the direct payroll costs of the firm. This percentage is then applied to the estimated payroll cost of the project as an increment in the total estimated cost exclusive of the engineer's net fee and out-of-pocket expenses for travel and the like. On this basis overhead costs for engineering and architectural firms vary from 50 to 100 per cent of direct technical payroll cost, depending on the type and size of the office and organization, and the accounting methods used.

The amount to be allowed in the estimate as the engineer's net fee will vary somewhat with the size and complexity of the work and the degree of professional skill required. The American Society of Civil Engineers concludes that the net fee should be not less than 50 per cent of the total of the estimated direct technical payroll plus overhead costs. For projects of average complexity, somewhat lower net fees are justified with perhaps 25 per cent representing an average, particularly when the time of partners or principals spent on the work is carried as a direct payroll charge.

Direct out-of-pocket expenses for travel, subsistence, cablegrams, long-distance telephone calls, and the like depend upon the distance from the site of the work to the engineer's central office. Such estimated costs are usually entered directly into the estimate and are not used in computing the overhead costs or the net fee. As an alternative they may be made reimbursable by special provision in the contract.

Lump-Sum Fees Based on the Number of Drawings to be Prepared. An alternate method of estimating the amount of lump-sum fees for design and preparation of plans and specifications is based on a careful analysis of the work and an estimate of the number of drawings required for contract and construction purposes. From the engineer's own cost records a unit cost per drawing is derived and applied to this total to obtain the estimated cost of the work. The cost per drawing should cover all direct technical payroll, overhead, and other expenses and the net fee. This method at best is satisfactory only when an engineer is engaged in specialized and more or less routine work on projects similar in size and character. Because the cost per sheet of drawings may vary from, say, $500 to $2,000 per sheet, this method is not recommended except as a rough check on estimates made by other methods and for relatively minor change orders.

Fixed Fees for Cost-Plus Type Contracts. The fixed fee may be determined as a percentage of the estimated construction cost or as a percentage on the estimated engineering cost. When based on the estimated construction cost an allowance of 2 to 4 per cent, depending upon the magnitude and complexity of the work, is considered reasonable for design and preparation of plans and specifications. The comparable allowance under the alternate method should be 25 to 50 per cent of the estimated engineering cost. By either method the fee should not be less than the engineer's estimated direct payroll cost. Under both methods it is understood that the fixed fee covers the engineer's profit, and sometimes the services of the principals of his firm. Once determined and agreed upon, the fixed fee remains unchanged irrespective of changes in the actual cost of construction as compared with the original estimate.

In addition to the fixed fee the engineer is reimbursed for all direct payroll costs, overhead expenses, and for travel, subsistence, and other direct expenditures. Payroll and other direct costs can be audited readily and offer no administrative difficulty. Overhead costs, on the other hand, are indirect and difficult to determine precisely for any particular project when more than one assignment is being undertaken concurrently in the engineer's office. For this reason the general practice is to allow a percentage of the direct payroll costs to cover overhead expenses. This allowance is of the order of 50 to 100 per cent depending upon the engineer's cost records from past assignments. Under an alternate method, a negotiated lump sum is determined to cover all overhead costs.

In writing the contract great care is required in defining specifically all reimbursable direct costs and the factors to be covered by the overhead allowance. Likewise, it is necessary to define those costs which are considered to be covered by the engineer's fee and are therefore non-reimbursable as items of cost. Time of partners or principals may be of this classification, particularly for large undertakings. In smaller engineering firms and for projects where an unusually large part of the work requires the personal services of a principal, it is customary to allow a negotiated imputed salary to cover his time on the work, as an item of direct cost, in addition to his fee. The relationship of such imputed salaries to the overhead allowance is a matter for negotiation.

Fees for Supervision of Construction. Standard schedules of fees generally include provision for consultation and advice during construction and the checking of shop drawings prepared by contractors. When the client desires the engineer to furnish detailed inspection and supervision services throughout the work, standard schedules are impracticable because of the wide range in the scope of such services. Complete supervision of construction may include property and layout surveys in the field, inspection of materials and workmanship at the site, supervision of shop and mill inspections, and supervision of materials testing. The cost of all borings and tests is generally excluded from the supervisory responsibilities and paid by the contractor or the client as provided for in the construction contract. Layout surveys are sometimes handled in a similar manner.

In order to arrive at a reasonable fee for supervisory services, a detailed estimate should be made of the engineering costs involved including overhead and travel and subsistence expenses for the resident engineer, inspectors, and surveyors. An overhead allowance of 50 per cent of the direct payroll cost and a net fee of 25 to 50 per cent of the total engineering costs is considered equitable. The cost of supervision on average projects will usually amount to about 4 to 6 per cent of the construction cost.

When supervision of construction is to be accomplished on a cost-plus basis, an allowance of 100 per cent of the engineer's estimated direct payroll cost is considered reasonable to cover his overhead expense and net fee but not including necessary travel and subsistence costs which are reimbursable. On this basis the engineer's net fee will usually range from about $1\frac{1}{2}$ to 3 per cent of the construction cost, and these percentages are recommended as a guide in arriving at the amount of the fixed fee, when all actual direct and overhead costs are reimbursable.

When the construction contract is on a cost-plus basis, the cost of complete supervisory services will be considerably higher than for lump-sum and unit-price contracts. This is because some of the construction management responsibilities must be assumed by the engineer, and he will need accountants and auditors in addition to the usual technical staff.

Per Diem and Retainer Fees. Per diem rates may either be charged on the basis of the time devoted to the engagement personally by the consulting engineer, with the salaries of technical assistants and general office salaries and overhead costs charged against the per diem fee, or they may be based upon the time of all technical staff members of the consulting engineer's organization at rates varying with experience and qualifications. In either method the per diem rates generally represent gross payments which cover all direct salaries, overhead costs, and net fee, exclusive of travel, subsistence, and other out-of-pocket direct expenses which are reimbursed separately as expense account items. Per diem rates vary from $150 per day to much higher amounts, depending upon the magnitude and importance of the work. For members of the consulting engineer's technical staff, the corresponding rates vary from $50 to $100 or more per day, depending upon the experience and qualifications of the staff member.

Retainer fees are intended to cover the availability of the engineer and his staff and not entirely as payment on account of charges at per diem rates. Therefore, the retainer usually is determined on the basis of an estimate of the probable total number of days to be billed at the per diem rates and is then considered as a minimum fee for the engagement if the actual number of days devoted to the engagement should be less than that estimated.

Under per diem and retainer fee agreements, time usually is charged in hourly increments as spent on the work, the hourly rate being determined on the basis of the number of hours in the standard working day. Time spent in necessary travel during working hours is charged as time spent on the work. Where the services include expert testimony, any part of a day spent in court or in testifying is considered a full day.

Services of the Architect. Although architectural services closely parallel those of the engineer, there are some differences in scope and practice. In addition to the architect, modern building construction generally requires the services of engineers in structure, foundations, sanitation, heating, ventilation, and electricity. Specialists are also sometimes required for unusual problems in airconditioning, acoustics, power, waste removal, water supply, and the like. These services are usually retained by the architect at the owner's expense and with his consent. The work is performed under the architect's general direction. The New York Chapter of the American Institute of Architects classifies the services of the architect as follows.

1. *Preliminary Services.*
   a. Determination of the Owner's Program: By conferences and communications, the architect informs himself fully regarding the purposes or requirements to be served by the new construction. In order to assist in this, the owner should provide the architect with an accurate boundary and a topographical survey, subsoil data, information regarding location of utilities and adjacent property conditions, and other necessary data.
   b. Schematic Drawings: These are drawings prepared by the architect to describe a recommended solution of the problem in an overall sense rather than in detail. These drawings are revised until there is an agreement on a fundamental solution.
   c. Final Studies: These drawings are prepared by the architect at small scale to show the owner graphically how his requirements will be served in the new building.
   d. Outline Description of Materials and Methods of Construction: The architect records in writing a brief description of the type of construction and character of materials to be incorporated in the building and calls the owner's attention to any unuual conditions affecting its cost.
   e. Preliminary Estimate of Cost: The architect will assist in securing an approximate estimate of the cost of the proposed work and may consult one or more general contractors, inviting them to compute the cost of the proposed construction on the basis of final studies and outline description of materials and methods of construction. At this stage, a general contract may be let by negotiation.
      On the basis of work so far executed the architect will require from the owner decisions and approvals before proceeding with the next division of the work.

2. *Working Drawings.*
   These are the drawings which are utilized by the general contractor and subcontractors in purchasing material and equipment and in constructing the buildings. They include all information required for these purposes.

3. *Specifications.*
   This is a very detailed document describing the requirements for all trades involved and the type or standard of materials and workmanship to be incorporated in the building. One important section, called "general conditions," binds the general contractor and his subcontractors in general basic requirements for carrying out their work.

4. *Determination of Cost.*

When all the above described in 1, 2, and 3 has been completed, general contractors can compute accurately the cost of the work. Competitive bids can be received or a contract negotiated on an accurate budget basis. In the case of competitive bidding, no contractor should be invited to bid unless the owner is prepared to award the contract to him.

5. *Contract Documents.*

The architect prepares or assists in preparing all contract documents. These documents consist of the working drawings, specifications, general conditions, and the construction contract. The architect receives the bids and assists the owner in negotiating the contract. All drawings and other contract documents should be signed by the owner and general contractor. The owner should make himself familiar with all of the contract documents. The construction contract should contain a clause providing for arbitration as the method for settlement of disputes.

6. *Services during Construction.*

During the course of construction, the architect performs further services as follows.

a. Full-Size and/or Large-Scale Details: The architect will supplement his working drawings with necessary full-size detail drawings or large-scale drawings in order that work may be shown in more detail for construction purposes.

b. Checking of Shop Drawings: Fabricators, equipment firms, and many trades prepare shop drawings showing how their own work is to be executed in their shops and installed in the building. Before such drawings can be utilized they are checked by the architect.

c. General Supervision: While the architect will endeavor at all times to guard the owner against defects and deficiencies in the work of the contractor, he does not guarantee the performance of the contract by the contractor, nor insure the owner against defective work thereunder. His supervision must be distinguished from continuous superintendence which may be secured by the employment at the owner's expense of a clerk-of-the-works or an architect's superintendent of construction.

d. Requisitions and Certificates of Payment: The owner pays the general contractor periodically in proportion to the work accomplished on the basis of certified requisitions submitted by the contractor and forwarded to the owner by the architect after his examination and issuance of this certificate.

e. Change Orders: If changes involving additional costs or credits are duly authorized and approved by the owner, the architect issues a form called a "change order" which all parties concerned should sign.

f. Final Inspection: Before the general contractor receives his final payment, the architect makes a final inspection of the premises and certifies to the completion of the work.

Fees for Architectural Services. The amount of the percentage fee for architects is determined in a similar manner to that for engineering services. The New York Chapter of the American Institute of Architects has adopted the following principles for the determination of percentage fees.

1. The amount paid the architect for his services under this form of agreement is governed by the total construction cost of the work.

2. The architect's fee should include the services of all engineers and other experts normally required for the project. Where the contemplated work involves unusual engineering and design features (or special consulting service) the extra cost of such special services should be borne by the owner, either by increase in the basic fee or by direct payment of such special costs by the owner.

3. Various types of work involve varying amounts of cost to the architect. This makes the establishment of a single basic rate impracticable and would result in excessive fees in some cases or losses in others. The minimum basic rates stated below should be considered as the usual percentage fee to be paid the architect, unless before executing a contract the owner is made aware of conditions which by agreement with the architect warrant the establishment of a higher fee.

4. Work of great magnitude and on the other hand of small construction cost may involve architectural expenditures which are not in proportion to the average size project for which the minimum basic rates are established. Because of this variance, the architect may, by agreement with the owner, graduate his fees downward for the large project and upward for the smaller work.

5. *Minimum Basic Rates:* (*a*) A minimum basic rate of 7 per cent should apply to structures of a simple utilitarian character and will usually apply to work without complications of design. (*b*) A minimum basic rate of 8 per cent should apply to structures requiring more than ordinary amount of detail, or where the project requires specialized knowledge or experience. (*c*) A minimum basic rate of 10 per cent should apply to structures of a monumental character. (*d*) A minimum basic rate of 12 per cent should apply to residential work, not including multiple dwellings. (*e*) A minimum basic rate of 15 per cent should apply to interior design. (*f*) For alterations and additions to existing structures, the minimum basic rates under each classification should be increased by 50 per cent. Special conditions may require adjustment of recommended percentage increase.

6. In addition to the agreed percentage fee, the architect should be compensated for: (*a*) Necessary traveling expenses; (*b*) blue prints and other reproductions, models, perspectives, and special disbursements; (*c*) additional services required by an operation to be constructed under separate construction contracts rather than under a single general contract, or by changes in the scope of the initial program; (*d*) services rendered in the purchase of articles which are not designed by him.

For lump-sum, cost-plus-fixed-fee, and other types of contracts for architectural services, the amount of the fee is determined by methods similar to those previously described for engineering services.

Changes and Extra Work. Frequently during the life of contracts for engineering and architectural services, changes in the original scope are desired by the client and the engineer is required to perform additional work over that originally contemplated. Alternative designs and bids received on alternative types of construction may be necessary before the client can make decisions as to the type to be constructed. Redesigns may be required after the original drawings had been completed in accordance with previous understandings. Litigation may develop out of the work, and the engineer may be required to make lengthy investigations and spend many hours in court as a witness or as an advisor to the client's attorneys. These and similar services, unforeseeable at the time the agreement was negotiated, may be required, and the contract should make provision for an adjustment in the engineer's fee when they occur.

As for construction contracts, changes and extra work are accomplished under change orders which should be received by the engineer in writing before the extra work is undertaken. In addition to defining the scope of the extra work, the change order should also state the method for determining the adjustment in the engineer's fee. For ordinary changes the adjustment in the fee is based upon either the actual or the estimated engineering costs involved plus about 100 per cent to cover the engineer's overhead expenses and profit. Basing the change in the fee upon a percentage of the change in construction cost usually is not an equitable procedure, because, for instance, the additional engineering work ordered could result in a reduction of the construction cost.

Unforeseeable conditions which develop during the life of a contract occasionally indicate the need for the services of a specialist consultant on a problem for which the contract engineer is not technically qualified. The contract should provide for such an eventuality by specifying that the cost of such specialized services will be paid by the client directly or that they will be paid for under a change order to the engineer's contract.

It is particularly important that the contract be clear as to method for the settlement of a contract canceled before the completion of the work when the fee is based upon a percentage of the construction cost. If the contract is not specific in this respect, the engineer may find himself unable to collect payment for the preparation of plans and specifications, for instance, if the project should be abandoned after completion of the design but before the beginning or in the early stages of construction.

Termination of the Contract. Under ordinary circumstances a contract for engineering or architectural services is terminated when all of the required services have been completed and the total fee has been paid in accordance with the terms of the agreement. In addition, specific provision should be made for termination at any time by force majeure or other reasons beyond the control of either party which prevent performance of the obligations of the contract. Other conditions which justify termination are the failure of the client to pay promptly on the account of the fee, unreasonable delay or incompetence on the part of the engineer, and the like.

The contract should be specific as to the methods for settlement in the event that the contract is terminated for valid reasons. This is sometimes accomplished by applying the percentage of the completed portion of the work to the total fee. A more equitable method, however, is based upon the actual engineering costs involved including the engineer's overhead expenses and a reasonable allowance for his profit.

Arbitration. Many contracts for engineering and architectural services include a clause providing for the arbitration of disputes, and this procedure is recommended in lieu of litigation. The discussion of arbitartion of construction disputes in Chapter 6 is equally applicable to contracts for professional services.

Ownership and Reuse of Plans. Although the engineer or architect has a common law right of ownership to plans he has drawn, court decisions are conflicting where the question has arisen. Therefore, the contract should be explicit as to the ownership and reuse of the plans after the completion of the work. It is a custom and a professional principle that the engineer or architect shall retain ownership of the plans and that they may not be reused without his permission and the payment of a reasonable fee therefor. This intention may not be enforceable upon a client, however, in the absence of a specific provision in the contract. Some additional protection against the copying or reuse of plans may be obtained by copyright of plans and designs. It should be noted in this connection that agencies of the U.S. Government always require a clause in the contract vesting ownership of plans in the Government upon completion of the work. The drawings then may be reused, sometimes being considered as standard drawings, without additional compensation to the engineer.

Responsibilities of the Engineer. The engineer's rights and responsibilities as the agent of his client should be definitely stated. Under the ordinary contract the agency exists whether stated expressly or implied. Therefore, it is recommended that the terms be definite in order that both parties may understand clearly the basis of the agreement. This is advisable particularly with respect to the engineer's authority to make changes in the plans during the progress of the work, to issue change orders to the construction contract, to receive bids and award contracts, to approve the contractor's shop drawings, and other similar functions required on all construction work. Otherwise, the engineer may find himself financially liable for the cost of changes or the cost of mistakes discovered in contractor's shop drawings after they have been approved in general by the engineer. In the latter connection the engineer's approval of shop drawings should never imply responsibility for their accuracy afterwards.

When the engineer receives bids and awards contracts or approves subcontracts he should always state expressly that this is being done in the name of his client; otherwise, he may be held liable if the client should default or breach the contract. In this connection reference is made to the laws concerning the responsibilities of an agent acting for an undisclosed principal. If it is intended that the engineer shall not be authorized to act as an agent of the client in these matters, the contract should state that the engineer is acting as an independent contractor rather than as an agent. Under this condition the engineer should refuse to take any direct action unless he is expressly authorized to do so. All his recommendations should be made to the client who should be required to authorize in writing any necessary orders or approvals to the engineer or the construction contractor.

Litigation Arising from Supervision of Construction. An engineer or architect engaged in the supervision of construction work incurs risks of becoming involved in litigation and arbitration proceedings which are not contemplated in the usual methods for determining fees. In disputes between the owner and the contractor, the engineer is expected to assist in preparing the case for his client and to appear in court as his client's witness. If the dispute arises from an interpretation of the contract documents, the engineer may be sued jointly with the owner or a separate suit may be filed against him by the contractor. Similarly, in a dispute between the contractor and a subcontractor in which the owner may not be involved, the engineer may be made a party to the lawsuit or he may be sued separately. In either instance the engineer is forced to expend much time, work, and money which would cause him to suffer a large loss unless his contract with his client afforded him protection against such risks. Therefore, it is recommended that every contract for supervision of construction contain a clause similar to the following.

If the engineer or his employees shall be made parties to any action or proceeding or be called as witnesses or be required by any party to assist in the preparation of any action or proceeding, arising out of this contract or the work or their acts or omissions while in the performance of their duties under this contract, the owner will indemnify them and hold them harmless from and against any claims or actions or proceedings, will indemnify them against their expenses, including reasonable attorneys' fees, and will pay the engineer reasonable compensation for services performed by him and his employees. This paragraph shall not apply to matters as to which it shall be adjudged that the engineer is liable for negligence or misconduct in the performance of his duties.

U.S. Government Contracts for Professional Services. Service contracts between the United States and professional engineers or architects follow private practice more closely than do construction contracts. Inasmuch as engineering and architectural agreements are always negotiated, legal restrictions relative to bidding and awarding of contracts are not involved. However, the requirements in connection with disputes and the settlement of claims are the same as for construction contracts. Lump-sum and fixed fees are limited, usually to 6 or 7 per cent of the estimated construction cost of the work. This limitation is imposed either by administrative action or by the Congressional Appropriation Act which provides the funds for the work, and its amount is variable, depending upon the wording of the act. Moreover, some Government agencies have decided the fee should be limited to 10 to 25 per cent of the engineering costs. This results in considerably lower fees in Government projects than are generally allowed for private work. Cost-plus-percentage contracts are prohibited by law on U.S. Government work.

It is a custom in private practice that tracings, drawings, and other permanent records in connection with the design are owned by the engineer and may not be used again without his permission, whereas U.S. Government service contracts always stipulate that such drawings are to become the property of the government.

Contracts with U.S. Government agencies and all state, county, city, and other public bodies as well are invalid if the contracting officer is not legally authorized to enter into the contract for his agency. Likewise, such contracts are not enforceable if funds have not been appropriated for the project even though the engineer has proceeded with his work and expended funds therefor in good faith. However, the courts will take exception to this principle, under the doctrine of *Quantum Merit,* if the plans are accepted and used by the public agency, and they will award a reasonable fee to the engineer notwithstanding the illegality of the contract.

Contracts for Engineering Services outside the United States. Whereas there is considerable latitude as to types of contracts for domestic engineering work, the cost-plus contract usually is a practical necessity for American engineers working in foreign countries. It is seldom possible to determine working conditions, tax and labor laws, social security assessments, and the like to the extent necessary for firm cost commitments in advance of the work, and this precludes use of the lump-sum form of agreement.

Engineering costs and fees are considerably higher for American firms working abroad, and standard schedules of fees are not applicable except as a guide in arriving at fair compensation. The differential may be as much as 25 to 50 per cent greater than for corresponding work in the United States. This results from the necessity of paying travel and subsistence costs for engineers and their families while away from their homes for protracted periods. Substantially higher salaries also are justified for American staff members on foreign assignments. The increase may amount to 25 to 50 per cent over United States salary levels, depending upon the hardships of the assignment. Also many countries have higher social security and payroll assessments than are paid in the United States.

Engineering costs in foreign countries must be classified into dollar reimbursements and local currency reimbursements. Because of world currency restrictions, free currency conversion is possible in only a few countries. Therefore, the contract should provide for all dollar engineering costs, salaries, and fees to be reimbursed in dollars, with local currency reimbursement for costs and salaries of indigenous personnel and travel and subsistence expenses for American employees.

Because of unsettled world conditions there are frequent changes of government in many countries. It is not uncommon for a new government to refuse to recognize the commitments and obligations of its predecessor. If this should happen, the engineer's only chance for compensation would be through the courts of the country in which he is working, since one country cannot be sued before the courts of another without its permission. Obviously the engineer would be at a disadvantage in this situation because of a lack of familiarity with foreign litigation and legal procedures. To guard against losses in foreign litigation, it is advisable to require advance payments to cover engineering costs and fees before the costs are incurred. This can be accomplished by periodic cash advancements, an irrevocable letter of credit against which costs and fees may be billed monthly, or the establishment of an irrevocable revolving fund in an United States bank which is replenished monthly after approval by the client of the expenses of the preceding month.

As an additional precaution against foreign litigation the contract should provide for the arbitration of all disputes in lieu of lawsuits. Each party should be authorized to select one arbitrator, and the two so named should select a third. If agreement cannot be reached on the third arbitrator, provision should be made for his selection by a recognized international agency.

State Registration of Professional Engineers, Architects, and Surveyors. With but a few exceptions all states have registration and license laws to regulate the practice of engineering, architecture, and land surveying. Although these professions overlap to a considerable extent, each is covered by a separate act which defines the conditions and qualifications under which registration is granted before anyone may be considered eligible to practice in the particular profession. A registered professional engineer, for instance, may not practice land surveying, as defined in the act, unless he is specifically licensed to do so. State registration is intended to protect the public against practitioners unqualified by training and experience to perform professional services and to protect qualified members of the professions against unethical competition by unqualified practitioners. A violation of registration laws is usually classed as a misdemeanor punishable by fine, imprisonment, or both.

State registration laws are administered by state boards appointed for that purpose. Registration or license is granted upon the basis of minimum experience requirements and the satisfactory completion of written examinations. Qualification requirements are not uniform in the various states, and some are ambiguous. Progress is being made towards standardization of state registration requirements, and it is hoped that all states may eventually adopt similar measures. Many states now have reciprocity agreements whereby registration in one state automatically qualifies the recipient to registration in the others which are parties to the agreement.

In most states which have registration laws, a contract for the performance of professional services by unregistered practitioners is invalid and not enforceable. This applies not only to unqualified individuals who may be attempting to practice illegally but it could also be important to a qualified member of one profession performing services of a type in which engineering, architecture, or land surveying overlap and the jurisdiction of each is difficult to determine.

Another noteworthy requirement in some states prohibits registration and professional practice of engineering, architecture, and land surveying by corporations on the grounds that personal services are necessary for the responsible practice of these professions. Other states permit corporations to practice but require the corporate officers to be registered or licensed. Partnerships are acceptable in all states, but requirements vary with regard to the registration of partners. In some states the registration of one partner qualifies the firm whereas in others all partners are required to be registered.

EXAMPLES OF CONTRACTS FOR ENGINEERING AND ARCHITECTURAL SERVICES

Typical Letter for "Proposal and Acceptance" Contract (Lump-Sum Fee)

Board of County Commissioners
Cuyahoga County
1926 Standard Bank Building
Cleveland, Ohio

Attention: Mr. Albert S. Porter
County Engineer

GENTLEMEN:

In accordance with our recent conversations relative to a Comprehensive Traffic Survey and Arterial Highway Planning Study for the Cleveland metropolitan area, we submit herewith our proposal to perform that portion of the work which you expect to assign to consulting engineers. We understand that certain parts of the project will be accomplished by the engineering staffs of the Ohio State Highway Department and the Cuyahoga County Engineer and that the remainder of the work will be accomplished by your consulting engineers.

Our understanding as to the portions of the work to be undertaken by the State Highway Department, the County, and the Consulting Engineers, and the approximate timing of critical phases of the field work is as follows:

I. Work to be Undertaken by the State Highway Department.

1. Collaborate with the Consulting Engineers in the subdivision of the Cleveland metropolitan area into traffic zones which will facilitate the control of the field work, the collection of data, the analysis of traffic movements, and the use of the traffic data in the planning of highway improvements.
2. Collaborate with Cuyahoga County in the preparation of a traffic zone index for use in determining the zones of the residences of registered vehicle owners and the zones of origins and destinations of vehicle trips.
3. Collaborate with Cuyahoga County in developing the cooperation and response of the registered vehicle owners in the area.

4. Prepare and mail questionnaire postcards. Although the procedure may be varied, the following steps are anticipated:

   a. Designate on machine record cards the names and addresses of vehicle owners and the registration numbers of the vehicles.

   b. Sort and gang-punch residence zone designations on the cards described in a above.

   c. Transfer to business reply questionnaire postcards the data contained on the machine record cards described in a and b above.

   d. Arrange with post office officials for the release of questionnaire postcards on a Tuesday or Wednesday early in October. The timing would be such as to avoid possible conflict with major events which would interfere with the normal traffic pattern of the area or tend to offset the effect of information programs for enlisting the cooperation of the public.

5. Undertake screen line traffic counts required to provide a check on the accuracy of the postcard survey and volume and turning movement required to prepare traffic flow diagrams for arterial routes.

6. Follow up with personal interviews as required to obtain data for vehicle trips of taxi and truck fleets.

7. Supplement the origin and destination survey with a cordon type roadside interview survey to obtain data on through traffic not covered by the postcard questionnaire survey.

8. Prepare machine record cards for the trips correctly reported. Sort the cards by origins and destinations, and expand the data thereby obtained to represent an average 24-hour period during the survey.

9. Prepare a report containing tabulations of origins and destinations by zones and showing traffic flow on existing arterial streets. This report would not include graphical presentations of desired lines of travel.

II. Work to be Undertaken by Cuyahoga County.

1. Collaborate with the State in securing the needed publicity for the survey.

2. Collaborate with the State in preparing a traffic zone index.

3. Secure traffic volume and turning movement data where required to supplement the data to be obtained by the State or now available.

4. Make any test borings and soil analyses that may be required.

III. Work to be Undertaken by the Consulting Engineers.

1. Consultation and collaboration with participating agencies in the planning for the collection of survey data to insure that specific needs are covered and that the work of the survey is expedited.

2. Analyses of traffic patterns.

3. Studies of existing street capacities and conditions.

4. Graphical analysis of traffic data.

5. Studies of related transportation problems.

6. Speed and delay surveys.

7. Assignment of traffic to alternative alignments.

8. Recommendations for the general selection of routes and locations for a comprehensive freeway system in the Cleveland metropolitan area.

9. Recommendations for interim traffic relief.

10. Recommendations for the sequence of construction of proposed highway improvements.

11. Preparation of a report on the traffic surveys and recommended highway program.

As the first phase of the work under Item III-1 above, we would analyze the specific needs for origin and destination data and for traffic volume turning movement data; consult with officials of the Ohio Planning Survey as to the procedures available; select, in collaboration with the Ohio Planning Survey, the traffic zones into which the metropolitan area would be subdivided; and otherwise assist in the planning and scheduling of the survey work.

Costs and Fees.

The costs of the work outlined under Part I, to be performed by the State, and under Part II, to be performed by the County, are estimated at approximately $............................. and $............................., respectively. For the work outlined in Part III, we propose a lump-sum fee of $............................., payable in monthly installments in accordance with the progress of the work.

We understand that office space and stenographic service will be provided for our staff by the County and that the County will furnish transportation in the form of car pool service for field trips, speed and delay studies and inspection trips by our engineers.

We would begin the work immediately and would proceed with the various phases as rapidly as feasible in conjunction with the scheduled field surveys. We anticipate that all of the required survey data to be obtained by the State and the County will be completed and available to us prior to July 1, 1953. In the event that the required field survey data to be obtained by others is not made available to us by August 1, 1953, we shall be compensated for the additional costs incurred as the result of such delays.

We will complete the surveys and report within six months after all of the survey data to be furnished to us by the State and the County are made available to us. We will furnish to the County 100 copies of the report printed by photo-offset methods. The report will include our recommendations, a summary of our studies, desired lines of traffic movements, schematic and preliminary alignments of recommended arterial highways, schematic and preliminary designs of principal traffic interchanges and structures, and preliminary estimates of cost of the improvements recommended.

Two copies of this letter are enclosed herewith. The return of one copy signed in the space provided will constitute the County's acceptance of its general intent.

Yours very truly,

....Consulting Engineers

By..Partner

Accepted..
(Date)

Board of County Commissioners
Cuyahoga County, Ohio

By...

STANDARD FORM OF AGREEMENT BETWEEN OWNER AND ENGINEER FOR PROFESSIONAL SERVICES

Issued by the National Society of Professional Engineers
(Percentage of Construction Cost, and Per Diem Fees)

1. THIS AGREEMENT *made at* ....this
....................day of............................in the year....................by and between................
....hereinafter called
the OWNER, and..
hereinafter called the ENGINEER, WITNESSETH, That whereas the OWNER
intends to construct..
                          (include statement of legal authority of the OWNER to proceed

....

in the case of municipal or other public or political subdivision officials)
    NOW, THEREFORE, in consideration of these premises and of the mutual
covenants herein set forth for the construction of above named improvement
as follows:

2. The ENGINEER agrees to furnish and perform the various professional
services required for the construction of above named improvement as follows:

    a. Preliminary investigations, studies and reports, preliminary general plan
or plans, approximate estimate of cost, and all necessary conferences with the
OWNER.

    b. Complete general and detail plans, specifications and detailed estimate of
cost.

    c. Prepare forms for construction proposals, advertisements, construction
contracts and bonds, subject to the approval of the OWNER.

    d. Receive and tabulate proposals, report same to the owner, and assist in
awarding contract for construction.

    e. Furnish general supervision of the work of the Contractor including line
and grade surveys as the construction progresses, to assist in a correct interpretation of the plans and specifications and to safeguard the OWNER against
defects and deficiencies on the part of the Contractor, but the ENGINEER does
not guarantee the performance of the contract by the Contractor. (The
general supervision of the ENGINEER is to be distinguished from and does
not include the resident personal supervision as hereinafter mentioned.)

    f. Furnish resident supervision and/or such other surveys or services as
may be mutually agreed upon between OWNER and ENGINEER, at the field
payroll actual costs to the ENGINEER plus field, traveling and "out of office"
expense.

    g. Furnish property, boundary, right-of-way or other surveys at the actual
cost to the ENGINEER plus field, traveling, and "out of office" expense.

    h. Furnish three (3) copies of reports, plans (on paper) and specifications
and furnish additional copies to the OWNER or Contractor at net cost.
(Original documents, survey notes and tracings are and shall remain the
property of the ENGINEER.)

    i. Meet with the OWNER or his representatives when requested or necessary
for consultation or conferences.

    j. Furnish and perform the supervision of work of inspection bureaus and
laboratories in the inspection and tests of materials entering into the construction of structure, receive, and pass upon by approval or rejection all

reports by such laboratories or bureaus on the material tested for use in the structure. (N.B. The cost of all such tests and inspection by laboratories or bureaus to be paid for by the OWNER.)

k. Furnish and perform the supervision of all test borings, sub-surface explorations or other investigations required for the determining of foundation conditions for the structure. (The cost of such borings, tests, explorations, or investigations to be paid for by the OWNER.)

l. Compute and determine the amount of special assessments if required by the OWNER. The cost of such service to be paid for by the OWNER at actual cost to the ENGINEER or as by mutual agreement.

m. Prepare necessary plans and applications for permits for the submission to and approval of local, state and Federal authorities (such as municipal building departments, state boards of health, etc.) as may be required for the initiation, prosecution and construction of the improvement. The cost of such permits to be paid by the OWNER but the cost of preparing such plans and applications shall be included in the fee paid the ENGINEER.

n. Assist the OWNER when requested in negotiations with the owners of property required for or affected by this improvement, preparing all surveys, maps, plans, and estimates required for said negotiations. (The cost of this service to be paid for by the OWNER at the same rate as hereinafter specified for the principal engineering service under this agreement computed on the amounts finally agreed upon or finally determined as the actual compensation and/or damages to the owner of the property.)

3. The OWNER AGREES to pay the ENGINEER as compensation for such professional engineering services................per cent of the entire cost of the construction of the structure, to be paid as follows:

a. ....per cent upon the completion of the preliminary investigation studies, preliminary general plan or plans and the approximate estimate of cost;

b. ....per cent additional when working or contract plans and specifications with detailed estimate of costs are completed;

c. ....per cent additional payable in............monthly payments, the first payment to be payable................months after the contract for construction has been awarded and the balance monthly thereafter until the aggregate of all payments shall equal the amount due under this agreement exclusive of any amounts that may be due for extra compensation provided for in this agreement. (Art. III, par. *d, e, f, h, i,* and *k.*)

d. For any additional service required by the OWNER due to changes ordered by the OWNER or due to the delinquency or insolvency of the OWNER or CONTRACTOR or as a result of fire or flood or causes beyond the control of the ENGINEER, the OWNER shall pay the ENGINEER the expense of such additional service.

Said expense to the amount of the payroll incidental to such additional service, plus one hundred (100%) per cent for overhead, readiness to serve and profit; this expense also to include at net cost all other expenses incidental to this item, such as traveling expense, telegrams, telephone toll calls, extra reproductions of prints or photos of drawings, specifications and other documents required for the proper execution of the services for the Client. These payments to be due and payable from time to time as the services are performed or as the expenses are incurred.

e. For agreement for compensation of services based on per diem instead of percentage rates the OWNER shall pay the ENGINEER the amount of the payroll chargeable or incidental to such service including a rate for the principal of $.................... per day plus one hundred (100%) per cent for overhead, readiness to serve and profit together with the amount of such special expenses incidental to this item as traveling, telegrams, telephone toll calls, extra points, etc., at net cost.

f. Traveling or other authorized expenses incurred by the ENGINEER or his assistants where traveling in the discharge of duties connected with these services other than expenses between the ENGINEER's office and/or OWNER's place of business and/or the site of the improvement.

g. As far as the work under this agreement may require, to furnish the ENGINEER with a complete and accurate survey of the site of the improvement giving the grades and lines of streets, pavements, and boundaries, right-of-way or other surveys. If the OWNER so elects, the ENGINEER shall furnish the services enumerated in this class at cost as stipulated in Par. 2-*f.*

h. To pay for test borings, sub-surface explorations or other investigations required for the determination of foundation conditions and to pay for chemical, mechanical, or other tests when required.

i. To pay for the computations, determination, and tabulation of special assessments if required by the OWNER at the actual cost to the ENGINEER or as by mutual agreement.

j. That if any work (covered by this agreement) designed or specified by the ENGINEER shall be suspended or abandoned, the OWNER shall pay the ENGINEER for the service rendered on account of it, the payment to be based as far as possible on the fee as established in this agreement or where the agreement cannot be applied, then the basis shall be a per diem or payroll cost plus 100% plus expenses. (Par 3-*d* and *e.*)

k. To pay for building or other permits, licenses, etc., as may be required by local, state, or Federal authorities.

l. To pay, if required by the OWNER, for the services of the ENGINEER in the negotiations with owners of property required for or affected by this improvement. The basis of payment to be computed on the amounts paid (or agreed upon) for the property and/or damages at the basic fee rate herein mentioned in this agreement for the principal engineering service ....per cent.

4. It is further mutually agreed by the parties hereto:

a. That the estimated cost shall be used as a basis for monthly, partial or final payments until the actual costs have been established by proposals or by contracts for construction, and

b. That partial payments may be made under Art. 3, paragraphs *a* and *b* if requested by the ENGINEER and approved by the OWNER, and

c. That the cost used as a basis for computation of payments means the cost to the OWNER of the entire construction, including all materials, labor, and use of equipment and without deductions on account of penalties, liquidated damages or other amounts withheld from payments to Contractors but such cost shall not include the ENGINEER's fee or other payments to the ENGINEER under this agreement. The "cost of construction" does not include the cost of land, right-of-way or compensation for and/or damages to property unless the agreement so specifies, in which case it is understood that the cost means the cost of the improvement.

d. That the drawings and specifications are instruments of service and as such the original documents, tracings, and field notes are and remain the property of the ENGINEER whether the work for which they were prepared be executed or not, and shall not be used by the OWNER for any other project or construction.

e. That all questions in dispute under this agreement shall be submitted to arbitration at the choice of either party hereto.

f. That the OWNER and ENGINEER each binds himself, his partners, successors, executors, administrators and assigns, to the other party to this agreement and to the partners, successors, executors, administrators and assigns of such other party in respect of all covenants of this agreement. Except as above, neither the OWNER nor the ENGINEER shall assign, sublet or transfer his interest in this agreement without the written consent of the other party hereto.

5. IN WITNESS WHEREOF, the parties hereto have affixed their hand and seals at..this....................day of..........................., 19...........

WITNESS:

....

....

....

....
                                                OWNER

....
                         ENGINEER

ATTEST OF RESOLUTION
    (or other authority)

....

CLERK OF BOARD
(or authorized secretary)

....

Certification of Funds

I, ..., Auditor, hereby certify that
            (or other authorized fiscal official)
the funds to wit to the amount of..
(........................) dollars, have been appropriated and are in the...........................
Treasury or are to be obtained from an issue of bonds..
                                                        (specify authority,
..............................which are in process of sale and collection, and that said
date, etc.)
funds are not and cannot be appropriated for any other purpose. Dated
at..this..........................day of

....
        month   year   ....
                                                    AUDITOR
                                        (or other authorized fiscal officer)

Certification of Legality

I, ..., hereby approve the
        (prosecuting attorney or other authorized legal adviser)
legality of this Agreement.

                                        ....
                                        (prosecuting attorney or other authorized adviser)
APPROVED

....
                    CLERK
....
                    ENGINEER
(County-city or other authorized engineering representative)
....

....

(Line out all clauses, subparagraphs, or sentences not essential to this agreement)

THE STANDARD FORM OF AGREEMENT BETWEEN
OWNER AND ARCHITECT *

ISSUED BY THE AMERICAN INSTITUTE OF ARCHITECTS FOR USE WHEN A
PERCENTAGE OF THE COST OF THE WORK FORMS THE BASIS OF PAYMENT,
AND ENGINEERS' FEES ARE REIMBURSED TO ARCHITECT BY OWNER.

SIXTH EDITION—COPYRIGHT 1948–1951 BY THE AMERICAN INSTITUTE OF
ARCHITECTS, WASHINGTON, D. C.

THIS AGREEMENT made the..day of........................in the
year Nineteen Hundred and...................................... by and between....................
....
....hereinafter called the Owner, and
....
....hereinafter called the Architect,
WITNESSETH, that whereas the Owner intends to erect....................................
....
...., hereinafter called the Work,
Now, THEREFORE, the Owner and the Architect, for the considerations
hereinafter named, agree as follows:

The Architect agrees to perform, for the above-named Work, professional
services as hereinafter set forth.

The Owner agrees to pay the Architect for such services a fee of....................
per cent of the cost of the Work, with other payments and reimbursements
as hereinafter provided, the said percentage being hereinafter called the Basic
Rate. ....
....
....
....

The parties hereto further agree to the following conditions:

1. *The Architect's Services.* The Architect's professional services consist
of the necessary conferences, the preparation of preliminary studies, working
drawings, specifications, large-scale and full-size detail drawings, for architectural, structural, plumbing, heating, electrical, and other mechanical work;
assistance in the drafting of forms of proposals and contracts; the issuance
of certificates of payment; the keeping of accounts, the general administration
of the business and supervision of the Work.

2. *Reimbursements.* The Owner is to reimburse the Architect the costs
of transportation and living incurred by him and his assistants while traveling
in discharge of duties connected with the Work, the cost of all reproductions
of drawings, the cost of the services of engineers for normal plumbing, heating, electrical, and other mechanical work and of special consultants, and
other disbursements on his account approved by the Owner.

3. *Separate Contracts.* The Basic Rate applies to work let under a single
contract. For any portions of the Work let under separate contracts, on
account of extra service thereby required, the rate shall be four per cent
greater, and if substantially all the Work is so let the higher rate shall apply
to the entire Work; but there shall be no such increase on the plumbing,

* A.I.A. Form A-102 (Rev. 9-1-51).
Reprinted with permission of The American Institute of Architects.

heating, electrical and other mechanical work or on any contracts in connection with which the Owner reimburses special consultants' fees to the Architect, or for articles not designed by the Architect but purchased under his direction.

4. *Extra Services and Special Cases.* If the Architect is caused extra drafting or other expense due to changes ordered by the Owner, or due to the delinquency or insolvency of the Owner or Contractor, or as a result of damage by fire, he shall be equitably paid for such extra expense and the service involved.

Work let on any cost-plus basis shall be the subject of a special charge in accord with the special service required.

If any work designed or specified by the Architect is abandoned or suspended, in whole or in part, the Architect is to be paid for the service rendered on account of it.

5. *Payments.* Payments to the Architect on account of his fee shall be made as follows, subject to the provisions of Article 4:

Upon completion of the preliminary studies, a sum equal to 25% of the basic rate computed upon a reasonable estimated cost.

During the period of preparation of specifications and general working drawings monthly payments aggregating at the completion thereof a sum sufficient to increase payments to 75% of the rate or rates of commission arising from this agreement, computed upon a reasonable cost estimated on such completed specifications and drawings, or if bids have been received, then computed upon the lowest bona fide bid or bids.

From time to time during the execution of work and in proportion to the amount of service rendered by the Architect, payments shall be made until the aggregate of all payments made on account of the fee under this Article, but not including any covered by the provisions of Article 4, shall be a sum equal to the rate or rates of commission arising from this agreement, computed upon the final cost of the Work.

Payments to the Architect, other than those on his fee, fall due from time to time as his work is done or as costs are incurred.

No deductions shall be made from the Architect's fee on account of penalty, liquidated damages, or other sums withheld from payments to contractors.

6. *Survey, Borings, and Tests.* The Owner shall, so far as the work under this agreement may require, furnish the Architect with the following information: A complete and accurate survey of the building site, giving the grades and lines of streets, pavements, and adjoining properties; the rights, restrictions, easements, boundaries, and contours of the building site, and full information as to sewer, water, gas and electrical service. The Owner is to pay for borings or test pits and for chemical, mechanical, or other tests when required.

7. *Supervision of the Work.* The Architect will endeavor by general supervision to guard the Owner against defects and deficiencies in the work of contractors, but he does not guarantee the performance of their contracts. The general supervision of the Architect is to be distinguished from the continuous on-site inspection of a clerk-of-the-works.

When authorized by the Owner, a clerk-of-the-works acceptable to both Owner and Architect shall be engaged by the Architect at a salary satisfactory to the Owner and paid by the Owner, upon presentation of the Architect's monthly statements.

8. *Preliminary Estimates.* When requested to do so the Architect will furnish preliminary estimates on the cost of the Work, but he does not guarantee such estimates.

9. *Definition of the Cost of the Work.* The cost of the Work, as herein referred to, means the cost to the Owner, but such cost shall not include any Architect's or Engineers' or Special Consultants' fees or reimbursements or the cost of a clerk-of-the-works.

When labor or material is furnished by the Owner below its market cost the cost of the work shall be computed upon such market cost.

10. *Ownership of Documents.* Drawings and specifications as instruments of service are the property of the Architect whether the work for which they are made be executed or not, and are not to be used on other work except by agreement with the Architect.

11. *Successors and Assignments.* The Owner and the Architect each binds himself, his partners, successors, legal representatives, and assigns to the other party to this agreement, and to the partners, successors, legal representatives, and assigns of such other party in respect of all covenants of this agreement.

Except as above, neither the Owner nor the Architect shall assign, sublet, or transfer his interest in this agreement without the written consent of the other.

12. *Arbitration.* All questions in dispute under this agreement shall be submitted to arbitration at the choice of either party, in accordance with the provisions, then obtaining, of the Standard Form of Arbitration Procedure of the American Institute of Architects.

The Owner and the Architect hereby agree to the full performance of the covenants contained herein.

In Witness Whereof they have executed this agreement, the day and year first above written.

Owner ....

Architect ....

A FORM OF
AGREEMENT BETWEEN OWNER AND ARCHITECT *

ON THE FEE PLUS COST SYSTEM

SIXTH EDITION, COPYRIGHT 1917-1926–1945–1951 BY THE AMERICAN INSTITUTE OF ARCHITECTS, THE OCTAGON, WASHINGTON, D. C.

This Agreement made the.................................day of.........................in the year Nineteen Hundred and.................................by and between........................

....hereinafter called the Owner, and

----

* A.I.A. Form 103 (Rev. 9–1–51).
Reprinted with permission of The American Institute of Architects.

....hereinafter called the Architect,

WITNESSETH, that whereas the Owner intends to erect..
(Add here brief description of scope and manner of execution of work.)

....

....

...., hereinafter called the Work,

Now, THEREFORE, the Owner and the Architect, for the considerations hereinafter named, agree as follows:

The Architect agrees to perform for the above-named Work, professional services as hereinafter set forth.

The Owner agrees to pay the Architect the sum of..

....dollars ($......................)

as his fee, of which..

dollars ($......................) is to be paid in .... equal installments

monthly beginning ...., the balance to be paid

on issuance of final certificate; and to reimburse the Architect monthly all costs incurred by him in the performance of his duties hereunder as hereinafter more fully set forth.

The parties hereto further agree to the following conditions:

1. *The Architect's Services.* The Architect's professional services consist of the necessary conferences, the preparation of preliminary studies, working drawings, specifications, large-scale and full-size detail drawings, for architectural, structural, plumbing, heating, electrical, and other mechanical work; assistance in the drafting of forms of proposals and contracts; the issuance of certificates of payment; the keeping of accounts, the general administration of the business and supervision of the Work.

2. *The Architect's Fee.* The fee payable by the Owner to the Architect for his personal professional services shall be named elsewhere in this Agreement.

In case of the abandonment or suspension of the Work or of any part or parts thereof, the Architect is to be paid in proportion to the services rendered on account of it up to the time of its abandonment or suspension, such proportion being 25% upon completion of preliminary sketches and 75% upon completion of working drawings and specifications.

If the scope of the Work or the manner of its execution is materially changed subsequent to the signing of the Agreement the fee shall be adjusted to fit the new conditions.

If additional personal service of the Architect is made necessary by the delinquency or insolvency of either the Owner or the Contractor, or as a result of damage by fire, he shall be equitably paid by the Owner for such extra service.

3. *The Architect's Costs.* The Architect shall maintain an efficient and accurate cost-keeping system as to all costs incurred by him, in connection with the subject of this agreement, and his accounts, at all reasonable times, shall be open to the inspection of the Owner or his authorized representatives.

The costs referred to in this Article comprise the following items:

(a) The sums paid for drafting, including verification of shop drawings, for specification writing and for supervision of the Work.

(b) The sums paid to structural, mechanical, electrical, sanitary or other engineers.

(c) The sums paid for incidental expenses such as costs of transportation or living incurred by the Architect or his assistants while traveling in discharge of duties connected with the Work, costs of reproducing drawings, printing or mimeographing the specifications, models, telegrams, long distance telephone calls, legal advice, expressage, etc.

(d) A proportion of the general expenses of the Architect's office, commonly called "Overhead," representing items that cannot be apportioned in detail to this work, such as rent, light, heat, stenographer's services, postage, drafting materials, telephone, accounting, business administration, etc.

It is agreed that the charge for such general expenses shall be .... per cent of item (a) of this Article.

4. *Payments.* On or about the first day of each month the Architect shall present to the Owner a detailed statement of the payment due on account of the fee and the costs referred to in Article 3 and the Owner shall pay the Architect the amount thereof.

5. *Survey, Borings, and Tests.* The Owner shall, so far as the work under this agreement may require, furnish the Architect with the following information: A complete and accurate survey of the building site, giving the grades and lines of streets, pavements, and adjoining properties; the rights, restrictions, easements, boundaries, and contours of the building site, and full information as to sewer, water, gas and electrical service. The Owner is to pay for borings or test pits and for chemical, mechanical, or other tests when required.

6. *Supervision of the Work.* The Architect will endeavor by general supervision to guard the Owner against defects and deficiencies in the work of contractors, but he does not guarantee the performance of their contracts. The general supervision of the Architect is to be distinguished from the continuous on-site inspection of a clerk-of-the-works.

When authorized by the Owner, a clerk-of-the-works acceptable to both Owner and Architect shall be engaged by the Architect at a salary satisfactory to the Owner and paid by the Owner, upon presentation of the Architect's monthly statements.

7. *Preliminary Estimates.* When requested to do so the Architect will furnish preliminary estimates on the cost of the Work but he does not guarantee such estimates.

8. *Ownership of Documents.* Drawings and specifications as instruments of service are the property of the Architect whether the work for which they are made be executed or not, and are not to be used on other work except by agreement with the Architect.

9. *Successors and Assignments.* The Owner and the Architect, each binds himself, his partners, successors, legal representatives, and assigns to the other party to this agreement, and to the partners, successors, legal representatives, and assigns of such other party in respect to all covenants of this agreement.

Except as above, neither the Owner nor the Architect shall assign, sublet, or transfer his interest in this agreement without the written consent of the other.

10. *Arbitration.* All questions in dispute under this agreement shall be submitted to arbitration at the choice of either party, in accordance with the

provisions, then obtaining, of the Standard Form of Arbitration Procedure of the American Institute of Architects.

The Owner and the Architect hereby agree to the full performance of the covenants contained herein.

IN WITNESS WHEREOF they have executed this agreement, the day and year first above written.

Owner ....

Architect ....

Questions

1. List the work to be included in a contract for complete engineering or architectural services in connection with a construction project beginning with preliminary surveys and extending to the preparation of "as-built" drawings.

2. What are the essential terms of a contract for engineering or architectural services?

3. Why is it advisable to establish a time limit beyond which an engineer would not be obligated under the contract?

4. Explain the importance of a definite description of the scope of the work under a contract for engineering or architectural services.

5. What provisions should be made for the determination of the engineer's fee when work under the contract is terminated before completion?

6. Why should change orders on construction work always be signed by the owner although they may have been initiated by the engineer?

7. Compare the letter "offer and acceptance" contract with the formal type with respect to (a) legality, (b) relations with the client, and (c) adequacy.

8. What are the relative advantages and disadvantages of the lump-sum-fee type of contract for engineering or architectural services as compared with a fee based on a percentage of the actual cost of construction?

9. Can engineering and architectural plans and specifications be copyrighted?

10. Who owns the plans and specifications at the end of the job if the contract is not specific in this regard?

11. Under what conditions could the engineer or architect become obligated to pay the costs of borings and tests required in connection with a construction project?

12. What precautions should be observed by an engineer or an architect in accepting stock in a completed construction project as total or partial payment of his fee?

14

SPECIFICATION WRITING

The specifications are written instructions which supplement the drawings to formulate the technical requirements of the work. Also they define the quality of materials and workmanship desired by the owner and serve as a standard and guide for the contractor. In general, the drawings show *what* is to be done whereas the specifications state *how* it is to be accomplished. The specifications are usually given greater legal strength than the drawings, and most contracts state that in case of a conflict between the two the provisions of the specifications shall govern. Both the drawings and specifications are integral parts of the contract documents, and allocation of subject matter between the two is largely a matter of expediency. Those requirements, such as arrangement, dimensions, and types of construction which are more readily expressed graphically, are shown on the drawings, whereas instructions that can be more clearly expressed in writing are placed in the specifications. It is understood that the combination of the drawings and specifications will define completely the physical, technical, and operating characteristics of the finished project.

Specifications are of necessity comprehensive and deal with many kinds of materials and many types of construction. Therefore, the composition should be logical and systematic; otherwise there may be repetitions, omissions, and conflicting statements which lead to confusion and disputes during construction.

The clauses of the specifications are of two classifications.

a. The *general provisions* which apply to the work as a whole.

b. The *technical provisions* which describe the technical details of each type of construction.

The specifications are usually written in sections, the first of which consists of the general provisions. This is followed by the technical provisions using a separate section for each type of construction.*

Principles of Specification Writing. The qualifications of a competent specification writer are a full knowledge and understanding of the work to be accomplished, definite ideas as to the materials and methods which should be used, and the ability to express these ideas in the specifications and drawings in a manner readily understandable by those who are responsible for the execution of the work. The engineer is assumed to possess skill in design and construction, but his technical knowledge is of little value in so far as the specifications are concerned unless he can make his ideas clearly understood not only by other engineers but also by manufacturers, construction superintendents, foremen, and workmen, many of whom may have had little or no technical training.

The technique of specification writing depends on the application of a few basic principles of grammar, word usage, subject matter, and composition. The mandatory and legal character of the specifications requires great precision in wording and punctuation as compared with most other types of writing. There are cases on record where an error in punctuation, a misplaced comma, for instance, has changed the entire meaning of a sentence and provided the basis for a lawsuit for extra compensation in which large damages were assessed against the owner because his engineer did not detect the error before the award of the contract. It is evident therefore that unless the specifications are properly written the objectives and precautions of the design may be endangered.

The introduction of legal phraseology in the technical specifications for the purpose of protecting more fully the rights of the interested parties and to counteract or comply with conditions imposed by decisions of previous lawsuits has resulted in confusing and sometimes misleading instructions. Legal phraseology is rarely necessary in the specifications if the other contract docu-

* In an alternative method of composition the specifications are divided into Part I, General Provisions, and Part II, Technical Provisions, the sections of each part being numbered independently.

ments are properly prepared. Although the specifications, being part of the contract, have legal implications they should contain only the technical instructions necessary for the proper accomplishment of the work. Legal requirements related thereto should be written into the agreement.

Specification writing practice has developed through a long period of evolution in which early methods have served as models for later work. In many instances these methods which have been handed down are used from force of habit in situations where they are of no benefit, and they may even obscure the meaning intended. Experience on a large number of construction projects indicates that certain stereotyped clauses in the specifications chronically cause controversies and disputes. The suggestions which follow are intended to formulate some of the principles of good specification writing and to point out some of the most common faults encountered in past practice.

Style. Specifications should be written in short, concise sentences, in the simplest style possible. The style and tense should be the same throughout. Unfamiliar words, words having more than one meaning, and unusual technical and trade expressions should be avoided. Hyphens, commas, and semicolons should be used sparingly, and, when they or other extensive punctuation appear to be necessary, consideration should be given to a revision of the sentence. Misplaced or omitted punctuation marks can sometimes change the meaning of a sentence completely. To avoid this possibility it is better to convert the clauses of a compound sentence into separate sentences each with its own subject and predicate. This may lead to a monotonous repetition of terms in short sentences, but the likelihood of typographical errors and misinterpretation will be reduced. Relative pronouns should not be used unless they are absolutely necessary. "Which" clauses almost invariably are indefinite and confusing. All sentences should be complete in accordance with the rules of grammar. Tabulations and schedules are frequently advantageous, but broken or telegraphic-type sentences are usually not clear.

Precision in the Use of Words. It frequently happens that questions relative to the intention of the specifications have to be settled independently of the original writer. For this reason it is important to use words in their exact meaning, otherwise more than one interpretation may be possible. Colloquialisms, unusual technical words, and trade expressions should be avoided, or, if required, they should be defined in detail where used. Where a word may be interpreted in more than one way it is necessary to restrict its meaning to that intended or to use a more definite substitute. The following is a list of typical words which are frequently used incorrectly in specifications. The correct use of these and similar terms will tend to reduce ambiguity.

Any, all. "He shall make good any defects," should read, "He shall make good *all* defects."

Either, both. "Dolphins shall be placed on *either* side of the pier," should read, "Dolphins shall be placed on *both* sides of the pier." *Either* implies a choice.

And, or. "It shall be free from defects of workmanship *and* material which would impair its strength *or* durability." Interchanging the two words in this sentence results in an entirely different meaning. Although commonly used in legal documents, *and/or* is considered undesirable in specifications because it is an indefinite expression and indicates lack of certainty. *Or . . . , or both* is preferred.

Do not use apparent synonyms, e.g., *must* and *shall,* as different degrees of their mandatory nature may be inferred. Adopt one or the other, and use it consistently.

Avoid omission of the articles, e.g., "Contractor shall paint ceiling of office." This should read, "The contractor shall paint the ceiling of the office." Omission of articles leads to awkward composition.

Will, shall. Use *will* in connection with the acts of the owner and *shall* with reference to the contractor. This is not based on considerations of grammar but is a helpful distinction between the commitments of the owner and the obligations of the contractor.

Resisting, resistant. Corrosion-resisting is preferred to *corrosion-resistant.*

Amount, quantity. Amount should be used in connection with money; quantity when referring to volume, number of objects, and the like. The *amount* of the estimate refers to the total cost, but the *quantities* in the estimate are the numbers of the various units.

Avoid the use of *same* as a pronoun, as in, "If bracing is required the contractor shall furnish same at no additional cost."

Similarly, *said* should not be used as an adjective, as in "Said bracing shall be furnished at no additional cost."

Subject Matter. The subject matter of the specifications should be confined to the information and directions necessary for construction work after the award of the contract. All information pertaining to bidding procedure and conditions of bidding needed before the award of the contract belong in the instructions to bidders, and the word "bidder" rarely needs to appear in the specifications. Likewise, all information of a legal nature, such as time, payment, changes, and liquidated damage clauses, belongs in the contract and not in the specifications.

The specifications are intended to amplify, but not to repeat, information shown on the drawings. For example, the drawings show studs to be 2 inches by 4 inches and their spacing to be at 16-inch centers whereas the specifications should state whether they are to be southern yellow pine or Douglas fir and the grade required. The drawings show how the materials are to be incorporated into the structure. The specifications state the quality of the materials and workmanship to be employed. No information should be given in the specifications which is not necessary for the preparation of bids and the accomplishment of construction work, e.g., unit stresses and loadings which should be placed more properly on the drawings.

The specification writer should have definite reasons in mind for every requirement specified, but these reasons or explanations should not be given in the specifications because the reasons may become the cause of controversy. The justification for a specification requirement is a matter of engineering before the award of the contract and has no place in the specifications.

Cross references should be kept to a minimum, and paragraph numbers alone should not be used for this purpose. If it is necessary to refer to a particular paragraph, do so by its title and number and the section under which it is to be found. Cross references of the types, "Painting of the Woodwork is covered under Painting" and "Painting is specified hereinafter" are usually unnecessary.

Balanced Composition. A specification in its entirety as well as in its component parts should be commensurate with the size and importance of the work. Matters of major importance should be covered in detail while lesser items may warrant only a brief explanation. There is a tendency among engineers to use standard specifications for such types of construction as concrete work, structural steel, and the like for all such work notwithstanding the fact that the standard specifications were probably prepared originally for large important projects and include many requirements and refinements which are not justified on small isolated structures. Cement for a large concrete dam, for instance, must be closely controlled in manufacture and subjected to mill and field tests throughout the work. The mixing, placing, and curing of the concrete likewise must conform to detailed technical requirements. The same procedure applied to a small culvert on a secondary road would, of course, result in an absurdity which would be a nuisance to everyone concerned and would increase the cost of the work. Similarly, standard-grade structural steel bought from stock is adequate for a minor structure whereas for a large bridge or building it will be desirable to control the composition, fabrication, and erection of the steelwork in great detail.

Broad general requirements should be reviewed carefully because they are applicable to all parts of the work and therefore will have a direct influence on the character and construction cost of the project. Special requirements, on the other hand, may apply to only a small detail and may have a minor effect on the job as a whole. The careless application of a broad general requirement may result in refinements unnecessary or even undesirable in some parts of the work. In general, it may be said that the most satisfactory specification contains a minimum of requirements consistent with the results desired.

Definite Requirements. The instructions in the specifications should be written in the form of specific directions and never as suggestions or explanations. Unless this is done there may be some doubt in the contractor's mind as to what is intended, leaving openings for him to propose substitutions and alternatives in his bid and during construction. He is within his rights in doing this if the specifications are not explicit as to what is required.

Such relative and indefinite expressions as *reasonable, best quality,* and the like, which are commonly found in specifications, should be avoided. They are difficult to enforce and have been the causes of many disputes because no fixed definitions are available. Similarly, the *or equal* clause frequently used in connection with specifications for proprietary commodities is indefinite. Although final decision in such questions is made by the engineer, the contractor has no way of knowing in advance what will be considered *reasonable, best,* or *equal,* and therefore he must protect himself against the most unfavorable interpretation likely to be made by the engineer. The expression *in accordance with standard practice* is somewhat more specific, but it also is too indefinite to be used other than in connection with minor items.

The expression *as directed by the engineer* is always indefinite because the contractor cannot predict in advance what the engineer will direct. Frequently it is necessary to leave some items to the discretion of the engineer, e.g., the color of paint or the exact location where the contractor may dump excavated material, and even then care must be taken that the arbitrary action of the engineer does not involve a change in cost. The specification writer indicates either laziness or incompetency when he takes refuge behind blanket clauses.

Indefiniteness results from certain words which crop up in specifications repeatedly. For example, when weights are given in *tons,* there is always a question as to whether 2,000 pounds or 2,240 pounds are intended. A distinction should always be made between *short tons* and *long tons.* Similarly when time is specified in *days* there is always a question as to whether *working days* or *calendar days* are intended. The latter is more definite and to be preferred. The abbreviation *etc.* at the end of a list of items is vague and indefinite and should never be used in specifications. It should be noted that, if more than one meaning can be taken from an expression in the specifications, the contractor is entitled to take the one most favorable to him inasmuch as the owner, not he, was responsible for the writing of the specifications.

These examples illustrate the necessity to be precise and definite in all respects when specification requirements are being formulated. Inaccurate wording and vague clauses almost always result in increased construction costs because contractors increase their bids to cover the possibility of arbitrary decisions during construction.

Accuracy. When information is given in the specifications it must be correct and complete. Unless special provision is made to the contrary, the owner will be held responsible for the accuracy of the specifications and misleading information will be sufficient grounds for the contractor to collect additional compensation or damages if extra costs result. For example, borings taken every 100 feet on a sewer line might show no quicksand. If quicksand pockets were encountered between the 100-foot boring sections the contractor could claim and collect the extra cost of excavation. Technical information should be carefully checked for accuracy and completeness.

Repetition is a common source of errors in the specifications; therefore a requirement should not be repeated after it is once stated. Though repetition may sometimes serve for emphasis, errors and uncertainties result when corrections become necessary and are made in one place but are overlooked in another where the same statement occurs. A similar source of errors is that resulting from revisions in the drawings which are not followed through in the specifications.

The use of symbols lead to errors. Instead, write out or abbreviate the meaning of the term, e.g., 150 lb. instead of 150# and 2 ft. 4 in. instead of 2' 4". The latter may be mistyped 214". Other common symbols to be avoided are % for per cent, — for minus, + for plus, and the like.

Incorrect spelling and punctuation marks and typographical errors in the specifications may have serious consequences. Mistakes of this sort may completely change the meaning of a sentence and cause disputes or extra cost.

Practicability. It should be recognized that laboratory technique is not economically feasible in ordinary construction work, and procedures which are necessary in scientific research would be prohibitive in time and cost on a construction job. Moreover, the improvements which result from such refinements in most types of construction are not warranted when compared with standard production methods. Therefore the specification writer should recognize the limitations of materials and workmanship, and he should not specify practical impossibilities. Tolerances, clearances, finishes, and the like should be controlled only to a degree consistent with the functional, architectural, and structural requirements of the completed work. It is true that some work, such as delicate parts of a machine, requires great precision, but much money has been wasted through unnecessary requirements in types of construction where such refinements are not essential.

In so far as possible construction requirements should be adapted to standard sizes and patterns. Special sizes are expensive and require extra time for manufacture and delivery. Odd lengths of lumber have to be cut from the next larger standard size, and the waste will be charged to the special order. Variation from the standard shapes for structural steel sections require special rolling and such an order will be charged with the extra cost of setting the rolls and special handling. If rivet spacings different from steel fabricator's standards are called for, rivet holes have to be punched by hand at extra cost rather than by multiple-punch machines. Many other examples could be cited in which the economic advantages of mass production are lost by variations from standard sizes and patterns. Although the designer makes many of the decisions relative to dimensions, sizes, and patterns, the specification writer may accomplish a great deal along this line by constant vigilance to detect variations from established standards.

Whims and pet requirements frequently result in improper and uneconomical use of materials and construction methods. These may gratify the specification writer but are not appreciated and perhaps not known by the owner. Also, they may be a thorn in the flesh of the contractor during construction and increase the cost of the work whereas a more appropriate substitute could be used at less cost.

Conflicting Requirements. Conflicts between the various clauses in the specifications are the cause of much trouble in construction work. Conflicts which result from typographical errors, mistakes, lack of coordination between the drawings and the specifications, and the like can be corrected usually by carefully checking the manuscript before the specifications are issued. Other types of conflicts are more difficult to detect, however. Detailed technical requirements in one paragraph may produce results incompatible with those stipulated elsewhere in the specifications. For example, specifying the chemical composition of a material may produce strength and durability characteristics completely different than those required under physical properties. Unless the specification writer is thoroughly familiar with materials technology, such a conflict is not improbable. In fact, specifying both methods and results at any time in construction work is likely to lead to conflicts of this sort and may relieve the contractor of responsibility for the final results.

Another illustration of conflicts is sometimes encountered in pile driving specifications where both the penetration of the pile and its bearing capacity are specified. In some types of soil the required bearing power might be developed long before the piling had been driven to the specified penetration. In others the reverse might be true.

Fairness. The best interests of the work require that the specifications simply set forth the desired standards of quality and workmanship without imposing harsh and unfair conditions on the contractor. The engineer's first responsibility is, of course, to safeguard the owner's interests, but this should not be done by saddling the contractor with vague and indefinite specifications subject to unilateral interpretation by the engineer or by forcing the contractor to assume responsibilities which more properly should be placed on the owner or on the engineer himself.

An outstanding example of unfairness is the common practice of compelling the contractor to assume responsibility for the possible inaccuracy of information furnished by the owner. At all times the owner should be responsible for the accuracy and sufficiency of the information he gives to bidders or contractors. The interpretation of such information is, of course, the responsibility of the contractor. It is expected that the contractor will assume full responsibility for all of the normal hazards of his business, such as weather, changing labor rates and material prices, and the like, but if he also has to evaluate the effect of future acts of the owner or the owner's representatives and guarantee the accuracy and sufficiency of these acts, construction costs are bound to be increased.

When specific difficulties or hazards are known to exist in connection with the work, all available information related thereto should be furnished to the contractor. The contractor, of course, should be made responsible for the interpretation of such information. Difficulties and hazards should never be concealed from the contractor because to do so prevents him from making an intelligent bid and may impose unfair losses on him in the performance of the work. On the other hand he may suspect their existence and include an excessive contingency item in the bid which would result in a greater cost to the owner than would have been the case if all facts were made known.

In every specification requirement the question arises as to the degree of responsibility to be imposed upon the contractor. If the quality of materials and workmanship are specified in detail the contractor can only be required to guarantee freedom from defects. With more general specifications, it is reasonable to require the contractor to assume more responsibility.

Brevity. Specifications should be written as concisely as possible without sacrificing clarity. Unless the provisions are brief and to the point with all non-essential wordage eliminated, their meaning may be obscured. A great deal can be accomplished along this line by avoiding verbose grammar and the repetition of stereotyped mandatory provisions in connection with each type of construction, whether needed or not. Literary style is desirable in many forms of writing, but it should be kept in mind that specification writing consists of a simple listing of directions for the contractor to follow in accomplishing construction work and simplicity and brevity are essential.

The following, quoted from actual specifications, illustrate how clarity and meaning may be lost by ponderous and verbose grammar.

Paint shall be of such character that it will protect the steel against corrosion without being injurious to the health of persons drinking the water after the latter has stood in the tank for three months.

Drain piping in and about the pump room to be supplied by the subcontractor whether entirely buried in concrete or not.*

The effective length of spans will be the distance between the centers of gravity of end posts, and the centers of end shoes or end bearing plates must coincide therewith.*

All materials which, subsequently to the tests at the mill and to its acceptance there, during manipulation, in the shops under shears, punch, etc., which shows it is not of uniform quality, as herein specified, and also hard spots, brittleness, cracks and other defects are developed; such material shall be rejected.*

Traffic is now being carried on over the existing gap to be replaced by these two steel superstructure spans by means of a pile trestle temporary bridge.*

* From *Civil Engineering*, February, 1943, p. 107.

The repetition of expressions similar to "The contractor shall provide all materials and perform all labor in connection with each type of construction" contributes nothing to the strength of the specifications and should be avoided. One statement in the general provisions to this effect applied to all the work is binding in all cases except those specifically noted. The following is an example of an introductory paragraph which is frequently used for each section of the technical provisions, thereby increasing unnecessarily the bulk of the specifications.

Under this heading, the contractor shall provide all materials and perform all labor for the installation of all concrete, including reinforcement, necessary to the construction and completion of the work in accordance with the drawings, and specifications, and the intent thereof.

The above sentence means little, for unless the intent of the drawings and specifications is indicated more adequately elsewhere, the sentence cited adds nothing. For a particular project to which this might be applicable, the statement should be changed to read, *"General requirements.* Concrete work includes column footings, wall footings, first floor of storeroom, and pipe trenches." In most cases, this paragraph can be omitted altogether as a superficial examination of the drawings indicates clearly the exact limits of the various classes of work, and a definition of their scope in the specifications is superfluous.

The expression "as indicated on the drawings" is seldom necessary because the fact that a thing is shown on the drawings is a sufficient indication, making mention of it in the specifications unnecessary. Similarly, "in accordance with these specifications" should be omitted in practically all cases in the body of the specifications.

The dimensions of all concrete work and amount of steel reinforcement shall be as indicated on the drawings. Where not indicated, the walls and slabs shall be proportioned . . . etc.

should read:

Where dimensions of concrete work and amount of reinforcement are not indicated, the wall and slabs shall be proportioned . . . etc.

Such statements as the following are unnecessary and should be omitted.

Expansion joints shall be provided in the concrete trenches where and as indicated on the drawings.

Manhole frames and covers shall be installed in the concrete manholes to provide access to the equipment. They shall be of the size indicated and shall be provided with cast iron frames and cast iron covers as shown.

The latter should be omitted entirely if marked properly on the drawings. If needed, it should be revised to read:

Manhole frames and covers shall be cast iron.

The expressions "to the satisfaction of the engineer," "satisfactory to the engineer," "subject to the approval of the engineer," "in an approved manner," and the like contribute nothing to the strength of the specifications and should be omitted for the sake of brevity. The contract should state once and for all that all work must meet the approval of the engineer, and repetition of this provision in the specifications is not necessary. Similarly repetition in the specifications of the expression "the contractor shall guarantee the work to be free from defects in materials and workmanship," means little. Where needed it should be revised to read "the contractor shall replace at his own expense any portion of the work found by the engineer to be defective." This requirement is enforceable but not needed any place in the specifications if the contract contains a general requirement that the contractor shall make good all defective work found any place on the job.

Principles of Specification Writing Summarized. The principles of specification writing discussed in the foregoing paragraphs may be summarized as follows.

1. Make a clear concise analysis of the job requirements as to general conditions, types of construction, and quality of workmanship.

2. Prepare an outline of the specifications, segregating topics to be covered in the general provisions and the technical provisions respectively and giving due consideration to the contents of the other contract documents.

3. Analyze each type of construction to be covered in the technical provisions to determine the type of specifications to be used, i.e., materials and workmanship, performance, or proprietary commodities.

4. Outline the requirements to be included in each section of the specifications, coordinating the subject matter with the drawings and the other contract documents.

5. Write each section concisely, using short sentences and simple composition.

6. Use words in their exact meanings.

7. Make each requirement definite and complete.

8. Give directions, never suggestions.

9. Do not give reasons for direct specification requirements.

10. Do not specify impossibilities.

11. Avoid conflicting requirements.

12. Give no information which is unnecessary for the preparation of bids and accomplishment of the work.

13. Specify standard sizes and patterns wherever possible.

14. Avoid whims and pet requirements.

15. Do not repeat information given on the drawings.

16. Do not repeat a thing once stated.

17. Do not repeat superfluous mandatory provisions when the same things are required in general by the contract.

18. Minimize the use of cross references.

19. Do not place upon the contractor the responsibility for the possible inaccuracy of information furnished by the owner.

20. Do not impose harsh and unfair conditions on the contractor.

21. Never conceal difficulties or hazards from the contractor.

Questions

1. Rewrite the following specification clause to eliminate ambiguity. "Timber piles shall be driven to a minimum penetration of 60 feet below cut-off and/or a final penetration sufficient to develop a safe bearing capacity of 50,000 pounds as computed by the *Engineering-News Formula.*"

Note. A controversy over the intended meaning of this clause in the specifications for the construction of a building resulted in the award of a large extra payment to the contractor.

2. Explain how the following specification clause is indefinite, and clarify the sentence. "Concrete shall be made of one part of Portland cement, two parts sand, and four parts gravel."

3. Show how it may be impossible to mix concrete in accordance with the following specification requirement. "Concrete shall be made of one part Portland cement, two parts sand, and four parts gravel. Seven gallons of water shall be used per bag of cement, and the slump, as measured by the standard slump test, shall not exceed 4 inches."

4. What are the advantages of specifying standard sizes and dimensions whenever practicable?

5. Give an example of unfair specifications.

6. Explain what is meant by "balanced composition" in specification writing.

7. What is the inherent danger in broad general specification requirements? Should such general clauses be used?

8. Give two commonly used examples of indefinite specification requirements.

9. What are the objections to the use of symbols in specifications?

10. Why should repetition be avoided in specifications?

11. Discuss the following phrases which are frequently used in specifications.

a. "As indicated on the drawings."

b. "To the satisfaction of the engineer."

c. "The contractor shall furnish all materials and labor required to construct. . . ."

d. ". . . shall be Type B as manufactured by Blank Company, or equal."

12. What are the advantages of simplicity and brevity in specification writing?

13. Explain why it may be difficult to detect conflicting requirements in specifications.

15

GENERAL PROVISIONS OF THE SPECIFICATIONS

The general provisions of the specifications cover such topics as the description of the work, reference to the contract drawings, work or materials to be furnished by the owner or by other contractors, work or materials to be covered by standard specifications to be found elsewhere, shop drawings, and control of materials. These are usually technical in nature and apply to the project as a whole; but non-technical requirements not covered in the other contract documents may also be included.

The general provisions of the specifications are similar to the general conditions of the contract, and there is no fixed rule for the allocation of subject matter between these two documents. The development of professional practice in this regard has been inconsistent and sometimes illogical. This is a natural result of the fact that contracts are generally drawn by lawyers who have the responsibility to protect their clients from the legal viewpoint, whereas the specifications are written by engineers who must specify the technical requirements while conforming to the legal conditions and implications of the agreement. Moreover, there is no unanimity of opinion among engineers and architects as to either contents or terminology of the two documents. Frequently the allocation of subject matter is determined by preference and emphasis. Although the provisions of the contract and the specifications are equally binding, the contract is the stronger document legally, and its articles undoubtedly receive greater emphasis. In many instances the owner will desire to use a standard form of contract, and this will have an important bearing on the contents of the general provisions of the specifications.

When the business and financial arrangements between the owner and the contractor, the obligations and responsibilities of each, the authority and duties of the engineer, and the intent of the plans and specifications are covered in the contract, the general provisions of the specifications will be devoted largely to the general requirements related to the scope, quality, and control of the work.   Special requirements and precautions will also be included.

Although many of the clauses of the general provisions can be more or less standardized for ordinary engineering and architectural construction work, it should be noted that these general requirements frequently establish policy with regard to the entire project, and therefore they may have important effects on cost as compared with detailed technical requirements which may be concerned with only minor elements of the construction.   Therefore, no general provision should be incorporated into the specifications until all of its implications are clearly understood.

Description of the Work.   Although the scope of the work is defined in general terms in the contract, the definition must be detailed in the specifications.   The description should be sufficiently complete to indicate the types of construction and the character of the finished product.   Any unusual features or requirements should be mentioned.   The geographical location of the work should be described briefly with mention also of the methods of access to the site such as railroad, highway, and water transportation facilities.

Work and Materials to be Furnished by Others.   If it is intended that any part of the work necessary for the completion of the project is to be furnished by the owner or by other contractors, complete information should be given.   This will apply when the owner expects to furnish materials and equipment to be installed by the contractor, when the owner plans to accomplish part of the work with his own forces, and when work is to be done under other contracts.   This clause should require the contractor to conduct his operations in complete harmony with others engaged in the work and not to interfere with or delay the work of other contractors.   When more than one contractor is to be employed, the work of each should be defined in detail, particularly if the same set of specifications is to be used by several contractors.

Reference to Other Contract Documents.   The agreement, drawings, and other documents upon which the contract is based should be designated by reference, and the requirements of each should be given equal force irrespective of the documents in which they are contained. This section should contain a clear manda-tory statement that the contractor shall furnish all materials, labor, and operations necessary to complete the work, as defined in the contract documents subject to the qualifications, conditions, or exceptions noted.

Contract Drawings. The drawings which accompany the specifications and which form a part of the contract should be listed by number, title, and date. It should be made clear that unless stated otherwise reference is made to these drawings when "as shown," "as indicated," "as detailed," or words of similar import are used in the specifications.

Reference Drawings. Occasionally it may be desirable to list for the convenience of the contractor supplemental drawings which contain information necessary for the conduct of the work but which are not included in the contract drawings. Utility location drawings and boring logs are typical of this classification.

Standard Specifications. Standard specifications, such as those published by the U.S. Government and the American Society for Testing Materials, are widely used for construction work, particularly for materials of construction. These standards may be incorporated into the specifications for any project by reference. Although such references are made in various sections of the technical provisions, it may be convenient to summarize them in the general provisions giving complete information as to title, serial number, and date of issuance or revision as in the following example.

Standard Specifications. Except where modified by this specification or its accompanying drawings, the standard specifications given in the following list shall govern in all cases where references thereto are made. Special care shall be exercised to refer to these specifications and to any modifications thereof in requests for quotations, purchase orders, and subcontracts.

STANDARD SPECIFICATIONS

No.	Title	Date or Edition
....	....	....
....	....	....

Detail Drawings and Instructions. Ordinarily contract drawings are sufficiently complete to define the scope of the work and to enable contractors to prepare estimates for bidding purposes. Frequently, however, they require development and elaboration before the various components of the project can be fabricated or constructed. Therefore, it is desirable that the specifications provide for the engineer to issue from time to time during the progress of the work detail drawings and instructions necessary for the proper execution of the contract. These drawings and instructions must be true developments of the contract drawings and readily inferable therefrom. If they are not consistent with the contract documents, it may be concluded that they call for work outside the scope of the contract, and the contractor would be justified in asking for an adjustment in the contract price.

Drawings Required of the Contractor. In order to fabricate and erect the structure shown on the contract drawings, it is necessary for the contractor to prepare shop drawings, working drawings, rivet lists, match-marking diagrams, camber diagrams, and erection drawings for structural steel construction; bar schedules and bending details for reinforced concrete construction; and details for cofferdams, bracing, falsework, forms, and the like which show the methods the contractor proposes to use in the construction work. This clause of the specifications should require the contractor to submit all such drawings for examination and approval by the engineer prior to the ordering of materials although the engineer's approval of shop drawings is not to relieve the contractor of any responsibility for furnishing materials of the proper dimensions and quality. The following example illustrates the requirements for drawings prepared by the contractor in heavy construction work.

Drawings Required of the Contractor. The contractor shall submit triplicate prints of drawings showing bar schedules and bending details of all steel reinforcement; complete drawings of cofferdams, bracing, falsework, and forms; and complete general drawings, shop drawings, erection diagrams, camber diagrams showing the position of the structure at various stages during the erection, match-marking diagrams, and rivet lists for structural steel and miscellaneous metal parts. Approval by the engineer of shop drawings shall not relieve the contractor from furnishing material of proper dimensions, quantity, and quality, nor will such approval relieve the contractor from responsibility for errors of any sort in the shop drawings or for the strength and sufficiency of cofferdams, bracing, falsework, forms, and other construction proposed or designed by him.

The estimated weight of each shipping unit shall be clearly marked on the shop drawings on which this unit is detailed.

The contractor shall bear all costs or damages which may result from the ordering of any materials prior to the approval of the shop drawings; and no work shall be done until the shop drawings therefor have been approved.

In case of correction or rejection, the contractor shall resubmit shop drawings until such time as they are acceptable to the engineer, and such procedure will not be considered as a cause for delay.

After approval of shop drawings, the contractor shall supply the owner with as many prints of the approved drawings as may be ordered.

Omissions and Conflicts in the Contract Documents. Discrepancies in the contract documents, particularly between the plans and specifications, are to be expected in construction work. These arise from changes and corrections made in one document and inadvertent failure to do so in others, the improper use of superseded drawings or specifications, the improper use of standard specifications where deviations are called for elsewhere in the plans or specifications, the inadvertent omission of necessary instructions, and the like. A clause in the general provisions should state that anything called for in one contract document and not in the others shall be of like effect as if shown or mentioned in all. The contractor should be required to perform all work necessary to carry out the intent of the plans and specifications, although these may contain some omissions or misdescriptions of details of the work. In the case of a discrepancy between the plans and the specifications, the specifications are usually given precedence over the drawings, although the reverse of this is sometimes specified. Details shown on large-scale drawings are given precedence over those shown on small-scale drawings. Some offices, the Department of Public Works of the City of New York, for instance, specifies that the contractor will be assumed to have based his bid on the most expensive alternative when a conflict is encountered in the contract document, and this principle is followed when adjustments in the contract prices are necessary as a result of a discrepancy.

The contractor should be required to check all dimensions and report any discrepancies or errors to the Engineer. When new construction is to be joined to existing structures, all dimensions involved likewise should be checked by the Contractor. Thereafter the Contractor is held responsible for all errors and discrepancies.

Data from Borings, Test Pits, and Test Piles. Responsibility for the accuracy and interpretation of subsurface information is the source of many disputes and lawsuits in construction work. Government agencies implicitly assume some degree of responsibility in this connection by permitting an adjustment in the contract price if subsurface conditions are found to be materially different from those anticipated. In general this policy is not advisable for private work because private owners do not have the great contractual strength which is given to government contracting officers in deciding such questions. Therefore, the specification should contain a provision relieving the owner of responsibility for the accuracy and interpretation of subsurface data.

Boring logs and test pile driving and load information should be shown by indicating only results of such tests without any interpretation of those results. The projection of soil and geologic profiles from borings particularly is to be avoided in the contract documents although they are useful and justified for design studies. Even the terms used in defining types of soil such as sand, gravel, clay, and hard pan have formed the basis of disputes and lawsuits. Therefore, it is advisable to exclude all subsurface data from the contract documents for ordinary projects and to make this information available elsewhere for the information of contractors.

A disclaiming statement similar to the following is frequently placed in the specifications and on the drawings which show subsurface data.

The subsurface information and data furnished herein are not intended as representations or warranties but are furnished for information only. It is expressly understood that the owner will not be responsible for the accuracy thereof or for any deduction, interpretation, or conclusion drawn therefrom by the contractor. The information is made available in order that the contractor may have ready access to the same information available to the owner and is not part of this contract.

Sequence of the Work. When the desires of the owner, the character of the work, or the public interests indicate the necessity for a definite sequence of operations or a particular schedule for the completion of various parts of a project, details should be given in the specifications. Otherwise, the contractor should be allowed to exercise option in these matters subject to approval of the engineer. Ordinarily the contractor will schedule his work, and accordingly the time of completion of the various parts of the project, in the manner best suited to his organization, equipment, and methods of operation. Interference or change in his normal schedule required by the specifications can have important effects on cost. Therefore, such requirements should be imposed only when absolutely necessary.

Progress of the Work. The contractor should be required to prepare an approved progress schedule before the beginning of the work and to provide such equipment and carry on work simultaneously at as many points as will, in the judgment of the engineer, enable him to complete the work within the time specified in the contract. The progress schedule should show a breakdown of the principal elements and types of construction in the project, the date upon which each type of construction or operation is scheduled to begin and when it should be completed. For lump-sum contracts the contract price should also be broken down with appropriate amounts budgeted to each of the items in the progress schedule. This is for the engineer's use in preparing progress estimates and forms the basis for progress payments to the contractor.

As soon as practicable after the first of each month the contractor should forward to the engineer a summary of the progress of the various parts of the work in the mills, shops, and in the field, giving the existing status, rate of progress, estimated time of completion, and causes of delays if any have occurred. The progress schedule and periodic reports provide the means for the direct control and coordination of all the work under the contract.

Control of Materials. This clause should be a general requirement calling for the examination and testing of all materials of construction before acceptance, it being understood that the details of the various tests will be covered by the technical provisions. Samples are to be taken under the engineer's direction of all sand, gravel, stone, cement, concrete reinforcing bars, steel fabric, bituminous materials, structural steel, and other materials used in construction. The contractor is often required to furnish all samples and bear all expense in connection with their sampling, transportation, and testing. Such samples are to be tested, and no material of which laboratory tests are required shall be used until accepted as the result of such test, and then only as long as the quality remains equal to that of the accepted sample.

When large quantities of cement, structural steel, and the like are to be used, provision should be made for mill inspection and the testing of samples taken at the mill. This work may be performed by representatives of the engineer, or specialized inspection and testing companies may be retained for this purpose. Ordinarily the construction contractor is required to retain an approved inspection and testing company and to include the cost of such work in the contract price. Under an alternative arrangement, mill inspection and tests may be accomplished under a separate contract between the owner and the testing company. With either method the engineer maintains general supervision of the work.

Rates of Wages at the Site. The Davis-Bacon Act requires that mechanics and laborers on U.S. Government contracts shall receive not less than the prevailing wage rates. These are predetermined for each contract by the Secretary of Labor and inserted in the general provisions of the specifications. Most states follow a similar procedure. For private work also some stipulation should be made as to wage rates in order to protect the owner against labor controversies and strikes. On smaller jobs it will be sufficient to require the contractor to pay wages not less than those prevailing in the locality. For more important work it may be desirable to form an agreement with representatives of the various trade unions on specific wage rates for all trades and crafts and to require these to be paid by the contractor under this heading in the specifications.

Lines and Grades. When construction surveys, lines, and grades are to be furnished by the engineer this clause in the general provisions should so state and require the contractor to give ample notice of the time and place where line and grade stakes are needed. Provision should be made for the contractor to suspend work when necessary to permit the carrying on of surveys, and he should be required to provide stakes and such ordinary labor as may be required by the engineer in this work without extra compensation.

When responsibility for the layout of the work is to be placed on the contractor, this clause should state that control points and elevations will be furnished by the owner and that the contractor shall be responsible for all other surveys and measurements in connection with the work. The contractor at his own expense is to furnish all stakes, templates, platforms, equipment ranges, and labor required in laying out the work and is responsible for the proper execution of the work to the lines and grades shown on the drawings or indicated by the engineer.

Space for Construction Purposes and Storage of Materials. When limitations of space for construction operations and the delivery and storage of materials are expected, specific details should be given. The contractor should be required to furnish all additional land necessary for the erection of temporary construction and storage facilities with right of access thereto at his own expense unless it is to the owner's best interests to furnish these facilities himself.

Water, Power, Light, and Other Utility Services. Information should be given as to the availability and location of the various required utility services and whether the expense for their use is to be the responsibility of the owner or the contractor. Details should be given as to the quantities and sources of supply of water and methods for the disposal of waste water. Characteristics and sources of supply of electric power should also be given.

Facilities for the Engineer. Under this heading the contractor is required to furnish at his own expense the facilities necessary for the engineer to carry on his work. Representative of such facilities are survey platforms, towers, etc., necessary for laying out and inspecting the work, and a field office for the engineer and his assistants furnished with drawing table, desk, chairs, light, heat, telephone, water, and toilet, and the like. For work near water a power boat may be required. When important structural steel work is included two tested and certified steel tapes guaranteed to agree with the shop standards of the manufacturer of the metal work should be furnished within one month after the signing of the contract.

Warranties by Contract Bond. A list should be given showing the mechanical and electrical equipment or other work to be furnished by the contractor to meet service or performance requirements which the contractor is to warrant to be free from defects of materials and workmanship for a specified period, usually one year, from the date of acceptance of the contract work. The contractor is required, upon notice, to make good at his own expense all defects developing during this period. These warranties may be covered by the contract bond, but they do not operate to defer final payment during the period specified, however.

The Owner's Right to Use Completed Portions of the Work. Frequently the contract will include a general clause permitting the owner to use completed portions of the project. The general provisions of the specifications should cover the owner's requirements in this connection in detail. Such questions arise when the owner wishes to use the lower stories of a building while the contractor is completing the upper stories. Similarly, it may be desirable to open completed portions of a street or highway before the entire project is ready for service. When the owner takes over a completed portion of the work, a final inspection should be made, and thereafter the contractor should be relieved of his responsibilities to maintain that portion of the project which is placed in service, pending the completion and final acceptance of the entire project.

Special Precautions during Construction. Under this and subsequent headings instructions should be given on any special requirements with regard to the maintenance of traffic during construction, interference with navigation, protection of adjacent property, protection or relocation of public utilities, and the like. Such requirements will vary with individual projects and will depend upon the conditions surrounding the work. Generally special requirements of this type have important effects on cost, and all their implications should be carefully analyzed before they are included in the specifications.

EXAMPLE OF THE GENERAL PROVISIONS OF THE SPECIFICATIONS

SPECIFICATIONS FOR THE CONSTRUCTION OF NORTH CENTRAL STATE AIRPORT PROVIDENCE, RHODE ISLAND

SECTION I

GENERAL PROVISIONS

1-01. STATEMENT OF WORK

A. LOCATION. The work is located in the town of Smithfield, approximately 7 miles northwest of Providence, in Providence County, Rhode Island.

B. DESCRIPTION.

1. *General.* The work consists of furnishing all plant, labor, equipment, materials, and services, unless otherwise specified, for the construction of North Central State Airport. The airport is provided with a flexible pavement runway and parking apron, seeded runway, administration building,

hangar building, utility building, transformer vault, pipe drainage systems, drainage structures, sewage disposal system including septic tank, well water supply system including 60,000-gallon storage tank, electric lighting and power systems, vehicular access roads and parking area, seeding and landscaping, fencing, and other incidental facilities.

2. *Principal Dimensions.*

   a. Landing strip width, 500 ft.
   b. Landing strip length, 5,200 ft.
   c. Runway width, 150 ft.
   d. Runway length, 5,000 ft.
   e. Airplane parking apron, 200 ft. x 750 ft.
   f. Two-story administration building, 80 ft. x 30 ft.
   g. Hangar building, 200 ft. x 62 ft.
   h. Utility building, 12 ft. x 18 ft.
   i. Secondary seeded runway, 300 ft. x 2,200 ft.

3. *Administration Building.* The administration building is a reinforced concrete and masonry two-story structure with structural steel frame, approximately 80 ft. long x 30 ft. wide and contains a 5-ft. reinforced concrete overhang at the first floor and roof slabs. The first story contains such rooms as the charter office, lounge, restaurant, boiler room, electric service room, and toilets. To the rear of the building and at first floor level is a one-story garage approximately 17 ft. wide built integrally with the administration building. The second floor contains the observation deck, operations offices, and lobby.

4. *Hangar Building.* The hangar building is a steel framed structure with reinforced concrete floor and fluted aluminum siding with a canopy type overhead door. The building is approximately 62 ft. wide x 200 ft. long divided into an open hangar section, an enclosed hangar section, and a section for shops, offices, and utilities. The hangar is designed with cantilever welded steel frames spaced at 25 ft. with footings and counterweight of concrete. The open and enclosed hangars across the front of the structure are, respectively, 125 ft. and 75 ft. long with a depth of 42 ft. The 60-ft. overhead type door for the enclosed portion provides a 20-ft. vertical entrance clearance. The entrance clearance for the open section is about 28 ft. The rear portion of the hangar consists of shop, offices, boiler room, and toilet facilities for the full length of the structure and has a width of about 19 ft. and a ceiling height of 12 ft.

5. *Utility Building.* The utility building is a 12-ft. x 18-ft. structure with concrete and brick walls, membrane waterproofing, reinforced concrete roof slab, and poured concrete foundation. The roof slab is covered with four-ply built up asphalt aggregate roofing. Membrane waterproofing is provided for walls below grade. The building contains one wood exterior door, and access will be provided through the roof by means of a hatch. A well pump and motor, fire pump and engine, domestic water pump, motor and pressure tank, overhead lighting fixtures, electric unit heater, and all necessary piping and controls are among the facilities in the utility building.

6. *Preparation of Site and Earthwork.* Removal of trees, stumps, structures, etc., will be required to prepare the site for construction. Earthwork will include earth and rock excavation, topsoiling, seeding, planting, embankment, backfilling, and borrow operations.

7. Fresh-Water System. The fresh-water system consists of a deep well with electrically driven pump discharging into an underground 60,000-gallon storage tank and cast-iron water mains. An electrically driven domestic water pump with automatic pressure storage tank provides normal water service to the administration and hangar building and fire hydrants. A gasoline-driven fire pump with automatic controls and individual gasoline storage system is included for emergency service.

8. Drainage System. The drainage system consists of reinforced concrete, cast iron, corrugated metal, and vitrified clay pipe culverts interconnected with brick and concrete catch basins and manholes, discharging through concrete headwalls.

9. Sewage Disposal System. The sewage disposal system consists of a reinforced-concrete septic tank, discharging into an open jointed vitrified clay pipe drainage field.

10. Paving. The pavement on the access roads to the airport is a run-of-bank gravel base course with a bituminous treated gravel surface course. The runway pavement consists of sub-base and base course of graded gravel with a bituminous concrete wearing course.

11. Electrical Work. The electrical work consists of picking up the utility company's 4,160-volt line, adapting the power for airport use by means of transformers, current regulators, oil fuse cutouts, and other equipment and installing the complete airport lighting, distribution, and power systems. The bulk of the electrical equipment is contained in an underground vault.

1-02. WORK AND MATERIALS TO BE FURNISHED BY OTHERS

A. Electrical Equipment. The State will furnish to the Contractor at the site of the work, the airfield lighting control console and the gasoline-electric emergency generator set and controls in accordance with the provisions of Section XII, Airfield Lighting.

B. Gasoline Storage and Dispensing System. The gasoline storage and dispensing system will be installed by others under separate contract.

C. Extension of 4,160-Volt Power Lines. Extending 4,160-volt power lines to a terminal pole of the Narragansett Electric Co., shown on the drawings, will be performed by the utility company.

D. Cooperation between Contractor and Others. The Contractor shall cooperate with others engaged in work at the site and conduct his operations with a minimum of interference and delay to the work of other contractors.

1-03. COMMENCEMENT AND COMPLETION

The Contractor will be required to commence work under this Contract within 10 calendar days after the date of receipt by him of notice to proceed and to complete the entire work ready for use not later than 400 calendar days after the date of receipt by him of notice to proceed. The entire landing strip including all earthwork and pavement shall be completed as soon as practicable for use by the State but not later than 230 calendar days from the date of receipt by the Contractor of notice to proceed. (See paragraph 1–17, State to Use Landing Strip Prior to Completion of Contract.)

1-04. DRAWINGS

A. Contract Drawings. The work shall conform to the drawings titled North Central State Airport subtitled and dated as follows.

Drawing No.	Subtitle	Date
318-1	Cover Sheet and Index	Dec. 15, 1949
318-2	Vicinity and Location Plan	Dec. 15, 1949
318-3	General Plan	Dec. 15, 1949
318-4	Clearing and Grading Plan	Dec. 15, 1949
318-5 *	Log of Borings	Dec. 15, 1949
318-6	Runways, Taxiways, and Apron—Plan and Profile	Dec. 15, 1949
318-7	Runways, Taxiways, and Apron—Cross Sections	Dec. 15, 1949
318-8	Access Roads and Retaining Wall	Dec. 15, 1949
318-9	Layout of Utilities	Dec. 15, 1949
318-10	Drainage Plan—Profiles and Details	Dec. 15, 1949
318-11	Drainage and Utility Details	Dec. 15, 1949
318-12	Airport Lighting—Details and Wiring Diagram	Dec. 15, 1949
318-13	Administration Building—Plans and Sections	Dec. 15, 1949
318-14	Administration Building—Architectural Details No. 1	Dec. 15, 1949
318-15	Administration Building—Architectural Details No. 2	Dec. 15, 1949
318-16	Administration Building—Structural Details	Dec. 15, 1949
318-17	Administration Building—Mechanical Details	Dec. 15, 1949
318-18	Administration Building—Electrical Details	Dec. 15, 1949
318-19	Hangar Building—Plan, Elevations, and Sections	Dec. 15, 1949
318-20	Hangar Building—Architectural and Electrical Details	Dec. 15, 1949
318-21	Hangar Building—Structural Details No. 1	Dec. 15, 1949
318-22	Hangar Building—Structural Details No. 2	Dec. 15, 1949
318-23	Hangar Building—Mechanical Details	Dec. 15, 1949
318-24	Miscellaneous Details	Dec. 15, 1949

B. DETAIL DRAWINGS AND INSTRUCTIONS. The Engineer will furnish with reasonable promptness additional instructions, by means of drawings or otherwise, necessary for the proper execution of the work. All such drawings and instructions shall be consistent with the Contract Documents, true developments thereof, and reasonably inferable therefrom. The work shall be executed in conformity therewith, and the Contractor shall do no work without proper drawings and instructions. The Contractor and the Engineer, if either so requests, shall jointly prepare a schedule, subject to change from time to time in accordance with the progress of the work, fixing the dates at which the various detail drawings will be required, and the Engineer shall furnish them in accordance with that schedule. Under like conditions, a schedule shall be prepared, fixing the dates for the submission of shop drawings, for the beginning of manufacture and installation of materials, and for the completion of the various parts of the work.

C. DRAWINGS REQUIRED OF THE CONTRACTOR. The Contractor shall submit with reasonable promptness for the approval of the Engineer duplicate prints of all shop and erection drawings and schedules required for the work, and as may be additionally required under the various sections of the specifications. The Contractor shall make any corrections required by the Engineer, file with him two (2) corrected copies and furnish such other copies as may be needed. The approval by the Engineer of such drawings or schedules shall not relieve the Contractor from responsibility for deviations, from drawings or specifications, unless he has in writing called the attention of the Engineer to such deviations at the time of the submission, nor shall it relieve the Contractor from responsibility for errors of any sort in shop drawings or

* Drawing No. 318-5 is for reference purposes only (see paragraph 1-08, Subsurface Information and Data).

schedules. In case of correction or rejection, the Contractor shall resubmit the shop and erection drawings until they are acceptable to the Engineer. After approval of the drawings, the Contractor shall furnish the Engineer with as many prints as may be ordered.

D. MECHANICAL, ELECTRICAL, DRAINAGE DRAWINGS—DIAGRAMMATIC. The mechanical, electrical, and drainage drawings are diagrammatic and indicate the general arrangement of the work. The Contractor shall consult the architectural and other appropriate detail drawings for the exact location of all fixtures and equipment. Where these are not definitely located in the drawings and details, the Contractor shall request this information from the Engineer in writing.

E. RECORD DRAWINGS. The Contractor shall, during the progress of the work, keep a careful record of all changes and corrections from the layouts shown on the drawings. Upon completion of construction and prior to final payment the Contractor shall submit to the Engineer two (2) copies of marked up prints showing the changes.

1-05. INTENT OF SPECIFICATIONS AND DRAWINGS

The intent of the specifications and drawings is to prescribe a complete work which the Contractor undertakes to do in full compliance with the Contract Documents. If there is any apparent contradiction or ambiguity between the drawings and specifications, the Contractor shall bring the fact to the attention of the Engineer and shall obtain his decision as to the true meaning or intention before proceeding with the portion of the work affected. All work that may be called for in the specifications and not shown on the drawings, or shown on the drawings and not called for in the specifications, shall be performed by the Contractor as if described in both, and should any work be required which is not denoted in the specifications or drawings, but which is nevertheless necessary for the proper carrying out of the intent thereof, the Contractor shall perform all such work as fully as if it were particularly described. Wherever in the specifications or drawings "directed," "required," "ordered," or words of similar import are used, it shall be understood that directed, required, ordered by the Engineer is intended. In like manner when "as shown," "as indicated," "as detailed," or words of similar import are used in the specifications, reference to the Contract drawings, listed in paragraph 1-04, Drawings, is intended.

1-06. STANDARD SPECIFICATIONS

In all cases where standard specifications, such as the American Society for Testing Materials and Federal Specifications, are referred to in these specifications, the latest revision of the pertinent specification, in effect as of the date of bid opening, shall be considered effective. Special care shall be exercised to refer to the standard specifications and to any modifications thereof in requests for quotations purchase orders, and subcontracts.

1-07. MATERIAL REFERRED TO BY NAME

Specific reference in the specifications or drawings to any article, device, product, material, fixture, form or type of construction, etc., by name, make, or catalog number shall be interpreted as establishing a standard of quality and shall not be construed as limiting competition, and the Contractor, in such cases, may at his option use any article, device, product, material, fixture, form or type of construction, etc., which in the judgment of the Engineer is equal to that named.

1-08. SUBSURFACE INFORMATION AND DATA

The subsurface information and data furnished in the drawings are not intended as representations or warranties but are furnished for information only. It is expressly understood that the State will not be responsible for the accuracy thereof or for any deduction, interpretation, or conclusion drawn therefrom by the Contractor. The information is made available in order that the Contractor may have ready access to the same information available to the State and is not part of this Contract.

1-09. QUALITY OF WORK

The work shall be executed in the best and most workmanlike manner by qualified, careful, and efficient mechanics. Unless otherwise specified all materials to be incorporated in the work shall be new, unused, and of recent manufacture.

1-10. PROGRESS SCHEDULE

A. SCHEDULE. The Contractor shall, within 10 days or within such time as determined by the Engineer, after date of commencement of work, prepare and submit to the Engineer for approval a practicable progress schedule showing the order in which the Contractor proposes to carry on the work, the dates on which he will start the various features, and the contemplated dates for completing the same. The schedule shall be in the form of a progress chart of suitable scale to indicate appropriately the percentage of work scheduled for completion at any time. The Contractor shall enter on the chart the actual progress at the end of each week or at such intervals as directed by the Engineer, and shall immediately deliver to the Engineer three (3) copies thereof.

B. REPORT. The Contractor shall also prepare and submit to the Engineer as soon as practicable after the first of each month a written progress report summarizing the progress of the various parts of the work in the mills, shops, and in the field, giving the existing status, rate of progress, estimated time of completion, and causes of delays if any have occurred.

1-11. APPROVAL OF MATERIALS AND EQUIPMENT

A. SUBMISSION OF SAMPLES AND DATA.

1. Samples, drawings, catalog cuts, and other data shall be submitted for approval of the Engineer as required by the various sections of these specifications. Items submitted shall be properly labeled to indicate the Contract number, project, Contractor, source of supply, manufacturer, and other data required by the specifications. All items shall be submitted in sufficient time to permit proper consideration and action thereon without delaying the approved progress schedule. Items sent for approval shall be shipped prepaid by the Contractor.

2. Where samples are required to be tested by the State, samples shall be taken and submitted under the direction of the Engineer in accordance with the specified methods for sampling.

B. APPROVAL. Only materials and equipment which have been approved in writing shall be used in the work. All materials and equipment may be inspected or tested by the Engineer at any time during their preparation and use. If after trial, it is found that approved sources of supply do not furnish a uniform product or if the product from any source proves unacceptable at any time, the Contractor shall furnish approved material from other sources. No materials which have in any way become unfit for use shall be used in the work.

C. TESTING.

1. Unless otherwise specified, all testing of materials and equipment shall be performed by the Contractor. When such tests are made, satisfactory proof of compliance of the materials and equipment with the requirements of the specifications shall be furnished to the Engineer. Satisfactory proof of compliance with the specifications shall be submitted as directed by the Engineer in one or more of the following ways.

a. *Manufacturer's Certificate of Compliance.* In the case of standard labeled stock products of standard manufacture which have a record of satisfactory performance in similar work over a period of not less than two years, the Engineer may accept a notarized statement from the manufacturer certifying that the product conforms to the applicable specifications.

b. *Mill Certificates.* For materials where such practice is the usual standard, the Engineer may accept the manufacturer's certified mill and laboratory certificate.

c. *Testing Laboratory Certificates.* The Engineer may accept a certificate from a commercial testing laboratory satisfactory to him certifying that the product has been tested within a period acceptable to the Engineer and that it conforms to the requirements of the specifications.

d. *Report of Actual Laboratory Test.* The Engineer may require that the Contractor make actual tests of any product and submit a report of the specified test. Such tests shall be made by a commercial testing laboratory satisfactory to the Engineer.

2. The cost of any additional laboratory tests required through the resubmission of samples shall be borne by the Contractor and will be deducted from any money due him on his Contract.

1-12. RATES OF WAGES AT THE SITE

The minimum wages to be paid laborers and mechanics on this project, as determined by the U.S. Department of Labor to be prevailing for the corresponding classes of laborers and mechanics employed on projects of a character similar to the Contract work in the pertinent locality, are as set forth below. Under no conditions shall the wages paid be less than those designated.

Classifications	Per Hour
Air tool operators (jackhammermen, vibrator)	$1.775
Asbestos workers	2.375
Boilermakers	2.76
Boilermakers' helpers	2.51
Bricklayers	3.25
Cable splicers	2.65
Carpenters, journeymen	2.175
Cement finishers	2.20
Electricians, journeymen	2.75
Glaziers	2.30
Iron workers, ornamental	2.425

Classifications	Per Hour
Iron workers, reinforcing	$2.325
Iron workers, structural	2.575
Laborers, building	1.55
Laborers, concrete	1.55
Laborers, unskilled	1.55
Mason tenders	1.10
Mortar mixers	1.25
Painters, brush	2.20
Painters, spray	2.275
Painters, structural steel	2.45
Caulkers, cast iron pipe	1.25
Pipe layers, caulkers, banders	1.90
Plumbers, journeymen	2.625
Power equipment operators:	
Apprentices (oilers, firemen)	1.94
Asphalt plant engineer	2.29
Boxmen or mixer box (concrete or asphalt plant)	1.99
Compressor	1.94
Compressor (more than one)	2.24
Concrete mixers (up to 1 yard)	1.94
Concrete mixers (over 1 yard)	2.24
Concrete pump or pumpcrete guns	2.24
Derrick	2.44
Dual drum mixer	2.34
Firemen in hot plant	1.94
Fork lift or lumber stacker (on construction job site)	2.19
Handi-crane (no oiler required)	2.34
Heavy duty repairmen	2.29
Heavy duty repairmen helper	1.94
LeTourneau pulls (Jeeps, Terra Cobras, LaPlant Choate, and similar types of equipment)	2.44
Material hoist	2.19
Mechanical finishers (concrete or asphalt) (airport, highway, or street work)	2.29
Pavement breakers, Emsco type	2.34
Power grader, power planer, motor patrol, or any type of power blade	2.44
Power shovels and/or other excavating equipment with shovel-type controls (up to and including 1 yard)	2.59
Rollers	2.29
Screedmen	1.94
Self-propelled elevating grade plane	2.34
Spreader machines (Barber Greene, Jaeger, etc.) (engineer and screedman used in operation)	2.29
Tractor (boom)	2.44
Tractor (tandem)	2.59
Tractor-type shovel loader (scale not to apply when used as blade of bulldozer)	2.44
Trenching machine	2.34
Truck crane	2.44
Truck-type loader	2.44
Roofers, composition	1.75
Roofers' helpers	1.30
Waterproofers	1.75
Sheet metal workers, journeymen	2.3125
Steam fitters, journeymen	2.625

Classifications	Per Hour
Tile setters	$2.90
Truck drivers:	
Dump trucks:	
Under 4 yd. (water level)	1.61
4 yd. and under 6 yd. (water level)	1.74
6 yd. and under 8 yd. (water level)	1.83
8 yd. and over (water level)	2.20
Transit-mix:	
3 yd. and under (mfr's mixing capacity rating)	1.69
4 yd. and 5 yd. (mfr's mixing capacity rating)	1.78
Pick-ups carrying under 1,000 lb.	1.58
Flat rack trucks carrying between 1,000 lb. and 4,500 lb.	1.60
Flat rack trucks carrying over 4,500 lb.	1.68
Heavy duty transports	1.95
Helpers, warehousemen, teamsters	1.58
Dumpster trucks—paid according to water level yardage (use dump truck scale)	
Water truck drivers	1.68
Road oilers	1.68
Automotive oilers or greasers	1.68
Euclid type (trac) trucks (use dump truck scale)	
Truck repairmen (job site construction)	2.28
Truck repairmen helpers (job site construction)	1.83
Welders	Prevailing rate

1-13. LINES AND GRADES

The Contractor shall lay out his work from base lines and bench marks established by the Engineer and shall be responsible for all measurements in connection therewith. The Contractor shall, at his own expense, furnish all stakes, templates, platforms, equipment, and labor, including surveyors, that may be required in setting and cutting or laying out any part of the work. The Contractor shall be held responsible for the proper execution of the work to such lines and grades as may be established by the Engineer.

1-14. RIGHT OF WAY

The State will secure the right of way for construction as shown on the drawings including an area shown as available for the Contractor's use. Nothing herein contained and nothing marked on the drawings shall be interpreted as giving the Contractor exclusive occupancy of the territory provided. The State and its employees, for any purpose, and other contractors of the State, for any purpose required by their respective contracts, may enter upon or occupy portions of it, as directed or permitted by the Engineer. Should the Contractor require additional area for his operations, he shall obtain such land at his own expense.

1-15. WATER AND ELECTRICITY

It shall be the responsibility of the Contractor to provide and maintain at his own expense an adequate supply of water and electricity required for the work. Existing sources of supply are shown on the drawings. The Contractor shall install and maintain supply connections and lines satisfactory to the Engineer, and prior to completion of construction the Contractor shall remove the supply lines at his expense.

1-16. FIELD OFFICE FOR ENGINEER

The Contractor shall provide a suitable field office in an approved location for the exclusive use of the State's engineering staff. The building shall consist of a room having a minimum floor space of 800 sq. ft. and a minimum inside height of 8 ft. and be provided with at least nine (9) suitably located windows. The building shall be entirely closed and waterproof and provided with adequate sanitary facilities, artificial light, two exterior doors with approved cylinder locks, suitable furniture, and fuel and equipment for heating to maintain a minimum temperature of 70°F. Lighting and telephone service shall be provided at the Contractor's expense. The Contractor shall pay for all local telephone calls relating to the project for the duration of the Contract. The field office shall be maintained by the Contractor for the duration of the construction period and for such additional time as may be required by the State [not exceeding 60 days], after which, and before final acceptance and payment is made, the Contractor shall remove and dispose of the field office in a satisfactory manner.

1-17. STATE TO USE LANDING STRIP PRIOR TO COMPLETION OF CONTRACT

In accordance with Article 10 of the Contract, titled State's Right to Use the Completed Portions of the Work, the State will make use of the landing strip as soon as it is completed. The entire landing strip area as indicated on the drawings will be used for emergency landings by distressed planes and for such other uses as the State may require. To this end, the Contractor shall carry on his work in such manner as to leave the landing strip free from hazards at all times. The State's forces will maintain the landing strip after completion by the Contractor.

1-18. GUARANTEE

The Contractor shall guarantee that all work installed by him is free from any and all defects in workmanship and materials and that all apparatus will develop the capacities and characteristics specified. He further guarantees that if, during a period of one year from the date of the certificate of completion and acceptance of his work, any such defects in workmanship, material, or performance appear, such defects will be remedied by him without cost to the State. Should the Contractor fail to remedy the defects as outlined above within a reasonable length of time, to be specified in a notice from the Engineer to the Contractor, the State may have such work done and charge the cost to the Contractor.

1-19. PROTECTION OF MATERIAL, STRUCTURES, AND WORK

The Contractor shall at all times protect and preserve all materials, supplies, and equipment of every description (including property which may be State furnished or owned), existing structures including utilities, and all work performed. All reasonable requests of the Engineer to inclose or specially protect such property shall be complied with.

1-20. TEMPORARY ROADS

The Contractor shall construct and maintain all temporary roads required for the completion of the work under this Contract. Temporary roads shall be limited to locations and type approved by the Engineer and, when necessary, shall be relocated to permit installation of various portions of the work in an orderly manner.

1-21. TRAFFIC CONTROL AND PARKING

It is anticipated that a large portion of the employees on the site will use automobiles for transportation to and from work. It shall be the responsibility of the Contractor to maintain, protect, and control traffic in the vicinity and on the site; to enforce restrictions against parking on roads within the site; and to provide necessary parking areas for all workmen in suitable locations as approved by the Engineer.

1-22. TEMPORARY HEAT

A. The Contractor whose work requires temporary heat shall provide at his own expense such heat as is necessary to prevent injury to work or material through dampness or cold. Heat shall be maintained as required for the installation of all items as specified in the technical sections of this specification and as required for maintenance of scheduled progress and for good workmanship.

B. The use of open salamanders or other types of heating which may smoke and damage the finish in the buildings will not be allowed.

1-23. PRESERVATION OF EXISTING VEGETATION

A. The Contractor shall preserve and protect all existing vegetation such as trees, shrubs, and grass on or adjacent to the site which do not unreasonably interfere with the construction, as may be determined by the Engineer. The Contractor shall be responsible for all unauthorized cutting or damaging of trees and shrubs, including damage due to careless operation of equipment, stock piling of materials, or tracking of grass areas by equipment.

B. Care shall be taken by the Contractor in felling trees authorized for removal to avoid any unnecessary damage to vegetation that is to remain in place. Any limbs or branches of trees broken during such operations shall be trimmed with a clean cut and painted with an approved tree pruning compound if required by the Engineer. The Contractor will be liable for or may be required to replace or restore at his own expense all vegetation not protected and preserved as required herein that may be destroyed or damaged.

1-24. CLEANING UP

The Contractor shall at all times keep the construction area, including storage areas used by him, free from accumulations of waste material or rubbish and prior to completion of the work remove any rubbish from and about the premises and all tools, scaffolding, equipment, and materials not the property of the State. Upon completion of the construction the Contractor shall leave the work and premises in a clean, neat, and workmanlike condition satisfactory to the Engineer.

Questions

1. Explain the differences between the general or supplemental conditions of the contract, the general provisions of the specifications, and the technical provisions of the specifications.

2. A sanitary district induced a contractor to take on a sewer job by concealing, in soil-condition information furnished the contracting company, the true condition as disclosed by test-boring records in the district's possession. The company was thereby misled to believe that the work could be done by the free-air method. The withheld information disclosed that compressed air would have to be used on part of the job.

In the company's suit for reimbursement against additional costs and damages caused by the fraud: (a) Was the district properly held to the same standards of honesty as if it had been a private corporation or an individual? (b) Could the district escape liability because the bidding specifications required the company to make its own test of soil conditions? (c) Was the district exonerated from liability because the claim for additional costs and damages was not filed within a time limited by the contract for filing claims? (See *Contractors and Engineers Monthly*, January, 1949, page 98.)

16

TECHNICAL PROVISIONS OF
THE SPECIFICATIONS

The technical specifications contain the detailed instructions necessary to obtain the desired quality and service in the finished product. In addition to the technical requirements it also is necessary to provide for inspection and tests during construction to make certain that the specified requirements are being obtained.

The writing of the technical specifications must be coordinated closely with the preparation of the drawings. As the design progresses it will be desirable to prepare an outline of the items of work to be covered, and the writing of the technical provisions can proceed concurrently with the completion of the drawings. In making up the outline the entire work must be analyzed carefully and all items and processes noted.

One section of the specifications should be devoted to each type of construction, such as earthwork, concrete, masonry, carpentry, and structural steel. Equipment and machinery will be covered likewise either in an individual section for each item or with groups of related items collected in one section. Similarly, it may be convenient to devote separate sections to each of the basic materials which are used in several types of construction. This will reduce repetition and the number of cross references otherwise required. For example, Portland cement may be used in concrete, masonry mortar, plaster, stucco, and cement grout, and one section in the technical provisions may be made to cover its use in each of those types of construction. There is no fixed rule for the order in which the various types of construction should appear in the specifications, but, in general, they should follow as closely as practicable the sequence of construction operations.

Types of Technical Specifications. The form of the technical provisions will depend to a considerable extent on the methods of inspection and testing to be applied during construction. A comprehensive project, such as a building or bridge, is made up of a large number of parts, and no practicable test or series of tests is available to prove that the finished structure will perform its required service throughout the desired period of time. It becomes necessary therefore to control the quality of materials and workmanship in the manufacture, fabrication, and assembly of the various parts and to rely on the sufficiency of the design to obtain satisfactory performance in the finished structure. Accordingly, specifications for materials and workmanship are used for most of the basic types of construction. In this type of specification responsibility is placed upon the contractor for furnishing materials and workmanship conforming to the requirements specified for each type of construction and for the assembling of the component parts of the structure, but if these are free from defects the responsibility for the overall performance of the assembled structure rests upon the owner inasmuch as he furnished the plans and specifications. An exception exists when the contract requires the contractor to guarantee the sufficiency of the plans in which case he becomes responsible for performance also. This requirement was once widely used, but now it is seldom encountered.

A second type of specification is based on the overall performance of the finished product. When the desired operating characteristics can be measured by tests this form of technical specification is indicated. Specifications based on performance are frequently used for equipment and machinery, such as pumps, motors, and other accessories. Performance tests, however, are not necessarily indicative of the life of the structure, and it is customary to require the contractor to furnish warranties to guarantee durability and the absence of defects in materials and workmanship.

A third type of specification frequently used for construction work is based on the selection of proprietary or "standard brand" products in the open market. These may be either items of materials or equipment to be used separately or in connection with other processes or components of the work. No control can be exerted over the manufacture of proprietary articles, and the specifications may merely identify a desired item known to be satisfactory for the purpose intended, or quality and performance tests or standards may be prescribed. In either event careful wording of the specification is necessary because of the legal and competitive aspects of the procurement of patented commodities.

Specifications for Materials and Workmanship. The selection of materials for specific purposes in construction is a matter of design and economy and requires expert knowledge of the technology of materials. In general, the selection will be made on the basis of the critical properties of the material and its cost as compared with available alternatives. Frequently appearance is also a consideration. A wide variety of materials is available for construction work, and each has many characteristics. Most of these materials are suitable for more than one purpose, and two or more materials may be suitable for the same purpose. Therefore, the specification writer is faced with the problem of obtaining those properties necessary for the particular purpose at hand and of limiting or prohibiting those considered unsuitable or detrimental. Those properties which have little or no bearing should not be mentioned. Specifications for some materials will be based on physical or chemical properties, and for others performance will be the desirable method. Frequently a combination of the two types will be desirable. In general, proprietary or patented materials should not be specified by brand name, but the specifications should be so worded as to permit their use if the composition and performance requirements are fulfilled and if they are otherwise acceptable.

In so far as practicable the properties specified should be capable of measurement and proof by tests. The testing of materials is a well-developed science, and only standard tests should be specified. Some properties of materials, such as appearance, texture, and the like, cannot be measured and tested, and their approval and acceptance must be left to the judgment of the engineer during construction. Samples should be required for this purpose before the materials are purchased. An experienced engineer can determine the quality of many other properties by inspection and in these instances tests may be omitted.

The principal properties to be considered in the preparation of specifications for materials in construction work are as follows.

a. Physical properties, such as strength, modulus of elasticity, durability, hardness, and ductility.

b. Chemical composition.

c. Electrical, thermal, and acoustical properties, such as conductivity and resistance.

d. Appearance, including color, texture, and pattern.

e. Transportation, handling, and storage. Some materials are subject to deterioration or breakage under certain conditions, and the specifications should cover the protection necessary in the interval between the manufacture or processing of the materials and their incorporation into the structure.

f. Inspection, tests, and analyses, which are the methods for the control of the manufacture or processing of the materials and the grade or quality of the finished product. The specifications should state whether inspection is to be at the shop, mill, or in the field.

The first principle in writing specifications for workmanship is that, in so far as practicable, results only should be specified, allowing the contractor maximum latitude as to the selection of construction methods. Exceptions to this rule are justifiable and permissible, however, in the case of certain types of construction in which it is necessary to designate methods and procedures in order to insure the satisfactory completion of the work. For example, it is customary to specify the method for compacting earth fill used in the construction of embankments, earth dams, and such structures. Precautions, restrictions, and limitations may also be imposed on construction methods for the purpose of protection and coordination of the work as a whole without affecting the independent contractor relationship. This also applies to the question of the order of construction operations when a definite sequence is made necessary by design conditions or to meet conditions contemplated by the owner. It should be noted, however, that such restrictions and limitations have a direct influence on the cost of the work and may not be imposed on the contractor after the bid is submitted without a change in the bid amount.

The writing of specifications for construction and workmanship requires expert knowledge of construction methods and the standards of workmanship attainable. The usual procedure is along the following lines.

a. Specify the results desired as to quality of workmanship and finish, giving due consideration to practical limitations in tolerances, clearances, and the like.

b. State any detailed construction methods or procedures necessary for the accomplishment of particular purposes.

c. Stipulate any desired limitations or restrictions to be placed on the contractor's methods in the interests of the coordination of the work.

d. Give any precautions necessary for the protection of the work or adjacent property.

e. Specify the methods of inspection and tests to which the work is to be subjected with particulars as to mill and shop inspection as well as field inspection.

EXAMPLE OF SPECIFICATIONS FOR MATERIALS AND WORKMANSHIP

STEEL ARCH BRIDGE *

SECTION 6. STRUCTURAL METAL WORK

MATERIALS

6.01. *Structural Steel.* All structural carbon steel, bolts, and steel billets shall conform to the requirements of the Standard Specifications for Structural Steel for Bridges of the American Society for Testing Materials, Serial Designation A7-36, and all silicon steel to the requirements of the Standard Specifications for Structural Silicon Steel of the same Society, Serial Designation A94-36, supplemented by the following paragraphs:

Test specimens of structural steel shall show a fracture having a silky or fine granular structure throughout with a bluish gray or dove color, and shall be entirely free from granular, black and brilliant specks.

Finished rolled material shall be free from cracks, flaws, injurious seams, laps, blisters, ragged and imperfect edges, and other defects. It

* Adapted from the specifications for the Rainbow Bridge, Niagara Falls; Waddell and Hardesty, Consulting Engineers.

shall have a smooth, uniform finish, and shall be straightened in the mill before shipment.

Material shall be free from loose mill scale, rust, pits, or other defects affecting its strength and durability.

Physical tests shall be required for billets bearing against, or functioning as, expansion rollers or rockers.

Rivet steel shall conform to the requirements of the Standard Specifications for Structural Rivet Steel of the American Society for Testing Materials, Serial Designation A141-38.

6.02. *Steel Castings.* Steel castings shall conform to the requirements of the Tentative Specifications for Carbon Steel Castings for Miscellaneous Industrial Uses of the American Society for Testing Materials, Serial Designation A27-36T, with the following additions:

Unless otherwise specified, all steel castings shall be Grade B-1, full-annealed.

Test specimens shall show a fracture having a silky or fine granular structure throughout.

Blow holes appearing upon finished castings shall be so located that a straight line laid in any direction will not cut a total length of cavity greater than one inch in any one foot, nor shall any single blow hole exceed one inch in any dimension or have an area greater than one-half square inch. Blow holes shall not have a depth injuriously affecting the strength of the casting.

Large castings, if required by the Engineers, shall be suspended and hammered all over. No cracks, flaws or other defects shall appear after such treatment.

No sharp unfilleted angles or corners will be allowed.

One tension test and one bend test shall be made from each melt in each heat treatment charge and from each casting weighing 500 pounds or over.

6.03. *Iron Castings.* Iron castings shall conform to the requirements of the Standard Specifications for Gray Iron Castings of the American Society for Testing Materials, Serial Designation A48-36, Grade 30, with the following additions:

Iron castings shall be true to pattern in form and dimensions, and free from pouring faults, sponginess, cracks, blow holes and other defects in positions affecting their strength and value for the service intended.

Castings shall be boldly filleted at angles and the arrises shall be sharp and perfect.

Castings having blow holes plugged or filled with putty or cement of any kind will be rejected.

6.04. *Steel Forgings.* All carbon and alloy steel forgings from which pins, rollers, and other forged parts are to be made shall conform to the requirements of the Standard Specifications for Carbon-Steel and Alloy-Steel Forgings of the American Society for Testing Materials, Serial Designation A18-30, with the following additions:

Structural forgings shall be Class C carbon steel, unless otherwise called for. All forgings shall be thoroughly annealed.

The yield point of Class C forgings shall be not less than 33,000 pounds per square inch.

The tensile requirements for forgings from 20 to 30 inches in diameter shall conform to the requirements for forgings 12 to 20 inches in diameter.

A Class C carbon steel specimen ½ inch by ½ inch in section shall bend cold 180 degrees around a diameter of ½ inch, without cracking on the outside of the bend. The bending may be effected by pressure or by blows.

6.05. *Wrought Iron.* Wrought-iron plates shall meet the requirements of the Tentative Specifications for Wrought-Iron Plates of the American Society for Testing Materials, Serial Designation A42-37T.

Wrought-iron shapes and bars shall meet the requirements of the Tentative Specifications for Rolled Wrought-Iron Shapes and Bars of the American Society for Testing Materials, Serial Designation A207-38T.

Terms relating to wrought iron are to be as defined in the Standard Definitions of Terms relating to Wrought-Iron Specifications of the American Society for Testing Materials, Serial Designation A81-33.

6.06. *Wrought-Iron Pipe.* Wrought-iron pipe shall conform to the requirements of the Standard Specifications for Welded Wrought-Iron Pipe of the American Society for Testing Materials, Serial Designation A72-38. All wrought-iron pipe shall be marked in raised letters with trade name or manufacturer's name.

6.07. *Steel Pipe.* All steel pipe shall be copper-bearing welded steel pipe conforming to the requirements of the Standard Specifications for Welded and Seamless Steel Pipe of the American Society for Testing Materials, Serial Designation A53-36.

6.08. *Malleable Castings.* Malleable castings shall conform to the requirements of the Standard Specifications for Malleable Castings of the American Society for Testing Materials, Serial Designation A47-33.

Malleable castings shall be true to pattern in form and dimensions, free from pouring faults, sponginess, cracks, blow holes and other defects in positions affecting their strength and value for the service intended.

The castings shall be boldly filleted at angles, and the arrises shall be sharp and perfect. The surface shall have a workmanlike finish.

Particular care shall be exercised so as not to embrittle malleable castings when galvanizing. All galvanized malleable castings shall meet specification requirements after galvanizing.

6.09. *Rolled Phosphor-Bronze.* Rolled Phosphor-Bronze shall conform to the requirements of the Tentative Specifications for Rolled Copper-Alloy Bearing and Expansion Plates for Bridge and Other Structural Uses of the American Society for Testing Materials, Serial Designation B100-38T.

FABRICATION

6.10. *Storage of Materials.* Structural material shall be stored above the surface of the ground upon platforms, skids, or other supports, and shall be protected as far as practicable from surface deterioration by exposure to conditions producing rust. It shall be kept free from accumulations of dirt, oil or other foreign matter.

6.11. *Straightening Material.* All deformed structural material shall be properly straightened by methods which are non-injurious prior to being laid off, punched or otherwise worked in the shop. Sharp kinks and bends shall be cause for rejection.

6.12. *Workmanship and Finish.* The workmanship and finish shall be first class and equal to the best practice in modern bridge shops. Shearing and chipping shall be neatly and accurately done, and all portions of the work exposed to view shall be neatly finished.

6.13. *Rivet Holes.* When reaming is not required, holes in material ¾ inch or less in thickness may be punched full size. Holes in material more than ¾ inch in thickness shall be sub-punched and reamed, or drilled from the solid.

Holes in carbon steel more than ⅞ inch thick, or in silicon steel more than ¾ inch thick, shall be drilled.

6.14. *Punched Holes.* Full size punched holes shall be $\frac{1}{16}$ inch larger than the nominal diameter of the rivet. The diameter of the die shall not exceed the diameter of the punch by more than $\frac{3}{32}$ inch. Holes must be clean cut, without torn or ragged edges. If any holes must be enlarged to admit the rivets, they shall be reamed.

6.15. *Accuracy of Punched Holes.* The punching of holes shall be so accurately done that, after assembling the component parts of a member, a cylindrical pin ⅛ inch smaller than the nominal diameter of the punched hole may be passed through at least 75 of any group of 100 contiguous holes in the same surface or in like proportion for any group of holes. If this requirement is not fulfilled, the badly punched pieces shall be rejected. If any holes will not pass a pin $\frac{3}{16}$ inch smaller than the nominal diameter of the punched hole, this shall be cause for rejection.

6.16. *Drilled Holes.* Drilled holes shall be $\frac{1}{16}$ inch larger than the nominal diameter of the rivet. Burrs on the outside surfaces shall be removed with a tool producing a $\frac{1}{16}$-inch fillet around the edge of the hole.

6.17. *Sub-Punched and Reamed Holes.* Sub-punched and reamed holes for rivets having diameters greater than ¾ inch shall be punched $\frac{3}{16}$ inch smaller than the nominal diameter of the rivet, and for rivets having diameters ¾ inch or less the holes shall be punched $\frac{1}{16}$ inch less than the nominal diameter of the rivet. The punch and die shall have the same relative sizes as specified for full-size punched holes. After punching, the holes shall be reamed to a diameter $\frac{1}{16}$ inch larger than the nominal diameter of the rivet. Burrs resulting from reaming shall be removed with a tool producing a $\frac{1}{16}$-inch fillet around the edge of the hole.

Reaming of rivet holes shall be done with twist drills or with short taper reamers. Reamers shall not be directed by hand. No oil or grease shall be used as a lubricant.

6.18. *Accuracy of Reamed and Drilled Holes.* Reamed or drilled holes shall be cylindrical and perpendicular to the member, and their accuracy shall be the same as that specified for punched holes except that, after reaming or drilling, 85 of any group of 100 contiguous holes in the same surface, or in like proportion for any group of holes, shall not show an offset greater than $\frac{1}{32}$ inch between adjacent thickness of metal.

6.19. *Drifting of Holes.* The drifting done during assembling shall be only such as to bring the parts into position, and shall not be sufficient to enlarge the holes or distort the metal. Holes that must be enlarged to admit the rivets shall be reamed.

6.20. *General Reaming.* General reaming shall be required for all work under this Contract.

All rivet holes in main members shall be sub-punched and reamed, or drilled from the solid. This requirement shall not apply to holes for shop rivets in lateral members, sway bracing, and secondary floor members, and to the lateral plates and connection angles connecting these members to the main members of the structure. Connection plates or other parts acting both as main member material and secondary (lateral, sway bracing, etc.) member material shall generally have sub-punched and reamed holes in locations engaging similar holes in main members.

Reaming shall be done after the pieces forming a built member are assembled and firmly bolted together. No interchange of reamed parts will be permitted.

6.21. *Reaming of Field Connections.* Holes for all field connections shall be sub-punched and reamed or drilled with the connecting parts assembled, or else reamed or drilled to a metal templet not less than 1 inch thick.

Splices in the arch ribs shall be reamed assembled wih their splice plates in place and the members lined up perfectly, the surfaces milled for bearing being drawn into tight contact.

6.22. *Shop Assembling.* All surfaces of metal to be in contact when assembled shall be carefully cleaned to remove all dirt, loose mill scale or other foreign matter.

The component parts of a built member shall be assembled, drift-pinned to prevent lateral movement, and firmly bolted to draw the parts into close contact before reaming, drilling or riveting is begun. Assembled parts shall be taken apart, if necessary, for the removal of burrs and shavings produced by the reaming operation.

The members shall be free from twists, bends or other deformations.

Preparatory to shop riveting full-size punched material, the rivet holes shall be cleared for the admission of the rivets by reaming.

End connection angles, stiffener angles, etc., shall be carefully adjusted to correct locations and rigidly bolted, clamped, or otherwise firmly held in place until riveted.

6.23. *Match-Marking.* Connecting parts assembled in the shop for the purpose of reaming or drilling holes in field connections shall be match-marked, and a diagram showing such marks shall be furnished to the Engineers.

6.24. *Rivets.* The diameter of the rivets indicated upon the Plans shall be understood to mean their diameter before heating.

Heads of driven rivets shall be of approved shape, concentric with the shanks, true to size, full, neatly formed, free from fins and in full contact with the surface of the member.

6.25. *Field Rivets.* Field rivets, for each size and length, shall be supplied in excess of the actual number to be driven to provide for losses due to misuse, improper driving, or other contingencies. Rivets shall be free from furnace scale on their shanks and from fins on the under side of the machine formed heads.

6.26. *Bolts and Bolted Connections.* Bolted connections shall not be used unless specifically authorized. Where bolted connections are permitted the bolts furnished shall be unfinished bolts (ordinary rough or machine bolts) or turned bolts, as specified or directed by the Engineer.

Unfinished bolts shall be standard bolts with hexagonal heads and nuts. The use of "button head" bolts will not be permitted. Bolts transmitting shear shall be threaded to such a length that not more than one thread will be within the grip of the metal. The bolts shall be of lengths which will extend entirely through their nuts but not more than ¼ inch beyond them. The diameter of the bolt holes shall be ¹⁄₁₆ inch greater than the diameter of the bolts used.

Holes for turned bolts shall be carefully reamed or drilled and the bolts turned to a driving fit by being given a finishing cut. The threads shall be entirely outside the holes, and the heads and nuts shall be hexagonal. Approved nut-locks shall be used on all bolts unless permission to the contrary is secured from the Engineer. When nut-locks are not used, round washers having a thickness of ⅛ inch shall be placed under the nuts.

6.27. *Riveting.* Rivets shall be heated uniformly to a light cherry red color and shall be driven while hot. The heating of the points of rivets more than the remainder will not be permitted. When ready for driving they shall be free from slag, scale, and other adhering matter, and when driven they shall completely fill the holes. Burned, burred, or otherwise defective rivets, or rivets which throw off sparks when taken from the furnace or forge, shall not be driven.

Loose, burned, badly formed or otherwise defective rivets shall be cut out. Caulking and recupping of rivet heads will not be allowed. In cutting out defective rivets care shall be taken not to injure the adjacent metal and, if necessary, the rivet shanks shall be removed by drilling.

Countersinking shall be neatly done and countersunk rivets shall completely fill the holes.

Shop rivets shall be driven by direct-acting riveters where practicable. The riveting machine shall retain the pressure for a short time after the upsetting is complete.

Pneumatic hammers shall be used for field riveting except when the use of other hand tools for riveting is permitted by the Engineer. A pneumatic bucker shall be used where practicable.

Special precautions shall be taken in the driving of rivets of extra-long grip, in order to secure the best possible results.

6.28. *Sheared and Flame-Cut Edges.* Sheared edges of carbon-steel material more than ⅝ inch in thickness, and of all alloy-steel material, shall be planed to a depth of ¼ inch. Re-entrant cuts shall be filleted before cutting.

Carbon steel or alloy steel may be flame-cut, provided a smooth surface is secured by the use of a mechanical guide. Flame cutting by hand shall be done only where approved by the Engineer, and the surface shall be made smooth by planing, chipping and grinding. The cutting flame shall be so adjusted and manipulated as to avoid overheating the metal or cutting beyond the prescribed lines. All flame-cut edges of alloy steel shall be planed, chipped, or ground to a depth of ¼ inch. Re-entrant cuts shall be filleted.

Fillets at re-entrant cuts shall have a radius of at least 2 inches, unless a different radius is specifically called for on the drawings.

6.29. *Planing of Bearing Surfaces.* Ends of columns taking bearing upon base and cap plates shall be milled to true surfaces and correct bevels after the main section of these members and the end connection angles have been fully riveted.

Caps and base plates of columns and the sole plates of girders and trusses shall have full contact when assembled. The plates, if warped or deformed, shall be hot-straightened, planed or otherwise treated to secure an accurate, uniform contact. After being riveted in place, the excess metal of countersunk rivet heads shall be chipped smooth and flush with the surrounding metal and the surfaces which are to come in contact with other metal surfaces shall be planed or milled, if necessary, to secure proper contact. Correspondingly, the surfaces of base and sole plates which are to come in contact with masonry shall be rough finished, if not free from warps or other deformation.

Surfaces of cast pedestals and shoes which are to come in contact with metal surfaces shall be planed and those which are to take bearing upon the masonry shall be rough finished.

In planing the surfaces of expansion bearings the cut of the tool shall be in the direction of expansion.

6.30. *Abutting Joints.* Abutting ends of compression members shall, after the members have been riveted, be accurately faced to secure an even bearing when assembled in the structure.

Ends of tension members at splices shall be rough finished to secure close and neat, but not contact-fitting joints.

6.31. *End Connection Angles.* End connection angles shall be flush with each other and accurately set as to position and length of member. All floorbeams, and all other built-up girders having the same type of end connections, shall be milled on the ends to correct length after the end connection angles have been riveted on. The end connection angles of such floorbeams and girders shall, unless otherwise ordered, be set and milled on a slight skew, the angle of skew being computed so that the finished ends will rotate to a true vertical when the girders take their full dead load deflections. To insure a perfect end bearing, metal shall be cut from the faces of the end connection angles and from the ends of all web plates, flange material, and filler plates in the milling operation, but the end connection angles shall be so accurately fitted that not more than $\frac{1}{16}$ inch will be taken off them at their roots. The abutting ends of cantilever beams shall be milled in the same manner. The end connection angles of each I-beam stringer shall be milled, or they shall be riveted in place while assembled with the whole stringer and clamped firmly in an iron frame which will give exactly the correct length of stringer and the correct position of the angles.

6.32. *Built Members.* The several pieces forming one built member shall be straight and close fitting. Such members shall be true to detailed dimensions and free from twists, bends, open joints or other defects resulting from faulty fabrication and workmanship.

6.33. *Lacing Bars.* The ends of lacing bars shall be neatly rounded unless otherwise indicated.

6.34. *Plate Girders.* Web plates of girders having no cover plates shall be detailed with the top edge of the web flush with the backs of the flange angles. Any portion of the plate projecting beyond the angles shall be chipped flush with the backs of the angles. Web plates of girders having cover plates may be $\frac{1}{2}$ inch less in width than the distance back to back of flange angles.

When webs are spliced, not more than $\frac{3}{8}$-inch clearance between ends of plates will be allowed.

End stiffener angles of girders and stiffener angles intended as supports for concentrated loads shall be milled or ground to secure a uniform, even bearing against the flange angles. Intermediate stiffener angles shall fit sufficiently tight to exclude water after being painted.

Web splice plates and fillers under stiffeners shall fit within $\frac{1}{8}$ inch at each end.

6.35. *Pins and Rollers.* Pins and rollers shall be accurately turned to detailed dimensions and shall be smooth, straight and free from flaws. The final surface shall be produced by a finishing cut.

Pins having a diameter of 7 inches or less shall be forged or rolled carbon steel. Pins having a diameter greater than 7 inches shall be forged and annealed.

Pins larger than 9 inches in diameter shall have a hole not less than 2 inches in diameter bored longitudinally through their centers. Pins showing defective interior conditions will be rejected.

6.36. *Boring Pin Holes.* Pin holes shall be bored true to detailed dimensions, smooth and straight, at right angles with the axis of the member and parallel with each other unless otherwise required. A finishing cut shall always be made.

The length outside to outside of holes in tension members and inside to inside of holes in compression members shall not vary from detailed dimensions more than $\frac{1}{32}$ inch. Boring of holes in built-up members shall be done after the riveting is completed.

6.37. *Pin Clearances.* The diameter of the pin hole shall not exceed that of the pin by more than $\frac{1}{50}$ inch for pins 5 inches or less in diameter, or $\frac{1}{32}$ inch for larger pins.

6.38. *Welding.* Except where shown specifically on the Plans, welding of steel shall be permitted only to remedy minor defects, and then only with the approval of the Engineer.

All fusion welding shall be performed by the electric-arc process, and shall conform in every respect to the requirements of the 1936 Specifications for Design, Construction, Alteration, and Repair of Highway and Railway Bridges of the American Welding Society, and the qualification of welding processes and welding operators shall be in accordance with the requirements of Appendix E of the same specifications. Covered electrodes shall be used.

Steelwork shall not be painted on any areas where shop or field welding is to be done, except that a thin coat of linseed oil without pigment may be used as a temporary protection. After welding, the unpainted and welded areas shall be spot-painted with the shop paint used on the remainder of the steelwork, prior to the application of the field paint.

6.39. *Screw Threads.* Screw threads shall make close fits in the nuts, and shall be U.S. Standard, except that for diameters greater than $1\frac{1}{2}$ inches, they shall be made with 6 threads to the inch.

6.40. *Pilot and Driving Nuts.* Two pilot nuts and two driving nuts shall be furnished for each size of pin, unless otherwise specified.

6.41. *Shoes and Base Plates.* Structural billets shall be finished top and bottom to the thicknesses and curved surfaces indicated by the drawings; all holes for bolts and anchor bolts shall be drilled; and all bolt holes shall be spot-faced unless the faces are finished.

6.42. *Arch Ribs.* The arch ribs are to be fixed at the skewbacks for all conditions of loading, and are to be without intermediate hinges. They shall be so cambered and erected that, under full dead load at a normal temperature of 50 degrees F., the axis of the rib shall conform to the outline shown on the Plans, the effect of rib shortening under full dead load being neutralized by the closure operations.

The ends of the rib sections shall be accurately faced to provide full bearing at the splices, the angle of facing being such as to secure radial splices under full dead load. The Contractor shall provide such methods of control as to insure that each rib section shall be of accurate length, with the ends milled to the exact angles required. The bases of the skewback and grillage sections shall be plane and parallel; the billets, the rib, the grillage beams, and the stiffener angles shall be accurately faced to provide full bearing; and the lengths along the rib axis shall be exact.

The length of each rib section and the angles of the milled ends shall be checked carefully after milling, and any errors corrected in the milling of adjacent sections, in order to insure that the total length and shape of each rib shall be accurate. The rib sections shall then be shop-assembled in lengths of about 5 panels with the milled joints in tight contact, sections being added to one end of the assembly as sections at the other end are removed. The positions of the various splice points shall be checked at each stage of the assembly, no errors exceeding $\frac{1}{32}$ inch being permitted. The method of assembly shall be such as will insure full bearing of the milled ends before reaming or drilling the rivet holes for the field splices, and will also permit inspection of the faced ends. If the joints are not in full bearing at each point of the assembly, the errors in milling shall be corrected. Since the assemblies must be made to cambered outlines, proper corrections shall be made from the dead-load outline of rib.

Flange angles and cover plates shall be bent to the rib outlines. The spaces between webs and cover plates at each end of each fabricated length shall be closed by caulking with oakum and lead wool after milling, all water and foreign materials being blown out with compressed air before closing the openings.

6.43. *Railings.* The sidewalk railing shall have posts of two 6-inch channels, top rails of 5-inch extra-strong steel pipe, bottom rails of 4-inch channels, and spindles of 1-inch square steel bars. The pipes and channels shall be true and straight, and each piece shall be cut to exact length; the spindles shall be straight and without twists or wind.

Each clear space between adjacent spindles, or between a spindle and a post, shall be made as nearly 5 inches as possible but shall not exceed this amount, the spacing being uniform for each panel of railing. All spindles and posts shall be set truly vertical, each panel of railing being fabricated on a skew conforming to the rate of grade at the location in which the panel is to be placed. The ends of the spindles shall have approximately round tenons which are to be set into holes drilled in the rails. The rails shall be drilled to receive the tenons of the spindles, the drilling being done with jigs which will insure an accurate spacing. The railing panels shall be assembled to templets, and they shall be firmly clamped preparatory to welding the joints. The faces of the spindles shall be set exactly on plane surfaces parallel to the center plane of the railing panel. Each end of each spindle shall be secured to the rail by a neat bead of electric-arc welding extending entirely around the spindle.

The roadway railings shall have posts of single 7-inch ship channels, top rails of 8-inch channels, bottom rails of 4-inch angles, and panels of ⅜-inch plates. Channels and angles shall be true and straight, and plates shall be free from waves, twists, or buckles. Posts shall be vertical, and panels shall be framed on a skew conforming to the grade at the point where the panel is to be located. All parts shall be of correct lengths. Connections shall be by rivets or bolts, except that the plates shall be fastened to the top rails by welding. The upper ends of posts shall be coped and chamfered as indicated.

All materials for railings shall be copper-bearing structural steel or copper-bearing steel pipe.

Adjacent railing panels shall be securely attached to the intermediate posts. The panels shall be removable without removing any post.

The railing shall be adjusted accurately to alignment and grade. The adjustment shall be made by means of field reaming or drilling the holes for the connecting rivets or bolts, these holes being either sub-punched or left blank in the shop.

6.44. *Floor Joints.* The steel for the floor joints at the ends of the arch span shall be fabricated by electric-arc welding and riveting. The two parts for each joint shall be assembled together in the shop to insure a perfect fit, with temporary fillers in the ¼-inch clearance spaces between adjacent fingers, and the fingers shall be attached to the angles and plates by spots of welding while they are so assembled. Finally the fillet of welding for each finger shall be extended entirely across the area of contact between the finger and the angle or plate. Header pieces shall be added at the rear ends of the fingers, and securely welded in place. Steel bars for anchoring the joint material to the concrete work shall be riveted to the joint material. The pieces shall be match-marked, and they shall be erected in accordance with the match-marks. The various pieces shall be adjusted to correct position and with uniform clearance spaces between the fingers before drilling or reaming the rivet and bolt holes, in the field, for the connections to the stringers, which holes shall be sub-punched or left blank in the shop. Shims shall be provided as required. The removable sections of the joints shall be fastened down by bolts with springs between the nuts and the steelwork. The joints shall fit together perfectly and they shall operate freely.

The floor joints in the roadways located about 100 ft. from the center of the span shall be formed by steel castings resting on shelf angles supported on the stringers and floorbeams. The castings shall provide an anti-skid pattern of a type approved by the Engineer. The holes for connecting rivets shall either be sub-punched in the shop and reamed to size in the field after erection and adjustment, or left blank in the shop and drilled in the field. The joints shall be lined up perfectly, so that the floor castings will have full and even bearing, and be quiet under traffic. Any finishing, adjustment, or shimming required to secure this result shall be done by the Contractor. The cast plates shall be fastened down by bolts with springs between the nuts and the steelwork.

The floor joints in the concrete approaches shall consist of plates and angles with copper gutters as indicated. They shall be set accurately to position, and shall be firmly anchored to the concrete.

Expansion joints in curbs shall be adjusted and lined up accurately.

All structural steel in floor joints, together with their supporting details, shall be copper bearing.

The completed floor joints shall present a smooth even surface for traffic.

6.45. *Scuppers.* Scuppers and inlets in roadways shall be of cast steel, and those in sidewalks of cast iron. Drain pipe shall be wrought-iron pipe.

6.46. *Lamp Post Sockets and Bases.* Lamp post sockets shall be 7-inch double extra strong copper-bearing steel pipe. Sockets on steel spans shall have cast steel bases.

MILL AND SHOP INSPECTION

6.47. *Inspection of Steel.* All material shall be subject to mill and shop inspection, except that, at the discretion of the Engineer, stock steel may be accepted for miscellaneous parts not subject to stress.

6.48. *Stock Steel.* No mill inspection will be required for stock steel. Stock steel shall be subject to surface inspection and cold bending tests only. Test pieces cut from stock materials shall endure bending cold, one hundred and eighty degrees, around a circle whose diameter is equal to the thickness of the test piece, without signs of cracking. One bending test shall be made upon at least one piece taken at random from every ten pieces of any particular size of plate, angle or other shape in stock; this requirement shall not apply to material ordered from the mill, which will be subject to the regular tests elsewhere specified. Full-sized rivets shall endure bending flat upon themselves without signs of cracking. The Contractor will be required to furnish the Engineer with two certified copies of the records of chemical analysis and physical tests of the stock steel to be used in the work.

6.49. *Notice of Rolling and Fabrication.* The Contractor shall furnish the Inspector with two copies of the mill order and give ample notice to the Inspector of the beginning of the work at the mill and shop, so that inspection may be provided. No material shall be rolled or fabricated before the Inspector has been notified where the orders have been placed.

6.50. *Cost of Testing.* Unless otherwise provided, the Contractor shall furnish, without charge, test specimens as specified herein, and all labor, testing machines and tools necessary to prepare the specimens and to make the tests.

6.51. *Rejection.* The acceptance of any material or finished members by the Inspector shall not be a bar to their subsequent rejection, if found defective. Any material accepted at the mills which, under the punches, shears, etc., shows hard spots, brittleness, laminations, piping, cracks, lack of uniformity in quality or other defects, shall be rejected and replaced by satisfactory material solely at the expense of the Contractor. Rejected material and workmanship shall be replaced promptly or made good by the Contractor.

6.52. *Marking and Shipping.* Members weighing more than 3 tons shall have the weight marked thereon. Bolts and rivets of one length and diameter, and loose nuts or washers of each size, shall be packed separately. Pins, small parts, and small packages of bolts, rivets, washers and nuts shall be shipped in boxes, crates, kegs or barrels, but the gross weight of any package shall not exceed 300 pounds. A list and description of the contained material shall be plainly marked on the outside of each shipping container.

The weight of all tools and erection materials shall be kept separate.

Anchor-bolts, washers, and other anchorage materials shall be shipped to suit the requirements of the masonry construction.

The loading, transportation, unloading and piling of structural material shall be so conducted that the metal will be kept clean and free from injury by rough handling.

ERECTION

6.53. *Field Inspection.* All work of erection shall be subject to the inspection of the Engineer, who shall be given all facilities required for a thorough inspection of workmanship.

Material and workmanship not previously inspected will be inspected after its delivery to the site of the work.

6.54. *Handling and Storing Materials.* All steel members shall be shipped, trucked and handled in a manner that will cause no danger of permanent deflection or other injury, and in a manner to be approved by the Engineer. In unloading girders or other heavy members from cars or trucks and in moving the same preparatory to raising into place, skids or rollers shall be employed and no material shall at any time be dropped, thrown, or dragged over the ground. Any bends, sags or kinks in material will be sufficient cause for its rejection. Material to be stored shall be placed on skids above the ground. It shall be kept clean and properly drained. Girders and beams shall be placed upright and shored. Long members, such as columns and arch rib sections, shall be supported on skids placed near enough together to prevent injury from deflection.

6.55. *Falsework.* Falsework shall be properly designed and substantially constructed and maintained for the loads which will come upon it. The Contractor shall keep all falsework in a safe condition, and shall provide such temporary stairways, gangways, staging, rope, railings, safety nets, etc., as the Engineer may direct, for the safety of the workmen and to facilitate thorough inspection and checking of the work during construction; but nothing herein contained shall relieve the Contractor from any responsibility for injury to persons or property. All timber used in connection with falsework shall be treated with a fire-retarding process approved by the Engineer. The Contractor shall avoid the use of timber falsework wherever possible.

6.56. *Methods and Equipment.* Before starting work, the Contractor shall inform the Engineer fully as to the method of erection he proposes to follow, and the amount and character of equipment he proposes to use, which shall be subject to the approval of the Engineer. The approval of the Engineer shall not be considered as relieving the Contractor of the responsibility for the safety of his method or equipment or from carrying out the work in full accordance with the Plans and Specifications. No work shall be done without the sanction of the Engineer.

6.57. *Protection of Masonry.* Before beginning erection, the Contractor shall effectively protect, by methods satisfactory to the Engineer, all masonry against damage or discoloration. The exposed surfaces of any timber near the masonry shall be treated with an approved fire retardant process. The Contractor shall be responsible for any damage to the masonry, and shall restore it to its original condition if injured or stained in any manner before final acceptance of the bridge.

6.58. *Preparation of Bearing Areas.* Bearing areas for arch skewbacks, column bases, and shoes are to be dressed to exact elevation by the substructure contractors. The Contractor is to make exact measurements across the river between arch skewbacks, which measurements are to be for the use of the substructure contractors in setting anchor bolts and finishing the arch skewbacks.

Bases and shoes shall have a full and uniform bearing upon the substructure masonry, and shall be rigidly and permanently located to correct alignment and elevation. They shall not be placed upon bearing areas which are improperly finished, deformed, or irregular.

6.59. *Setting Anchor Bolts.* Anchor bolts for arch ribs and for the columns of the steel approach spans shall be furnished f.o.b. the bridge site on the two sides of the river by the Contractor, and received, unloaded, stored, and set by the substructure contractors. When the Contractor is ready to set the bearings, he shall clean out the pipe sleeves; and after the bearing is lined up he shall fill the pipe sleeves with Portland cement mortar mixed in the proportion of one part cement to two parts fine aggregate. Other anchor bolts shall be furnished and set by the Contractor, in holes drilled by the Contractor, using Portland cement mortar. The mortar shall consist of one part cement to two parts fine aggregate, mixed sufficiently wet to flow freely. The anchor bolts shall be set as follows:

Anchor bolts shall first be dropped into the dry holes to assure their proper fit after setting. They shall then be set as follows: Fill the hole about two-thirds full of mortar and by a uniform, even pressure or by light blows with a hammer (flogging and ramming will not be permitted) force the bolt down until the mortar rises to the top of the hole and the anchor bolt nut rests firmly against the metal shoe or pedestal. In case the mortar does not fill the hole, additional mortar shall be added. Remove all excess mortar which may have flushed out of the hole, to permit proper field painting of the metal surfaces.

The location of the anchor bolts in relation to the slotted holes in expansion shoes shall be varied with the prevailing temperature. The nuts on anchor bolts at the expansion ends of spans shall permit the free movement of the span.

6.60. *Setting Arch Skewbacks.* The arch skewbacks shall be set to exact position by the Contractor on the bearing areas which have been prepared by the substructure contractors. The Contractor shall be responsible for locating the skewbacks correctly, and shall make such exact measurements as may be required to secure this result. Any necessary correction of errors in the masonry bearing surfaces shall be paid for as Extra Work, unless such errors shall have resulted from inaccurate measurements made by the Contractor during the construction of the arch abutments, in which case the said correction shall be at the expense of the Contractor.

In case the Contractor desires to have any metal work for erection purposes encased in the substructure concrete, he shall arrange with the substructure contractors, at his own expense, for the placing of such metalwork, which metalwork will not be included in the weight paid for.

After the skewback grillages have been set to final position and before the skewback sections have been placed, the center cells of the grillage shall be filled with $1:2:3\frac{1}{2}$ cement concrete. After the skewback sections have been erected and adjusted, the remaining concrete of the arch abutments shall be placed. In placing the concrete, particular attention shall be paid to the compacting of the concrete within and around the grillages, in order to insure that all spaces are completely filled and the concrete is everywhere in close contact with the steelwork. All joints between concrete and steel surfaces shall be sealed as hereinafter provided under the paragraph entitled Filling Spaces and Sealing Joints.

6.61. *Erection of Arch Ribs.* It is expected that the Contractor will erect the arch ribs with their lateral bracing by cantilevering out from both shores with the aid of tie-backs anchored into the rock cliffs. The location and details of the tie-backs and anchorages shall be such as will, in the judgment of the Engineer, render them safe for the loads imposed on them, and will avoid any danger of displacement of the rock or other damage thereto; but approval of the said tie-backs and anchorages by the Engineer shall not relieve the Contractor of responsibility for their strength and sufficiency.

The Contractor shall erect the arch ribs to accurate outline by methods that will avoid stressing them beyond the limits hereinbefore specified. In advance of beginning such erection, he shall submit to the Engineer, for his check and approval, his detailed erection procedure, with computations of stresses and deflections at various stages. Excessive deflections or distortions of the ribs will not be permitted, and the two ribs shall be equally loaded at all times.

The center closure shall be made by a method that will insure that the axis of the rib will be of correct outline under full dead load at a normal temperature of 50 degrees F., and that the effect of rib shortening under full dead load will be entirely neutralized. In securing this result, the amount and position of the crown thrust shall be measured, by jacks or other approved means, after the erection tie-backs have been released and before final closure has been effected, and such adjustments made as will insure the correct condition of stress. The final closure shall then be made by a method that will not disturb this condition. The entire closure procedure shall be subject to the approval of the Engineer.

Riveting of the arch ribs shall follow closely the erection, and shall be done only when the milled joints are in full contact. All field splices in the ribs, with the exception of the closure point, shall be fully riveted before closure operations are begun.

6.62. *Handling Members.* The field assembling of the component parts of a structure shall involve the use of methods and appliances not likely to produce injury by twisting, bending or otherwise deforming the metal. No member slightly bent or twisted shall be put in place until its defects are corrected, and members seriously damaged in handling shall be rejected.

6.63. *Alignment.* Before beginning the field riveting the structure shall be adjusted to correct grade and alignment and the elevations of ends of floor beams properly regulated.

Unsatisfactory holes shall, at the discretion of the Engineer, be seamed to the next larger size after erection, and larger rivets used.

6.64. *Straightening Bent Material.* The straightening of bent edges of plates, angles and other shapes shall be done by methods not likely to produce fracture or other injury. The metal shall not be heated unless permitted by the Engineer, in which case the heating shall not be to a higher temperature than that producing a dark cherry red color. After heating, the metal shall be cooled as slowly as possible.

Following the completion of the straightening of a bend or buckle, the surface of the metal shall be carefully inspected for evidence of incipient or other fractures.

6.65. *Assembling and Riveting.* Splices and field connections shall have one-half of the holes filled with bolts and cylindrical erection pins (half bolts and half pins) before riveting. Splices and field connections in arch ribs shall have from 60 to 100 per cent of the holes so filled, as may be required by the erection stresses, the erection pins being tight-fitting.

Fitting up bolts shall be of the same nominal diameter as the rivets, and cylindrical erection pins shall be $\frac{1}{32}$ inch larger.

Railings shall not be riveted or bolted until the dead load is on the structure, and after carefully setting as to grade and alignment.

6.66. *Adjustment of Pin Nuts.* All nuts on pins shall be thoroughly tightened and the pins so located in the holes that the members shall take full and even bearing upon them.

6.67. *Filling Spaces and Sealing Joints.* All spaces in steelwork shall preferably be arranged to insure draining, but any spaces that will not drain shall be filled with $1 : 1\frac{3}{4} : 3\frac{1}{2}$ cement concrete. The edges of this concrete shall be finished with slight grooves against the faces of the steel, and these grooves shall be filled with a watertight seal of asphalt, to meet the requirement herein specified for Waterproofing on Concrete Surfaces, placed hot. Similar watertight seals shall be provided at all points where concrete abuts against steelwork. There will be no direct payment for the grooves and asphalt seals, but any concrete used for filling spaces shall be classified for payment as Concrete in Slabs and Curbs.

Specifications for Performance. In the performance-type specification the contractor guarantees to furnish an article which will serve a specified purpose and warrants its future performance in service. The satisfactory use of this type of specification depends upon the degree to which service requirements can be defined, measured, and proved by practical short-time tests. If this can be accomplished performance specifications are very suitable. The contractor is permitted maximum latitude in the design, thereby providing wide competition among manufacturers and great flexibility as to materials and workmanship. Indeed if performance is strictly adhered to in the specifications, only limited control may be maintained over the details of materials and workmanship. Performance specifications are particularly suitable for equipment, machinery, and some types of construction materials.

Mechanical and electrical equipment will be required in connection with most construction projects. When these items are of the nature of accessories and appurtenances, they are usually furnished by the contractor under the construction contract although they are sometimes obtained by the owner under separate contracts. The latter arrangement is usually made for large machines and equipment, such as heavy cranes, turbines, and engines. When machinery is procured under a special contract, both the construction specifications and the machine specifications should be explicit as to the responsibility for its erection and installation and the facilities required therefor, such as foundations, anchor bolts, and services. Also, provision should be made for the coordination of the machinery installation with the rest of the work.

With regard to the mechanical and electrical equipment included under the construction contract, the contents of the specifications will depend on the extent to which details of their design are furnished on the drawings. If the equipment is specially designed and all details of its construction are shown on the drawings, complete detailed specifications covering materials and workmanship are required, as for any other element of the project in as much as the owner is responsible for the design and therefore for performance as well. Frequently, however, standard products of reputable manufacturers are adequate, or the manufacturer is required to furnish the detailed design of special equipment. In this case a somewhat different point of view is indicated in the preparation of the specifications. The usual practice is to specify type, power, capacity, efficiency, and the like, allowing the contractor latitude in the actual selection of the item and details of the design. The details of warranties as to sustained performance and durability, spare parts and tools to be furnished, together with installation instructions, should also be included. When performance is specified or the contractor is required to state and guarantee performance, the specifications should state explicitly how this performance is to be determined. Details of measurements and tests should be given. If the machine is unusual or complicated, consideration should be given to requiring the installation and tests to be performed under the supervision of the manufacturer. The specifications should be worded in such a way as to eliminate inferior equipment and the products of disapproved manufacturers.

Quality in the finished product must be guaranteed by warranty although some control in the design and manufacture is sometimes permissible by specifying the type and quality of the materials to be used in the various parts or elements of the article. It should be noted in this connection, however, that when standard products of recognized manufacturers are contemplated under performance specifications, any special quality requirements written into the specifications may be in conflict with manufacturing standards. Variation from manufacturing standards, of course, results in increased prices due to the special handling required.

With regard to the arrangement of performance-type specifications it will be advisable to follow an outline similar to the following.

    a. General description.
    b. Design and installation.
    c. Operating conditions.
    d. Performance requirements.
    e. Special requirements as to quality.
    f. Inspection and tests.
    g. Warranty.

EXAMPLE OF SPECIFICATIONS FOR PERFORMANCE

SECTION 9.   CENTRIFUGAL PUMPS

9.01. *General Requirements.* Four pumps of the centrifugal mixed-flow or modified-propeller type having 54-inch diameter discharge outlets shall be furnished for the pump-well installation. Suction to each pump shall be through a single opening in the bottom of the pump casing which shall be flanged or otherwise arranged to receive an intake bell. Each pump shall be direct-connected to a vertical-shaft induction motor specified elsewhere herein. Pumps shall be of one manufacture and like parts shall be interchangeable. Each pumping unit shall be complete with all accessory equipment specified, indicated, or necessary for efficient operation. Anchor bolts and nuts for securing pumps to their bases shall be machine-finished cadmium-plated steel. The pumps will be inspected during construction, and shall be tested before and after installation as specified hereinafter. The performance data furnished with the bids shall be guaranteed.

9.02. *Services of a Supervising Erector,* who is well qualified by long training and experience in the installation of pumps of the type specified, shall be furnished by the pump manufacturer to supervise the erection of the pumps and place them in operation.

9.03. *Design and Construction.* Pumps shall be designed in accordance with the most approved practice, shall be efficient, quiet and reliable in operation, shall have maximum accessibility for examination and repair, and shall be constructed of materials which will afford maximum resistance to the corrosive action of salt water and a minimum of wear of all operating parts. Each pump assembly shall be free from critical speeds and harmful vibrations under all conditions of service. All passages through the pumps shall be of liberal area and contours shall be designed to reduce friction and eddy currents to a minimum. The design characteristics of the pumps shall be such as to insure satisfactory operation under parallel operation. All working parts, bonnet edges, and nut bearing surfaces and the faces of other supporting surfaces shall be machine finished. Each pumping unit shall be provided with base flanges, lugs or feet, which preferably shall be cast integral with the pump casting, for attachment to the pump-well structure, with the bearing surfaces drilled for holding down bolts. Pump outlets shall be located to conform with the general piping systems indicated and shall be arranged to provide for convenient disconnection. All flanged connections of pipe outlets on the pumping units shall be faced and drilled in accordance with the 125-pound American Standards. Flanged outlets on the pumps shall be not less than 2.25 inches thick.

9.04. *Pump Operating Conditions.* The pumps will take their suction from a common suction chamber at the bottom of the pump well, to which water will be conveyed through culverts, approximately 8 feet wide and 8 feet high opposite each pump. The arrangement and design of the pump well, as indicated on the drawings, is diagrammatic and only the general arrangement of the pumps and suction bells in reference to the suction chamber is intended.

9.05. *Capacity and Efficiency.* Each pump shall have an average capacity of not less than 96,000 g.p.m. of sea water when pumping against a total dynamic head varying from 20 to 54 feet. The total dynamic head is that defined for centrifugal pumps by the latest standards of the Hydraulic Institute. The rated capacity of each pump motor shall not be exceeded under any load condition, including shut-off, that can occur in normal use of the pumps, and the speed of each pump shall not exceed 300 revolutions per minute. The pumps shall have characteristics such that the shut-off head shall be not less than 64 feet and the average overall efficiency, over the range of head between 20 and 54 feet, shall be as high as practicable. The average overall efficiency shall be defined as the ratio of the total work done to the power in-put to the motor, noted at increments of 5,000 gallons per minute, on the characteristic curve (between the two load-points mentioned) and averaged. The full load speeds of the pumps shall be less than ½ of their first critical speed.

9.06. *Pump Casings* shall be of the volute type without guide vanes or diffusers between the impeller and casing. Casings shall be divided in two or more parts in a manner which will best facilitate ready removal of the impellers and shafts without the necessity of breaking pipe joints or removing other important parts of the machinery. The inside surfaces of casings shall be smooth and the contours shall be designed to minimize friction and eddy currents. The parts of each casing shall be fitted with flanges which shall be accurately faced and drilled, shall be aligned with dowel pins or snug-fitted bolts, and shall be bolted together securely. Pump casings shall be close-grained high-test cast iron conforming in general to the Standard Specifications for Gray Iron Castings of the American Society for Testing Materials, Serial Designation A48-36, Grade 30, and having a tensile strength of not less than 35,000 pounds per square inch. Casings shall be fitted with renewable wearing rings of bronze having a zinc content not in excess of 5 per cent, and with suitable outlets at top for connection to a priming system. Bolts and nuts for casing joints shall be finished machine bolts and nuts conforming to the Standard Specifications for Heat-Treated Carbon Steel Bolting Material of the American Society for Testing Materials, Serial Designation A261-44T, except that bolts and nuts shall be cadmium plated after fabrication. Suitable inspection doors shall be provided in the casing covers to permit examination of the impellers without the necessity of opening the casing. Lifting eyes or other acceptable means shall be provided for handling the pump casings.

9.07. *Impellers* shall be cast in one piece and finished smooth; shall be designed to provide ample clearance and smooth passages between vanes so as to insure unobstructed flow of grit, solids and debris through the pumps; and shall be close-grained iron or semi-steel castings having a tensile strength of not less than 40,000 pounds per square inch (A.S.T.M. Spec. A48-36 Class 35). Impellers shall be fitted with renewable wearing rings of bronze as specified for the pump casings. Impellers shall be keyed on the shafts and secured thereto by lock-nuts or other approved means. The threading of the nut shall oppose the direction of impeller rotation. The lock-nut design shall be such as will eliminate the possibility of collecting debris. Each impeller, together with the rotating parts connected to it, shall be designed to insure that no objectionable vibration will develop at the operating speeds under any condition of service.

9.08. *Pump Shafts* shall be turned, ground, and polished to accurate dimensions at bearings, and shall be of ample size to resist bending, torsional, or other stresses to which the shafts may be subjected. The portion of the pump shafting outside the pumps shall be forged or rolled steel (A.S.T.M. Spec. A235-42 class G). Shafts inside pump casings shall be of nickel-copper-alloy, or rolled or forged steel with renewable, centrifugally-cast nickel-copper-alloy sleeves designed to exclude salt water from contact with the steel. The renewable jackets shall be forced on the shafts, each jacket being secured by key and lock-nut. Each pump shaft at the impeller shall be tapered, keyed and locked in an approved secure manner. Joints in shafts shall be connected by forged steel shaft couplings.

9.09. *Stuffing Boxes and Sleeve Type Bearings.* Stuffing boxes for pumps shall be of adequate design to permit the satisfactory packing of the glands with an approved flexible metallic packing, and to reduce leakage to a minimum under all conditions of operation. The design of stuffing boxes, which are subjected to pressures above atmosphere, shall be such as will trap the leakage therefrom by the use of lantern rings or other approved means and return leakage by drain piping to the suction side of the pump. The glands shall be the same material as specified for the internal shafts. Ample drip pockets or drains shall be provided under the glands in all cases to permit proper drainage and to insure that no leakage from the glands will reach the bearings. The glands shall be split to allow their removal and to provide convenient access to the packing. Sleeve-type bearings shall be sufficient in number to maintain properly the shaft alignments and prevent whip, and shall have renewable linings of approved anti-friction bearing metal or alloy. Bearings shall be of proper proportions and sufficient size to insure cool operation. Bearing brackets and caps shall be of the same material as specified for the pump casings.

9.10. *Thrust Bearings.* Each pump unit shall be provided with a single thrust bearing of the Kingsbury type, located at the upper end of the motor shaft and designed to take the weight and thrust of the rotating parts of the entire pumping unit. The bearings shall be designed for the extreme possible conditions of loading and shall be of size sufficient for self-cooling. They shall be absolutely dependable, shall impose a minimum amount of friction and shall be self-aligning and as nearly as possible self-adjustable for wear. They shall not be dependent upon any external means for their lubrication. All wearing parts shall be accurately machined, ground and hardened. Provision shall be made for easy examination and for the renewal of parts. The bearings shall be of a design which has been successfully used in installations of similar size and type.

9.11. *Guide Bearings.* Each pump shaft shall be provided with a guide bearing at the intermediate gallery. These guide bearings shall be of liberal proportions and ample to prevent any noticeable vibration or deflection of the rotating shafting. The bearings shall be of the sleeve or anti-friction type, easily adjusted for position. The design shall admit of accurate erection, easy dismantling and the renewal of wearing surfaces, which in case of sleeve bearings shall be of bronze or genuine babbitt metal. The bearings shall run cool under all conditions, and the lubricating appliances shall be self-contained, automatic, and such as to require only occasional attention.

9.12. *Rigid Couplings* shall be provided in each pump shafting above the pump casing to permit the removal of the pump shaft and impeller without disturbing the position of the motor and at other points necessary for the installation and removal of the line shaft. Couplings shall have tight bolts and the coupling flanges shall be marked for selective assembly. The hubs or heads of couplings shall be of forged steel and shall be keyed on the shafts and secured by lock-nuts.

9.13. *Lubricating Apparatus* shall be provided for the proper and efficient automatic lubrication of all bearings of pumps and shafts and shall include all necessary auxiliary equipment, oil piping, valves, and accessories including grease guns and brass lubricating fittings at greased ball bearings. The apparatus shall be of simple design consistent with maximum safety, easy accessibility, and minimum attention. Ball bearings may be arranged for grease or oil lubrication, unless approved specifically otherwise. In general, the maximum ambient temperature in service shall be assumed to be 120 degrees F.

9.14. *Designation and Marking of Pumping Units.* The direction of rotation shall be marked clearly by an arrow on the housing of each pump.

9.15. *Special Tools.* A complete set of special wrenches and special tools required for making adjustments, indelibly marked for the purpose intended and mounted on a suitable rack, shall be furnished with the pumps. (Ordinary adjustable wrenches are not required with the set furnished.)

9.16. *Painting.* All mechanical equipment shall be painted or enameled at the factory in accordance with the manufacturer's standard practice.

9.17. *Inspection and Tests.* Material and equipment shall be subject to inspection and tests at the place of manufacture and after installation. The testing code of the American Society of Mechanical Engineers, the standardization rules of the American Institute of Electrical Engineers, and the latest Standards of the Hydraulic Institute shall be followed in making the tests.

9.18. *Factory Tests* shall be made at the expense of the contractor. No painting or filling material shall be applied to castings until the latter have been inspected and passed. The pump casings shall be subjected to hydrostatic pressure tests of 50 p.s.i. and shall show no indication of leakage or other defects under the tests. Authorization from the Engineer shall be obtained before shipment is made.

9.19. *Tests after Installation.* Upon completion of the installation, the pumps shall be tested under the supervision of the pump manufacturer to demonstrate compliance with all the contract requirements and guarantees. The tests shall be conducted in the presence of the Engineer and shall be subject to his approval. The costs shall be borne by the contractor, including the services of supervising engineer and expenses incident to retests occasioned by defects and failure of equipment to meet contract guarantees at the first tests. Water, electric current, water level indicators and recorders, observers for taking readings, gauges and connections for measuring the velocity heads on the pumps shall be provided by the contractor. All test gauges shall be proven accurate to the satisfaction of the Engineer. All defects disclosed in the equipment by those tests shall be made good by the contractor or the equipment replaced by him without additional cost to the owner. Tests after installation shall include the following:

(*a*) *Operating tests* as in service shall be given all pumps. The alignment of each pump and motor unit shall be proven straight and plumb and the satisfactory operating of each pump unit shall be demonstrated.

(*b*) *Performance and efficiency tests of the pumps* shall include tests to prove the guaranteed capacity and approximate efficiency of one pump, as selected, and shall include:

(1) *Capacity and overall efficiency.* The work done by the unit shall be determined by multiplying the total weight of water pumped by the average total head. The overall efficiency shall be the ratio of the total work done to the power input to the motor, both expressed in foot-pounds. During the tests, readings will be taken from a depth gauge on the basin walls which will be marked to give the elevation of the water at 10-minute intervals, and with these readings the quantity of water pumped shall be

determined from the basin volume curve. The basin volume curve will be furnished by the Engineer.

(2) *Total dynamic head.* Readings simultaneous with the capacity readings shall be taken of the head in feet as determined by calibrated gauges in the suction and discharge connections of the pump; the gauge in pump discharge connection shall be located on the inboard end of the check valve.

(3) *Power input.* Readings simultaneous with the capacity readings and total-head readings shall be taken of the power input to the motor as determined by calibrated polyphase watt-meters and current and potential transformers placed in the circuit and near the motor.

(4) *Pump test performance curves.* Pump characteristic curves shall be prepared from the pump test data by the contractor and shall be furnished to the Engineer in triplicate.

9.20. *Warranty.* The contractor shall warrant the pumps and all accessory equipment furnished therewith to be free from defects of material and workmanship for a period of one year from date of acceptance; and he shall upon notice promptly make good at his own expense all defects developing during this period.

Specifications for Proprietary Commodities.

Many of the materials and appurtenances of construction are manufactured and sold under patents, trade secrets, and other proprietary conditions. These so-called "standard brand" products are utilized in practically all construction work. Inasmuch as no control over manufacturing methods can be enforced, it is necessary for the engineer to be familiar with the quality and character of the various standard products on the market and to word the specifications in such a way that suitable products will be obtained while those considered unsatisfactory will be eliminated. It is inappropriate, and in public works illegal, to specify trade names or brands without qualification because to do so confines the selection to one manufacturer whereas most of such commodities are sold competitively by several manufacturers.

When it is practicable the most satisfactory method for procuring proprietary commodities is by means of quality or performance-type specifications setting forth the required properties and characteristics and the tests which are to be applied to prove them. Under this type of specification any proprietary commodity which meets the specified requirements is acceptable.

In many instances quality and performance specifications are not feasible, and some other method must be adopted. Traditionally it has been customary to specify one selected brand or trade name followed by the qualifying "or equal" phrase. This has been a chronic trouble maker during construction because of the lack of suitable methods for determining equality when a substitute is proposed by the contractor and because of failure to designate who shall decide questions of equality. The latter objection can be overcome partially by providing for the engineer's decision to be final on such matters and by writing the qualifying phrase "or approved equal." In the absence of suitable tests, however, the determination of equality is always uncertain and subject to criticism by contractors and manufacturers alike. Moreover, the contractor has no means of knowing what will be considered equal when he is preparing his bid, and therefore he is forced to base his estimates on the brand named nothwithstanding the fact that a lower price might be obtained by using a competitive product. If both products are equally satisfactory the owner should receive the benefit of the lower price, but few contractors will assume the risk involved in the uncertain equality question. Therefore, the "or equal" clause is not recommended except for minor items.

When it is necessary to procure proprietary commodities by trade or brand names, probably the most satisfactory method is to include a clause in the general provisions to the effect that the brand names specified are to be used as the basis of the bid and furnished under the contract but if the contractor desires to use products other than those specified he should so state in his bid, giving the details of the proposed substitution and the adjustment to be made in his bid if the substitution is approved. A similar procedure may be followed if the contractor for any reason desires to make a substitution during construction, but the contract or specifications should state that after the execution of the contract the engineer's decision on substitutions shall be final.

An alternative method which has some advantages consists of a list in the specifications of all brands or models which would be acceptable for a given purpose with a provision for additions to be allowed if it is desired that the products of other manufacturers be approved as substitutes before the award of the contract.

The Use of Standard Specifications. Specifications are seldom written in total for any particular project. Like the other contract documents certain parts are capable of standardization for use on all work of a given type, and practically all engineering offices develop their own standard clauses. Some engineers find it convenient and efficient to catalog such clauses in a card index file or to prepare them in skeleton form with blank spaces provided for the insertion of variable requirements. In preparing specifications for a particular job the standard clauses are combined with those written especially for the work, and the complete assembly makes up the specifications.

The development of office standards is a continuous process in which it is necessary for the engineer to keep abreast of standard practice and improvements in construction materials and processes. Careful notes and records should be made during the construction of one job in order that revisions, corrections, and improvements may be made in the specifications for the next job. The repeated use of specification clauses from job to job by the so-called "scissors and paste pot" method is a notorious cause of difficulties during construction. No specification clause from a previous job should ever be used until a careful review has indicated its suitability.

Standardization of specifications for construction materials and processes has been the subject of much study and research by the various professional societies, government agencies, and manufacturers' associations. The recommendations of these organizations form the basis of current construction practice, particularly with regard to quality of materials. Any standard specification may be incorporated into a contract by reference, and this procedure is universally followed for construction work. Most of the common standards are so widely distributed as to be readily available to everyone in the construction industry. The following is a representative list of sponsors of standard specifications for various materials and types of construction.

PROFESSIONAL ENGINEERING SOCIETIES.
   American Society for Testing Materials.
   American Railway Engineering Association.
   American Society of Mechanical Engineers.
   American Concrete Institute.
   American Association of State Highway Officials.
   Society of Automotive Engineers.
GOVERNMENT AGENCIES.
   Federal specifications.
   Various department and bureau specifications.
   Various state highway department specifications.

MANUFACTURERS' ASSOCIATIONS.
> American Institute of Steel Construction.
> Portland Cement Association.
> National Lime Association.
> Various lumber associations.

The most widely used of all standard specifications are probably those of the American Society for Testing Materials and the federal government. The A.S.T.M. specifications are published individually and may be purchased from the American Society for Testing Materials, Philadelphia, Pa. They are also collected in book form under the following titles.

A.S.T.M. Standards:
> *Part 1, Ferrous Metals.*
> *Part 2, Non-Ferrous Metals.*
> *Part 3, Cement, Concrete, Ceramics, Thermal Insulation, Road Materials, Waterproofing, and Soils.*
> *Part 4, Paint, Naval Stores, Wood, Fire Tests, Sandwich Constructions, Building Constructions.*
> *Part 5, Fuels, Petroleum, Aromatic Hydrocarbons, and Engine Antifreezes.*
> *Part 6, Electrical Insulation, Plastics, and Rubber.*
> *Part 7, Textiles, Soap, Water, Paper, Adhesives, and Shipping Containers.*

Federal Government specifications are prepared by committees representing the various Government agencies and bureaus and may be obtained from the Government Printing Office, Washington, D. C.

Most individual standard specifications are given serial designations and are revised from time to time to reflect developments and improvements in techniques and processes. When a standard specification is to be incorporated into a specific contract by reference, care must be taken to make certain that the proper edition is designated. Otherwise uncertainty may result when revisions are made to reflect improved techniques. The reference should always include the title, serial designation, date of issue, and name of sponsoring organization. Any desired changes or modifications in the standard specification should be covered in detail.

In addition to the organizations listed herein, many individual manufacturers maintain research staffs which prepare specifications for their particular products and make them available to specification writers. Although such specifications may be biased and written with the view of promoting sales, they usually represent the best current practice when prepared by reputable firms. The products covered are usually patented or otherwise proprietary, and care should be taken to see that competitive products are not eliminated by the wording of the specifications.

Specification clauses are sometimes prescribed by law. All major cities have building codes and safety codes designed to enforce construction methods which will be safe for the probable loadings and occupancy and which will result in structures resistant to fire, wind, and other hazards. Many states have similar codes which are also enforced. Specification clauses must conform to such codes. Likewise, the recommendations and specifications of the National Board of Fire Underwriters must be followed closely to avoid excessive insurance rates. Also specification clauses may be influenced by local court decisions in previous lawsuits. Legal interpretations of ordinances and laws may vary from state to state or even from city to city, and the specification writer must have some familiarity with local precedents in this respect. Within their jurisdiction all of these requirements assume the aspects of standard specifications.

When appropriate, the use of standard specifications has many advantages. Standardization of products and production methods is encouraged, resulting in reductions in manufacturing costs and, therefore, in lower costs to the consumer. Constant improvements in products may be expected through cooperative research. Field control is simplified through the use of standard methods of inspection and tests. On the other hand, it should be kept in mind that standard specifications are written for average conditions and frequently require supplemental clauses to fit them for a particular purpose. Furthermore, special features of a particular project may make them entirely unsuitable. Such possibilities emphasize the necessity for a complete understanding of standard specifications before they are adopted for contract purposes.

Selection of Specification Type. For most of the basic types of construction, such as timber, structural steel, concrete, and masonry structures, the materials and workmanship type of specification is almost always used on important work because an over-all performance test usually is not feasible. For these types of construction, rigid control of the methods of manufacture, composition, character, and physical properties of the components is practicable. It should be noted that the cost of inspection and tests required under this type of specification may not be justified on smaller projects particularly with regard to the procurement of materials. Most materials are obtainable from reputable well-known manufacturers, and for minor work it is sufficient to permit such materials to be procured from stock thus dispensing with mill inspection, chemical, metallurgical, and physical tests, and the like. This will usually result in a reduction in costs.

Large specially designed and specially built machines will also require materials and workmanship specifications. Other equipment and machinery manufactured in standard models can usually be obtained best under performance and quality specifications. Likewise, this type of specification is indicated for special-purpose materials, such as waterproofing compounds and fire-resisting materials.

Many standard brand products, such as those desired for character, texture, appearance, and the like, are not adaptable to performance specifications. For commodities in this classification a list of acceptable brands is indicated in combination with a clause in the general provisions providing methods for approval of substitutions.

Basis of Measurement and Payment. When the work is to be accomplished under a unit-price contract it is necessary to specify the base units for each of the bid items and the details of the methods to be used in arriving at the actual quantities of units completed under the contract. The latter procedure usually requires a combination of field measurements and mathematical calculations and is a common cause of controversy. For instance, earthwork is normally paid for on a cubic yard basis, the quantity being computed from cross sections taken before and after the completion of the work. The quantities are computed by the average-end-area method or more accurately by the average-end-area method combined with prismoidal corrections. The latter method is laborious and tedious to apply and is seldom used in current practice, but unless the specifications cover this point uncertainty will exist inasmuch as the two methods never agree exactly.   Likewise, structural steel is usually paid for by the pound or ton, and payment may be based on the actual weights shown on the shipping lists or computed from the dimensions shown on the drawings and the weights given in the steel handbooks.   Theoretical and actual weights seldom agree exactly, and the specifications should be explicit as to which method is to be used.   These examples are cited because they have been the sources of many disputes and lawsuits.

Alternative methods of measurement are sometimes permitted but almost always lead to controversies because of discrepancies between the results given by the alternative methods.   Therefore this practice is not recommended when it can be avoided.   In earthwork, measurement of quantities is sometimes permitted on the basis of truckloads in lieu of volumes computed from cross sections. Truck measurement is, of course, inaccurate because of the bulking of excavated soil and variable quantities loaded into trucks. Dredging contracts frequently permit barge measurement of the excavated material in lieu of cross-section measurements.   Conversion factors of from 20 per cent to 40 per cent, depending on the material, are usually specified, however, to make allowance for the swelling of dredged material in the barge as compared with cross-section measurements in place.   To avoid these discrepancies earthwork quantities should always be measured in excavations, such as cuts and borrow pits, rather than in fills because disturbed earth usually requires years of consolidation before it settles to its original volume.

Precise definition of the units of measurement is essential particularly when one type of construction is subdivided into several classifications.   Examples are found in concrete construction where one class of concrete may be used for foundations, a second class for columns, and perhaps a third class for beams and slabs.   Each class will consist of a different mixture and will be paid for at its own price.   Similarly, excavation may be subdivided into earth excavation and rock excavation each at its own unit price.   Earth excavation usually includes all soil and loose rock which does not require special equipment for handling, whereas rock excavation is usually defined as solid ledge rock, requiring blasting for removal, and large boulders exceeding a certain specified size, usually $\frac{1}{2}$ cubic yard.   Unless careful distinction is made between these classifications controversies are bound to occur because of the wide range in unit prices.

With regard to the basis of payment to the contractor, the specifications should state explicitly how payment is to be made for each element of the work. The work included under each bid item should be carefully described. Particular care is necessary to make certain that provision is made for payment of every part of the work in one place or another in the specifications and that no element of the work is paid for more than once. For work required but not listed as a bid item, reference should be made to the item under which payment is included. For example, in highway construction on a unit-price basis, the material from excavations is utilized for the construction of embankments, and the price bid for excavations includes payment for placing the excavated soil in fills. Although the specifications contain specific instructions covering the methods for constructing embankments no direct payment is made therefor. Likewise, payment for fine grading and preparation of the subgrade might be included in the unit-price bid for pavement. Also in some classes of earthwork it is impracticable to excavate to the exact theoretical outlines shown on the drawings, and the specifications should state the maximum variation from the theoretical limits for which payment will be allowed, it being understood that greater variations than the prescribed maximum will be at the contractor's expense.

The foregoing examples illustrate questions which may arise in connection with almost every type of construction. Therefore, the specifications should prescribe clearly the basis of measurement and payment for each item in the bid. Measurement and payment instructions should be inserted in the specifications in a titled paragraph at the end of each section of the technical provisions, or they may be collected for all bid items and placed in a special section of the technical provisions at the end of the specifications.

Changes in the Specifications. It frequently happens in construction work that changes must be made in the plans and specifications after they have been issued to contractors. Such changes result from revisions in the requirements of the owner, changes in the design because of new information as to foundation conditions, and the like. A change in the contract plans is accomplished by the issuance of revised drawings. The corresponding change in the specifications is accomplished by the issuance of an *addendum* to the specifications.

When addenda to the specifications are issued during the bidding period but before the opening of bids, contractors are directed to acknowledge their receipt and to confirm that the effect of the changes has been considered in the preparation of their proposals. After the award of the contract an addendum usually results in a change of the contract price and requires the issuance of a change order under the contract. For these reasons it is always necessary for the engineer to obtain the owner's approval before an addendum can be authorized.

ABBREVIATED OUTLINE SPECIFICATION FORM

The foregoing discussions and examples of specification writing were based on the use of traditional style, composition, and grammar. In practice, it has frequently happened that such specifications have been so filled with legal phraseology as to become primarily legal instruments and only secondarily construction specifications. Furthermore, many specification writers have adopted the practice of applying certain stock phrases and stereotyped expressions to each requirement in the specifications. In most instances such repetition adds nothing to the strength of the specification and may obscure its meaning. To avoid this situation Mr. H. W. Peaslee has proposed a new form without sentence structure. This form subdivides the specification into two separate specialized writing operations.*

Part 1.   For Lawyers—to distil the essence of the contract and to draft a single iron-clad condition to the effect that *everything* listed thereinafter, material or operation, shall be put into the job, subject to qualification, condition or exception noted.

Part 2.   For Technicians—to distil the construction essentials, boiling down to a clear concise analysis of materials and methods— An Outline of Requirements. With the body of the specifications left free for technical details only, the specification writer may then express his requirements in clear concise form in headings and sub-headings, without sentence structure, using phrases in preference to clauses, with only essential adjectives or adverbs, with no articles, definite or indefinite, unless positively required.

* From "Streamlined Specifications," by H. W. Peaslee, *F.A.I.A. Pencil Points,* August, 1939.

The outline form is best illustrated by the following examples of actual specifications which are quoted from Mr. Peaslee's original article:

CASE EXHIBIT A

APPLICATION OF MEMBRANE WATERPROOFING

A typical old line form using a sentence structure with incessant repetition of mandatory provisions in each sentence. Cross references by italicized letters show where descriptive requirements of this form recur in the tabulated requirements of the streamlined form.

A streamline specification with a single governing mandatory provision followed by headings and subheadings only in phrase form.

"Mention herein or indication on drawings of articles or materials, operations, or methods, requires that the Contractor shall provide each item listed—of quality, or subject to qualifications noted; perform according to conditions stated each operation prescribed; and provide therefor all necessary labor, equipment, and incidentals."

a. Surfaces to which waterproofing is to be applied shall be free from holes, cracks, projections, and conditions that would prevent complete adhesion of waterproofing.
See a'

b. Waterproofing shall not be applied at temperatures lower than 50 degrees F.
See b'

c. The application of waterproofing material shall be done in the most workmanlike manner and only by firms and workmen thoroughly familiar and experienced in this class of work.
See c'

d. Surfaces to receive 3-ply membrane waterproofing shall be uniformly coated with hot coal-tar pitch. Over this coating place 3 layers of 32-inch wide coal-tar pitch saturated felt, lapping each sheet 22 inches over the preceding sheet. Lap ends not less than 6 inches. (If 36-inch wide felt is used, lap sheets 24½ inches.) Mop each of the sheets full width of lap, with hot coal-tar pitch, using not less than 25 pounds

a'. Condition of surface to be waterproofed: free from holes, cracks, projections, and conditions preventing adhesion.

b'. Temperature for application: 50 degrees F. minimum.

c'. Prequalification of firms and workmen: experience statement required.

x'. Materials: coal-tar pitch: and coal-tar-saturated felt.

d'. Method of Application:
Uniform coating of surface with hot pitch (40 lb. per sq.).
3-Ply Membrane: Over base coating
   3 layers, 32" felt, lapped 22"
   or 36" felt, lapped 24½"
5-Ply Membrane: Over base coating
   5 layers, 32" felt, lapped 26"
   or 36" felt, lapped 29"
end-laps, not less than 6"

per square for each mopping. Over the entire surface apply a uniform mopped-on coating of hot coal-tar pitch, using not less than 50 pounds per square.   *See x' & d'*

e. Surfaces to receive 5-ply membrane waterproofing with hot coal-tar pitch, using not less than 40 pounds per square. Over this coating place 5 layers of 32-inch wide coal-tar pitch saturated felt, lapping each sheet 26 inches over the preceding sheet. Lap ends not less than 6 inches. (If 36-inch wide felt is used, lap sheets 29 inches.) Mop each of the sheets full width of lap, with hot coal-tar pitch, using not less than 25 pounds per square for each mopping. Over the entire surface apply a uniform coating of hot coal-tar pitch, not less than 50 pounds per square shall be used.

See x' & d'

each sheet mopped, full width of lap, with hot pitch (not less than 25 lb. per sq. for each mopping).

Entire surface mopped with hot pitch (not less than 50 lb. per sq.).

CASE EXHIBIT B

This shows the repetition and waste of words in an actual specification which was originally supposed to be a brief one. Yet it contained:

489 repetitions of mandatory expressions.

4 to 5 pages of superfluous wordage.

34 descriptions of named trade products; and

22 descriptions of materials covered by reference in federal specifications (for each of which it is necessary to send a nickel to Government Printing Office!).

The bold face type indicates superfluous words that are not essential to the clarity of the instructions. The italics are repetitious mandatory expressions which are covered in the streamlined system by a single paragraph at the beginning of each specification section or in the General Conditions. The Roman type indicates the words remaining which are essential to the meaning of the instructions! Four to five pages could thus be eliminated.

SECTION 5. SEWAGE DISPOSAL

5-01. *Sewage Disposal.* **A** complete sewage disposal plant *shall be* **furnished and installed by the contractor** and *shall* consist of **the** following parts:

For each house: 1 Super-Septic Tank, No. 113; 1 Syphon and Syphon Compartment, No. 151.

5-02. Absorption Field: **The contractor** *shall* **provide** four 4-inch lines of open drain 30 feet in length from the "Y" in **the** soil line. **The** pipe of absorption field *shall be* common farm tile laid in gravel-filled trenches **as detailed.**

5-03. Vitrified Pipe: **The** sewer line from **the** septic tank to **the** absorption field *shall be* 4-inch pipe.

SECTION 6. EXCAVATION

6-01. Excavation. **The kind of material to be encountered in excavation is** believed **to consist of** shale. Excavation *shall be* **made** for **the** cellars, garage and porches to **the** depth shown **on the drawings** and for **a** distance of 6 inches outside **of all** exterior footing lines. **All** trenches *shall be* excavated to **a** neat size **and** each is to be leveled to **a** line on **the** bottom ready to receive foundations. Any excavations made below grade for **the** footing or walls *shall not be* filled before **the** footings or walls have been constructed on **the** undisturbed earth.

6-02. Shoring. **Shoring** and bracing *shall be* **furnished and placed** wherever **there is any** likelihood of **a** slip occurring **and the** cost **thereof** *shall be* included in **the** contract price.

SECTION 7. CONCRETE FOOTINGS, FLOORS, WALKS, AND CEMENT BLOCK WALLS

7-01. Footings. **All footings** *shall be* **constructed of** concrete to **the** elevations shown **on the drawings. Footings** *shall* **rest** on undisturbed earth or other foundation satisfactory to **the** contracting officer.

7-02. Concrete Floors. **The floors in the** basement *shall be* **constructed of concrete** 4 inches thick, monolithic, smooth finish with uniform slopes to floor drains **as indicated on the drawings. A** 6-inch cinder fill *shall be* **made for the concrete basement floor. The** porches *shall* **have a** 5-inch concrete slab reinforced with triangular wire mesh **similar and equal to. . . .**

Selection of Specification Form. Much space in the specifications may be saved by the use of the outline form, particularly in specifications for building construction. The objectives of this form, however, may be accomplished largely with the older type of composition if the same general principles are applied, that is, by the elimination of extraneous, indefinite, and repetitive statements throughout the body of the specifications. If this is done the specification, though still somewhat lengthier than the outline form, will be clearer and more readily understood. Under these conditions it appears that the older type of specification is more desirable although individual requirements or personal preference may lead to the selection of the alternative form. Perhaps the most important element in the selection of form is the fact that any form or style whatsoever is acceptable if the requirements are definitely expressed in a clear, direct manner.

Questions

1. Define three types of technical specifications.

2. Select the type of specification to be used for each of the following construction items, and give reasons for each selection.

a. Concrete work.

b. Excavation.

c. Electric refrigerators.

d. Bridge cranes.

e. Electric elevators.

f. Brick masonry.

g. Asphalt roofing.

h. Floor-hardening compound.

i. Painting.

j. Pump motors.

k. Automatic sprinkler system.

3. What are the advantages in using standard specifications for materials and what precautions must be observed in their use?

4. Outline the general properties to be covered in the preparation of technical specifications for construction materials.

5. Outline the general procedure to be followed in writing the technical specifications for workmanship.

6. Outline the procedure to be followed in writing technical specifications based on the performance of the finished product.

7. Write the specifications for electric refrigerators to be installed in a high-grade apartment house. Each refrigerator is to have a capacity of 6.5 cubic feet. Only first-class standard models of recognized manufacturers are to be acceptable. Consult catalogs for manufacturers, models, and grades.

8. Write the basis of measurement and payment clause of the specifications for reinforced concrete pavement. Would actual measurement of the pavement on slopes be necessary or could areas be computed from the stations shown on the plans? If the pavement is 30 feet wide and extends from station $0 + 00$ to station $10 + 00$ on a 5 per cent grade, what would be the difference between the actual area measured on the slope and that computed from stations shown on the plans?

9. Describe two methods of measurement and payment for vitrified clay pipe sewer construction in which the diameter of the sewer in various locations is 8, 10, 12, and 15 inches and the depth of invert varies from 6 feet to 12 feet for the 8-inch pipe, 8 to 14 feet for the 10-inch pipe, 12 to 18 feet for the 12-inch pipe, and 12 to 20 feet for the 15-inch pipe.

10. In an excavation contract on a unit-price basis, rock excavation was defined as ledge rock which would require blasting for removal and was priced at $5.00 per cubic yard. All other excavation was classed as earth excavation, priced at $1.00 per cubic yard. The site of the work was in a built-up section of a city. Because of the proximity of buildings and streets and fears as to what blasting would do to them, the contractor resorted to removal of the rock by the jack-hammer method, himself absorbing the considerable difference between the cost of blasting and the method used. The owner refused to recognize the rock excavation and insisted it should be paid for at the price bid for earth excavation because there was no blasting. The contractor sued the owner for the higher price for the material removed by the jack-hammer method. What was the decision?

11. An engineer required a contractor to place larger rock than called for by the specifications. Was the contractor entitled to an allowance for increased cost of constructing an embankment wall?

17

ASSEMBLING THE SPECIFICATIONS

The first step in assembling the specifications for a construction project is to make a careful analysis of the job requirements and a check list of the topics to be covered. The various items should be segregated into three classifications, namely, those which properly belong in the agreement, the general provisions of the specifications, and the technical provisions of the specifications, respectively. This involves a close study of the drawings to determine the aspects of the work which are not shown completely thereon and the supplemental instructions required in the specifications. The resulting check list or outline headings will then represent a breakdown of the project into basic operations and types of construction, and the specification writer may concentrate on each topic successively until all are prepared.

As each section of the specifications is completed, it should be checked by the designer of that particular phase of the work, and when revised or corrected it is ready for assembly into the combined volume of the specifications. When all sections are completed, a final review of the plans and specifications should be made to detect any remaining errors, omissions, and duplications.

For use by contractors for the preparation of bids and for construction purposes in the field a considerable number of copies of the specifications will be required. The most suitable methods of reproduction will depend largely on the character and magnitude of the work. For large important projects the specifications are usually printed and bound with either heavy paper or cloth covers. Many offices use the mimeograph or hectograph processes in which the pages are reproduced, punched, collated, and bound in paper covers. For minor work a few typed carbon copies may be sufficient.

Title and Table of Contents. The title should be short but definite and descriptive. It should give not only the location and character of the project but should also specify the services to be performed, as in the following example.

SPECIFICATIONS FOR THE CONSTRUCTION OF PINE FLAT DAM AND APPURTENANT WORKS, KINGS RIVER, CALIFORNIA

Though not absolutely necessary a table of contents listing section numbers, section headings, and page numbers is a convenience for reference purposes, particularly when the specifications are of considerable length. For some types of construction it may be desirable to list paragraph or subsection headings in the table of contents as well.

Outline of the Specifications. The preparation of the specification outline for an important project is a lengthy process, and many general check lists have been prepared as guides for this phase of the work. Some of these are helpful expedients for the specification writer, but it should be emphasized that no standard check list ever written can be considered a satisfactory substitute for a careful analysis of the individual job. Likewise, specification handbooks are helpful but dangerous if used indiscriminately without a full understanding of the special requirements of the individual job.

The general provisions may be set up as Section 1 of the specifications, to be followed by the various sections of the technical provisions in numerical order. Arabic numerals are used in contrast to the Roman numerals frequently adopted for the articles of the contract. In another method of composition which is also widely used, the general provisions comprise Part I of the specifications, and the technical provisions make up Part II, with the sections of each part numbered independently. The former method is recommended in this book because it is believed to result in a more unified treatment of the specifications as a whole although the general provisions do not receive as much emphasis in the documents under this arrangement as in the alternative method. The decimal reference method is used for numbering paragraph or subsection headings.

Whereas the general provisions of the specifications follow a similar form for all kinds of work, the technical provisions are different for each type of structure. The job analysis will indicate the various section headings required in the outline. After the section headings are determined a careful summary of the subject matter should be made for each section in which the outline is broken down into paragraph and subparagraph headings. This can be done, of course, only after a thorough study and analysis of the detailed technical requirements for each type of construction. Students may gain much by a study of specifications which have been used for actual projects. Here again, however, it is necessary to warn that there is no guarantee that a specification clause is sound and appropriate simply because it has been printed and used.

For some kinds of work, notably building construction, it is desirable to include an introductory paragraph to each section of the technical provisions defining in detail the scope of the work under each type of construction. This is for the convenience of contractors in obtaining bids from subcontractors who usually specialize in one trade or type of construction. The description of the work of each trade therefore serves to define the work to be included in each subcontract. It also results in economy when a large number of subcontracts is contemplated. The specifications can be taken apart and each subcontractor furnished with only the section covering the work to be performed by him. Otherwise a complete set of specifications would be required for each subcontractor. For engineering construction work which usually employs a relatively small number of subcontractors there does not appear to be any advantage in this practice, and the writing of the specifications will be simplified by its omission. If a general description is required it should be placed in the first paragraph, followed in turn by the specifications for materials, workmanship, and location if the latter is not clearly evident on the drawings. For unit-price contracts the methods of measurement and payment should be covered in detail for each type of construction listed in the bid items.

The outlines which follow indicate the form and composition of the specifications for representative engineering and architectural construction projects. Section headings are given together with a brief discussion of the contents of each. The comments and suggestions are intended to illustrate the procedure of analyzing job requirements and assembling the specifications, but the outlines should not be considered complete check lists.

HIGHWAY CONSTRUCTION

1. GENERAL PROVISIONS

2. CLEARING AND GRUBBING

Specify the requirements in connection with the clearing of objectionable materials from the right of way, approaches, inlet and outlet ditches, and the grubbing of all stumps, large roots, buried logs, and other objectionable material below the ground's surface. The methods for removal and disposal of waste material and timber and the ownership of material cleared from the site should also be covered.

3. EARTHWORK

Give the general instructions for excavating earth and rock from the roadway prism, ditches, channels, and borrow pits. Also include the instructions for constructing embankments at the proper elevations, backfill of ditches and depressions caused by the removal of obstructions, construction of approaches to new structures, and the like. Provide for the removal of all unsuitable materials and the backfilling of the resultant space with approved materials, and describe the methods for finishing slopes. Various classes of excavation should be accurately defined.

4. EXCAVATION FOR STRUCTURES

Specify the instructions for excavation and backfill of trenches for pipe culverts, head walls, cut-off walls for slope paving, concrete culverts, retaining walls, and other appurtenant structures. Backfill specifications should include the thickness of backfill layers before compaction, sprinkling, tamping, rolling, and other compaction methods required.

5. DITCH AND CHANNEL EXCAVATION

Include specifications for the excavation of all water channels above and below all culverts and bridges, and the use of the excavated material, when suitable, in road embankments and dikes.

6. EMBANKMENTS

Embankment specifications should include the preparation of original ground by plowing or scarifying in order to allow new fill material to bond with the original earth. Specify the thickness of horizontal layers to be required in the placing of embankment fill and the methods for sprinkling and compaction by tamping or rolling. The contractor should be made responsible for maintaining the planned grades and cross sections of all embankments until the completion and acceptance of the work. He should also be made responsible for the stability of constructed embankments and should replace any portions which shrink or otherwise become displaced during construction.

7. SURPLUS AND BORROW EXCAVATION

All surplus excavation should be used to widen embankments uniformly or to flatten slopes when desired. Provisions should be made for the disposal of all waste material not required for these purposes. When roadway excavation is not sufficient for the construction of embankments, provision should be made for local borrow from the excess widening of cuts or by excavating from sources within the right of way. The location and extent of local borrow should be placed under the control of the engineer. When imported borrow is necessary, state whether borrow areas are to be furnished by the owner or by the contractor, and the conditions under which they are to be used in lieu of roadway excavation.

8. OVERHAUL

State the conditions, if any, under which the contractor will receive special payment for the overhauling of excavated material and the distance within which the cost of hauling is included in the unit price paid for excavation.

9. DRAINAGE STRUCTURES

Paved gutters, drainage pipe, catch basins, drop inlets, manholes, culverts, and the like are specified under this heading. The desired quality of materials and workmanship must be set forth clearly. Many features of drainage structures are shown on the plans, and the coverage in the specifications will depend to a large extent on the information given on the drawings.

10. SLOPE EROSION PROTECTION

Cover the placing of top soil, seeding, and other methods to be used to protect roadway slopes from erosion. Instruction should be included for the preparation of the slopes to receive the protection material and the methods to be used in placing, spreading, and planting.

11. SUBGRADE

Specify the procedure for the finishing of the subgrade to true grade and cross section and the methods to be used in placing it in readiness to receive the roadway surfacing. The specific requirements for subgrade finishing will depend upon the type of pavement to be used.

12. PAVEMENT

Describe the type of pavement to be used, the quality of materials, the side forms, and the mixing and placing of the pavement. For reinforced concrete paving, specifications for reinforcement steel and dowels must be included. Give the details for construction joints, contraction and expansion joints, and the methods for finishing the pavement's surface. Also include the curing of the fresh pavement and its protection against damage during the curing period.

13. SHOULDERS

Provide for the placing of shoulders on both sides of the pavement or surfacing and the quality of materials to be used in their construction. Methods for placing, shaping, sprinkling, and compaction should be included.

14. FINISHING ROADWAY

Give the requirements for finishing the entire roadway after the earthwork has been completed and the surfacing or pavement has been placed. This includes the trimming of shoulders, slopes of embankments, excavation, road approaches, ditches, borrow pits, and the like and the removal of large loose rock from the finished roadbed.

15. GUARD RAILING

If the railing is to be of the cable type the number of strands, the number of wires to the strand, and the tensile strength of the cable must be specified. The method of galvanizing, the methods of splicing, and types of fittings must be covered. If the railing is to be of the timber, reinforced concrete, or steel beam type, the quality of materials, bolting or riveting, fittings, and connections must be specified. Posts will be of timber, steel, or reinforced concrete. The quality of the materials and methods of driving or setting must be given.

16. BASIS OF MEASUREMENT AND PAYMENT

It is usually desirable to include a description of the basis of measurement and payment for each of the bid items where each of these items is specified in the technical provisions rather than in a separate section. Payment is usually made on the contract unit price per cubic yard of excavation, and the specifications require that excavated material from the roadway be used for the construction of embankments. Borrow-pit excavation when required is likewise paid for on a cubic-yard basis except when it is elected optionally by the contractor in lieu of the placing of roadbed excavation. In the latter case, the roadbed excavation would otherwise be used on the job or disposed of at the expense of the contractor. No direct payment is made for embankment construction. Excavation may be classified as, for instance, earth, rock, and top soil with a separate unit price for each class; or it may be unclassified, in which case all materials are paid for at the same unit price.

BRIDGE CONSTRUCTION

1. GENERAL PROVISIONS

2. EXCAVATION

Define the classes of excavation for which unit-price payments are to be made. Provide for the use of suitable excavated materials in the construction of approach embankments; for the removal of all objectionable materials, such as trees, stumps, boulders, or other obstructions; and for the disposal of all unsuitable materials elsewhere. When pier foundations are to rest upon rock, the surface of the rock should be removed to expose sound material and the surfaces leveled off to provide horizontal or stepped bearing surfaces. Objectionable seams in the rock should be grouted under pressure. When piles are used and masonry is to rest on excavated surfaces other than rock, care should be taken not to disturb the original soil at the bottom of the excavation. All excavation should be completed to the bottom of the pier bases before any piles are driven. Any disturbances at the bottom of the excavation during the driving of piles shall be corrected. After the concrete is placed, excavation should be backfilled where necessary to the established grades, the filling material being thoroughly compacted and surfaced.

3. DEPTH OF PIER BASES

The elevations of the bottoms of piers and abutments shown on the plans are usually considered as approximate only, and it may be necessary to require the contractor to dig test pits or make test borings during the progress of excavation in order to determine the exact character of foundation material and the final depth to which piers are to be constructed.

4 OPEN COFFERDAMS

Where piers are to be constructed in open cofferdams, the specifications should state that the structural pier bases will be constructed in the dry. Before the pier bases are poured a concrete seal should be provided in the cofferdam by "tremie," or other underwater concrete placing methods. Details of materials and workmanship for concrete in pier bases and pier shafts will be specified under the heading "Concrete Work." Design of the cofferdams and the method of their construction should be the responsibility of the contractor subject to approval by the engineer of the design and methods of bracing and sealing. Allowable tolerances in location and alignment should be fully described. Also provide for the removal of cofferdams, sheeting, and bracing to the level of the stream bed after the completion of the substructure.

5. PNEUMATIC CAISSONS

The responsibility for the design and construction of pneumatic caissons should be placed upon the contractor, but complete working drawings should be furnished for examination and approval by the engineer before any construction work is started. Under an alternative procedure pneumatic caissons are designed by the engineer and set up as a separate item in the bid and the contract. Drainage wells, working shafts, air locks, and other facilities and equipment must usually conform to the requirements of state safety laws. Contractors should be required to furnish such facilities as hospital locks, lockers, and shower baths, as may be required for work under compressed air, when such facilities are not otherwise available. The method of sinking caissons should be optional with the contractor, and he should be permitted to predredge or to sink the caissons in so far as practicable by open dredging, provided that the final operation of preparing the rock bed, inspecting, and filling the working chambers is done under compressed air, reinforced concrete decks being placed in the dredging wells to permit the completion of the pier bases by the pneumatic process. Specifications should state the tolerances which will be allowed as to location, alignment, and level. Particular care should be taken in concreting the working chambers to obtain full bearing under

the entire cutting edge and under all bulkheads as well as against the roof of each chamber. Provision should also be made for the removal of the portion of the caisson above the stream bed after the completion of the work, if this is required.

6. PILING

When piers and abutments are to be founded upon piling, specifications should state the type of piles to be used, quality of materials, methods of driving, required bearing capacity, and methods by which the bearing capacity is to be determined. Provisions should be made for test piles when required, the removal of damaged and misdriven piles, and the tolerances which will be permitted with regard to location and alignment.

7. CONCRETE WORK

Define the classes of concrete work, and specify the quality of materials and workmanship required for each. Such features as standard specifications for cement, aggregate, water, and reinforcing steel and the requirements as to mixing, placing, vibrating, and curing concrete should likewise be covered. The requirements for the materials and workmanship in formwork should be given. When under-water tremie seals are contemplated in the open-cofferdam method of pier construction, detailed instruction should be included for the mixing and placing of tremie concrete. It may be desirable to devote a subparagraph to the particular requirements for formwork, mixing and placing concrete, and waterproofing for each of the types of work, as, for instance, tremie seals, pier bases, pier shafts, abutments, roadway deck, and the like. Anchor bolts, steel anchorages, sleeves, nuts, and other embedments must be set in concrete to the proper location and grade.

8. MASONRY

Where bridge piers and abutments are to be faced with sawn granite or other masonry materials, the specifications for this work should be given under this heading.

9. STRUCTURAL METAL WORK

See Example on page 360. Specifications for painting steel work may be included in this section, but it will usually be desirable to devote an entire section to the various classes of painting.

10. BRIDGE FLOORING

Where reinforced concrete flooring is used, specifications for this work may be included under the heading "Concrete Work." For other types of decking a separate section will be required. In addition to the specifications for materials and workmanship, this section should also cover such items as roadway curbs and curb openings, joints, ducts to be placed in the floor system, special wearing surfaces, and the like.

11. ELECTRICAL WORK

Under this heading give the specifications for metal conduits, fibre ducts, pull boxes, and splicing chamber frames and covers, lighting standards and brackets, traffic signals, marine and aviation lights, roadway lighting units,

electrical equipment, and wiring. Control equipment, such as switches, magnetic contactors, circuit breakers, and the like should also be included.

12. APPROACH EMBANKMENTS AND ROADWAY SURFACING

In general, the specifications for bridge approach embankments and roadway surfacing will follow those previously outlined for highway construction, and the extent to which they are detailed will depend on the magnitude and the importance of this element of the work.

13. RIPRAP AND REVETMENT

Under this heading give the specifications for any stone riprap, revetment work, or bank protection for the piers, abutments, and channel at the bridge site.

14. MISCELLANEOUS WORK

Under this heading give the specifications for miscellaneous items not otherwise classified, such as toll collection booths, traffic barriers, bronze tablets, and the like. For large toll bridges, the toll collection facilities, consisting of the toll plaza, collection booths, and administration building, will require separate sections in the specifications.

15. BASIS OF MEASUREMENT AND PAYMENT

Occasionally bridges are built under lump-sum and cost-plus contracts in which case no mention in the specifications is necessary relative to measurement and payment. More frequently, however, this type of work is constructed under unit-price contracts, and the basis of measurement and payment must be established in the specifications for each of the various bid items, as, for instance, excavation, concrete in the pier bases, concrete in the pier shafts, reinforcing steel, structural steel, etc. When pneumatic caissons are contemplated, a separate lump-sum bid item is sometimes set up for each; otherwise their cost would be included in other bid items.

BUILDING CONSTRUCTION

1. GENERAL PROVISIONS

2. PREPARATION OF THE SITE

This section covers the demolition and disposal of existing structures and provides for the ownership or re-use of materials cleared from the site. It may also include the protection of all conduits, drains, sewers, services, shrubs, and lateral support of adjacent buildings.

3. EXCAVATION

Give directions for the excavation for foundations, basements, trenches, tunnels, elevator pits, and the like. Give particulars as to caissons, shoring, bracing, sheeting, underpinning and protection of adjacent buildings, cribbing, etc. Also specify the quality of materials, the methods to be used for backfilling and grading, the disposal of surplus excavated materials, and the storage of top soil or other materials to be replaced upon completion of the work.

4. FOUNDATIONS

If pile foundations are to be used, give the type, quality, bearing capacity, and methods of driving. Specify the methods for determining pile bearing capacity, and give the details of any tests required. For pile caps, spread footings and foundation walls, and basement floors, cover the quality of materials and workmanship.

5. CONCRETE WORK

Specifications for all concrete work on the job should be covered in this section. Describe the various classes of concrete, and give the details as to quality of materials, proportions, and workmanship for each. Construction of forms, mixing, placing, and curing of concrete should also be included. Under subheadings, detailed instructions should be given for the concrete work for the various elements of the structures, such as footings, foundation walls, floor slabs, framing elements, concrete fill, and the like. Mill and field tests should also be described. Reinforcing steel may be specified either in this section or under "Structural Steel Work."

6. MASONRY

Specify the quality of materials, the proportions and the methods of mixing masonry mortars, and any tests required therewith. Also specify the quality of common brick, face brick, structural clay tile, concrete blocks, stone, glass blocks, architectural terra cotta, and other masonry materials. Standards of workmanship should be specified in detail. For small buildings, specifications for all types of masonry construction may be included in one section, but, for large important structures, a separate section may be required for each type.

7. WATER-PROOFING AND DAMP-PROOFING

Schedule the materials and methods to be used for the water-proofing of elevator pits, pump sumps, basement walls, tunnels, floors, spandrels, and ceilings. Roof water-proofing and flashing would ordinarily be specified in a separate section. Integral water-proofing will usually be included with the concrete specifications, but all other types may be placed in this section. This includes the membrane types and their protective coatings, plaster and iron coats, and damp-proofing mixtures that are brush, spray, or trowel applied.

8. STRUCTURAL STEEL AND IRON WORK

This section covers the specifications for the fabrication and erection of all structural steel and iron in the building frame, including the columns, beams, trusses, hangers, billet plates, stiffeners, anchor bolts, nuts, rivets, and connections. The quality of materials is usually made to conform with the Standard Specifications of the American Society for Testing Materials, and the fabrication erection standards are those recommended by the American Institute of Steel Construction, Inc. Miscellaneous items, such as lintels, elevator guides, steel door bucks, steel joists, and steel anchors for other trades, may be included in this section or specified separately under "Miscellaneous Metal Work." Ornamental metal work also is usually specified in a separate section.

9. MISCELLANEOUS METAL WORK

Under this heading are included the specifications for materials and workmanship for such miscellaneous items as gratings, coal chutes, package chutes, curb guards, trench covers, door bucks, lintels, flag poles, smoke stacks, steel stairs, and the like.

10. METAL DOORS, WINDOWS, AND TRIM

Fire-resistant metal doors and windows are generally furnished in standardized proprietary brands. The specifications therefore will indicate various approved models and manufacturers. Performance requirements may also be incorporated as well as the required quality of workmanship in fabrication and installation. Window screens may be specified under this heading, or they may be covered in a separate section.

11. CARPENTRY

For ordinary buildings this section will be subdivided in two parts: rough carpentry and finished carpentry. For larger projects each type may require a separate section in specifications. Rough carpentry includes framing, posts, bearing partitions, platforms, girders, basement steps, roof trusses, stud partitions, and the like. It also includes cornices, water tables, nailing strips, blocking, centering for masonry openings, timber door bucks, roof and wall sheathing, shingles, rough flooring, and the installation of building papers and insulation. Lumber specifications should describe the grades, species, and seasoning.

Finished carpentry includes moldings, ballustrades, doors and door frames, window frames, sash and trim, panelling, stair steps, cabinet work, and the like. Methods of nailing, joining, finishing, and other qualities of workmanship should be specified.

12. ROOFING AND SHEET-METAL WORK

Under this heading cover the quality of materials and workmanship for the roof surfacing, giving detailed instruction for construction at changes of roof slope, pipes and chimneys through the roof, dormers, roof vents and sky-lights. Some types of roof surfacing are proprietary and when these are desired, specifications should be based on guaranteed performances and approved types should be identified. Sheet-metal requirements, such as flashing at joints and openings, roof ventilators, louvers, ventilating ducts, radiator enclosures and gutter leaders and outlets should be specified as to the type, thickness, and quality of materials and workmanship. Methods for ribbing, seaming, and painting of sheet-metal roofing should be given.

13. FURRING, LATHING, AND PLASTERING

Type and quality of furring, lathing, metal beads, grounds, and parting strips should be described in the specifications when not shown on the drawings. For plastering the quality of the ingredients and the methods of mixing and placing the various coats should be specified. The desired wall finishes should be described, and, for special finishes, samples should be required in advance of the work.

14. GLASS AND GLAZING

State the type, thickness and quality of glass for windows, doors and mirrors, scheduling the types to be placed in steel sash, wood sash, transoms, wood doors and metal doors respectively. Types of glass will include clear window glass, polished plate glass, figured and processed glass, wire glass, and corrugated, prism, and safety glass, among others. Composition of putty, lead, mastic or other materials for setting glass should also be given. Specify the details for bedding, back puttying and face puttying, and for securing glass in the sash. Make provisions for the cleaning and replacing of the broken panes before the completion of the work.

15. PAINTING AND FINISHING

This section should contain the specifications for both outside and inside painting. Outside work consists of exterior woodwork and trim, siding, metal work, roofs, cornices, and gutters. Interior work includes floors, walls, ceilings, trim, exposed radiators, pipes, and the like. The American Society for Testing Materials has prepared standard specifications for the quality of paints, varnishes, stains, and shellacs, and these may be used in combination with paint materials' specifications, based on the furnishing of proprietary products by approved manufacturers. For large projects it will be advisable to specify mixing formulas as well. Workmanship specifications should contain detailed instructions for the preparation of surfaces and the manner in which priming and finished coats are to be applied to all surfaces, such as wood, metal, plaster, and masonry materials. It is advisable to provide for the approval of samples before any paint materials are ordered, and detailed instruction should be included for the protection of painted surfaces and cleaning up after the completion of the work.

16. HARDWARE

Rough hardware includes stair rail brackets, shelf supports, drawer guides, door stops, sash weights, folding partitions, and the like whereas finishing hardware includes door butts, knob locks, metal sash, transoms, etc. Nails, spikes, bolts, anchors, screws, etc., should be required in connection with hardware work. Usually the hardware specifications will contain schedules of approved proprietary brands for each type of hardware. These schedules will also supplement the drawings by indicating the location of hardware installations in the various parts of the building. It is advisable to require all hardware to be installed in accordance with the manufacturer's directions and to require approval of samples before ordering.

17. PLUMBING

Give the instructions for the installation of storm water and floor waste drainage systems, sanitary drainage system, hot and cold water systems, automatic sprinkling systems, gas piping and plumbing fixtures. Specify the type, size, weight, and quality of cast iron, wrought iron, brass, galvanized, and lead pipe, fittings, and sleeves. Pipe installation and painting should also be included. Such fixtures as water closets, bath tubs, sinks, laundry tubs, lavatories, and the like will usually be selected from the available standard brands, and the specifications should state the model, color, size, style of trimming and should list approved manufacturers for each fixture. Tabular schedules are very convenient for this purpose. It is usually advisable to require a warranty from the contractor that the plumbing installation will be free from defects of materials and workmanship for a stipulated period of years and he should be required to replace all defective work within that period at his own expense. "Roughing-in" work for plumbing installations requires close coordination with other types of construction and other trades, and the specifications should be explicit in this regard.

18. HEATING AND VENTILATING

Specify the installation of furnaces, boilers, registers, radiators and piping, ducts, stacks, louvers, control equipment, and other appurtenances. Ventilation equipment includes fans, motors, and ducts. Air-conditioning equipment should be installed in accordance with the manufacturer's instructions. All heating, ventilating, and air-conditioning installations should be warranted by the contractor to be free from defects of materials and workmanship.

19. ELECTRICAL WORK

This section covers the complete electric light and power system, including telephone, burglar alarm, fire alarm, and radio reception systems. All work should be in accord with the National Electrical Code and the local utility regulations. Specify the type and quality of wires and cables, conduits, outlet boxes, switches, outlet receptacles, and fixtures. Where not shown on the drawings, the location of all outlets and switch boxes should be described in the specifications. A warranty from the contractor should be required in connection with the installation of all electrical work.

20. ELEVATORS AND DUMB-WAITERS

Schedule the number of cars and locations for passenger, freight, and service elevators and dumb-waiters. Hatchway construction will be included with the structural specifications, but extraordinary precautions must be taken to make certain that the requirements as to dimensions and vertical alignment are properly specified. The materials, finish, and equipment for cars must be covered, likewise accessories and appurtenances, such as guides, hatchway entrances, signals, controls, and buffers. Provide for load, levelling, and speed tests and tests of the control and signal systems. Elevator specifications will usually be based on performance, and warranties should be required.

21. INSULATION

Under this section specify the requirements for heat insulation, such as mineral wool, felt, and reflective sheet metal. Acoustical insulation, such as acoustic tiles, plaster, and felts, may also be included in this section. Specifications for insulation materials will usually be based on performance, and they should be installed in accordance with the manufacturer's instructions. The specifications should contain a list of approved standard products.

22. SEWERS AND SEWAGE DISPOSAL

When outside sewers and sewage-disposal equipment, such as septic tanks, sludge pits, sand filters, and the like, are included in the building contract, specifications should contain instructions as to materials and workmanship.

23. LANDSCAPING

Specify the requirements as to finished grading, seeding, and planting after the completion of the building structure.

24. MISCELLANEOUS WORK

Specifications for minor items not otherwise classified may be grouped under this heading. Typical of these are driveways, sidewalks, curb and gutter, fences, gates, and the like. Individual items of this character are usually small in scope, but for larger projects some may justify individual sections in which specification requirements are treated in more detail than would be necessary for ordinary work.

25. BASIS OF MEASUREMENT AND PAYMENT

Practically all building construction in the United States is accomplished under lump-sum or cost-plus contracts, and no mention of measurement and payment is necessary in the specifications. Sometimes, however, it is desirable to handle some parts of the building, such as foundation work, on a unit-price basis, with the remainder of the structure paid for as a lump sum. When this is done the specifications must be explicit as to the bid units and methods of measurement and payment for the unit-price work.

EARTH DAM

1. GENERAL PROVISIONS

2. DIVERSION AND CARE OF WATER DURING CONSTRUCTION

Define the various classes of excavation for the diversion channel, and give instructions for backfilling after completion of the work. Specifications for the construction of the cofferdam or other diversion works and the maintenance of stream flow during construction should also be included.

3. CLEARING AND GRUBBING

Describe the areas to be cleared, the requirements as to the removal of all surface material and the grubbing of stumps, roots, buried logs, and the like below the ground surface, the disposal of waste material, and the ownership of timber and other materials cleared from the site.

4. DRAINAGE OF EARTH-EMBANKMENT DAM STRUCTURE

Classify the materials to be excavated for foundation drains, and give instructions for trench sheeting and slope protection to prevent slides. Specify the quality of materials and construction details for the various types of drainage pipe, and give the methods for placing backfill of both pervious and impervious types.

5. EXCAVATION AND EMBANKMENT

Cover the stripping of vegetation and top soil, the classification of excavated material, shoring, blasting, hauling, testing, and the like. Give the limits of excavation of such structures as the spillway, core trench, embankment, and tailwater channel. Specifications for embankments should include instructions for the preparation of foundations, control of spring areas, selection of materials for the various zones of the core and embankment, methods of placing selected impervious and pervious materials, moisture control, compaction, soil tests, and the like. Also include specifications for the placing of top soil on completed embankments, seeding, planting, and landscaping.

6. FOUNDATION DRILLING AND GROUTING

Cover the drilling and grouting procedure, records of drilled material, types of holes, grout outlets, and drainage. Also included are grouting pipe, fittings and valves, and the methods to be used in pressure grouting.

7. REVETMENT AND GRAVEL BACKING

Specify the quality and sources of material to be used for revetment and backing of slopes and channels and the instructions for placing truck-dumped and derrick-placed types of revetment.

8. CONCRETE CONSTRUCTION

Cover the quality of concrete ingredients, proportioning, mixing, placing and finishing of concrete mixtures, including classification of the different mixes to be used on the job. Also specify the use of admixtures and special mixes, construction joints, forms, expansion and contraction joints, embedded items and inserts, steel reinforcement, and testing.

9. STRUCTURAL METAL WORK

Specify the quality of materials and workmanship for such items as structural steel work, reinforcement bars, steel plates, forged steel, cast steel, cast iron, bronze castings, trash racks, penstocks, bulkheads, hand rail, pipe, pipe fittings, and valves. Give special attention to tolerances, finish, adjustments, and the like. For a small project the specifications for metal work, piping, and painting may be combined in one section. For larger projects they should be separated with a full section devoted to each.

10. GATE HOUSE

Specifications for the gate house will follow the general procedure indicated for building construction. For small projects the entire structure could be handled in one section, but for more complex installations the specifications should be subdivided with each type of construction covered in detail.

11. MACHINERY

Under this heading give the specifications for the cranes, movable gates, hoisting machinery, gate position indicators, and control equipment. These specifications will normally be based on performance and acceptance tests, and warranties should be required.

12. ELECTRICAL WORK

Give the details of the materials and workmanship for conduits, boxes, conductors, grounding system, receptacles and plugs, switches, lighting fixtures, flood lights, panel boards, motors, transformers, automatic recording gauges, and the like.

13. MISCELLANEOUS WORK

Specifications for minor miscellaneous items not otherwise classified, such as small water-supply systems, stand-by power units, gauge-recording houses, and the like may be grouped under this heading.

14. BASIS OF MEASUREMENT AND PAYMENT

Practically all contracts for earth-dam construction are on a unit-price basis. Methods of measurement and payment must be specified for each of the bid items, either in a separate section at the end of the specifications or in a separate paragraph in each of the sections containing the technical provisions for the various types of construction.

WATER DISTRIBUTION SYSTEM

1. GENERAL PROVISIONS

2. EXCAVATION, TRENCHING, AND BACKFILLING

Define the various classes of excavation. Give directions for the disposition of waste materials. Give the widths and depths of trenches for various pipe sizes in different kinds of material. Specify the different types of material and the methods of placing to be used in backfill. State restrictions as to blasting of rock material.

3. PIPE LINES

Specify material which should be used in manufacturing pipe and the standards to which it should conform, together with the tests to be used. State the manner in which the pipe should be laid, giving the minimum cover, method of support, condition of trench, weather conditions which are unsuitable for such work, protection of pipe, and testing of pipe after sections of the line are completed. Also state methods of handling and storing pipe and precautions to be observed to prevent damage to special coatings. Give directions for cutting pipe. State maximum tolerances for deflections from alignment or grade and bracing required to prevent such deflections when the pipe line is under pressure. Give precautions to be taken to protect pipes from the entrance of dirt or water from the trench.

4. SPECIALS AND FITTINGS

Specify type and class of fittings, standards to which they should conform, and tests to which they should be subjected. Give directions for cutting threads, arrangements of fittings, methods for connecting with mains, and precautions to be taken in handling.

5. JOINTS

State the type of joint compound to be used, the standards to which they shall conform, and the ingredients of patented compounds which may be used. The method of making the joint should be described in detail including the manner in which the jointing material should be prepared for use and the precautions to be taken to prevent the use of jointing material which has become dirty, cold, or otherwise undesirable. If gaskets for flanged connections are required, material of which they may be composed should be specified, and detailed dimensions should be given. Specifications for bolts should indicate the type of heads, the standards to which they should conform, and whether they shall be corrosion resistant, bituminous coated, etc. Where applicable, the method of tightening bolts and making screwed connections should be specified.

6. VALVES AND VALVE BOXES

The type of valves should be specified together with details of their design, the standards to which they should conform, and the tests to which they should be submitted. Directions for placing the valves in the lines, methods of bracing and backfill, and the use of bridle rods where required should be stated. The type of material, dimensions, design, and instructions for the installation of valve boxes should be specified.

7. HYDRANTS

The type of hydrants, the pressure for which they shall be suitable, and the standards to which they shall comply should be shown. Details of the design should be indicated, giving minimum allowances for valve openings, friction losses, and the number and size of hose connections. The kind of material and type of thread for the hose nipples should be stated. The manner in which the hydrants are to be placed in the trench, their foundation, method of drainage, bracing, and ties with the pipe line should be specified. The manner in which the hydrants should be painted should be shown, giving the type of paint and number of coats. Operating test for hydrants should also be specified.

8. SERVICE CONNECTIONS

Type of corporation stops, couplings, and joints between the mains and the house laterals should be specified. The method of making connections in the mains should be described, and the tests to which the house connections must be subjected should be stated.

9. HYDROSTATIC FIELD TESTS

The pressure to be used in the tests should be stated, giving the length of the section to be tested and the manner of blocking off sections of the pipe. The manner of making the tests, the determination of leakage, and the manner of correcting the leaks should be described.

10. STERILIZATION OF COMPLETED LINE

Specify the agent to be used for sterilization, the method of application, the results desired, and the test for the effectiveness of the process.

11. MISCELLANEOUS WORK

Specifications for minor miscellaneous items not otherwise classified may be grouped under this head.

12. BASIS OF MEASUREMENT AND PAYMENT

Unit-price payments cover all service, labor, equipment, material and supplies specified and indicated for the work complete and ready for use. Incidental work including shop drawings, paint, couplings, joint material, testing and sterilization necessary in connection with each item is generally included with these items. Payment for excavation for pipe trenches is based on the theoretical yardage to be removed from the trench having the dimensions specified, and any over-cutting is the responsibility of the contractor or separate prices are given per linear foot of water pipe for various depths of trench and pipe sizes. Payment for the pipe is based on the length of the pipe as laid, exclusive of the fittings. Fittings are usually paid for on a tonnage basis computed from manufacturer's figures. Valves, hydrants, and service connections are generally paid for on the basis of a lump sum for each.

SEWAGE COLLECTION SYSTEM

1. GENERAL PROVISIONS

2. EXCAVATION, TRENCHING, AND BACKFILLING

Define the various classes of excavation. Give directions for the disposition of waste materials. Give the widths and depths of trenches for various pipe sizes in different kinds of material. Specify the different types of material and the methods to be used in backfill. State restrictions as to blasting of rock. Give instructions for dewatering, sheathing, and shoring trenches, and describe the manner in which sheathing or shoring should be removed.

3. PIPE LINES

Specify types of pipe which should be used and the standards of materials and manufacture to which each shall conform, together with the tests to which it should be submitted. State the manner in which the pipe should be laid; specify condition of the trench, holes for bells, method of support for the barrel of the pipe, exclusion of water and dirt from the pipe during laying, protection of the pipe, and testing of the pipe after sections of the line are completed. State methods of handling and storing pipe and precautions to be observed to prevent damage to special coatings.

4. SPECIALS AND FITTINGS

Specify type and class of fittings, standards to which they should conform, and tests to which they should be submitted. Give directions for cutting pipe, arrangements of fittings, and precautions to be taken in the handling and laying.

5. JOINTS

State the characteristics of the joint materials to be used, the standards to which each shall conform, and the ingredients of patented compounds which may be used. The method of making joints should be given in detail including the manner in which the jointing material should be prepared for use. Gasket material, couplings, and other incidental items where applicable must be specified. Where cement mortar joints are to be used, specifications should be given for the cement, lime, sand, and water and the method of mixing and application to the joints.

6. MANHOLES AND CATCH BASINS

Describe the types of manholes and catch basins required and the quality of materials and workmanship to be used in their construction. The manner of making excavation, mixing concrete, laying bricks, etc., should be covered in detail. Specify the results to be attained in regard to grade, alignment, and infiltration.

7. MANHOLE CASTINGS

Cast iron castings for manhole frames, covers, gratings, and ladder rungs should be specified as to quality of materials, and instructions should be given for their installation. Details of design and dimensions are usually shown on the plans.

8. MISCELLANEOUS WORK

Specifications for miscellaneous minor items not otherwise classified may be grouped under this heading.

9. FIELD TESTS

Describe the tests to be made for alignment, grade, and watertightness.

10. MEASUREMENT AND PAYMENT

Unit-price payments cover all service, labor, equipment, material, and supplies specified and indicated for the work complete and ready for use. Incidental work including jointing material, cutting pipe, and testing necessary in connection with each item is generally included with these items. Payment for excavation for pipe trenches is based on the theoretical yardage to be removed from a trench having the dimensions specified or separate prices are given per linear foot of sewer for different depths of trench and for the various pipe sizes. Payment for the pipe is based on the length of the pipe as laid including the fittings. This price also includes the cost of dewatering the trench and sheathing left in place. Fittings are paid for on a unit-price basis. Manholes are paid for in accordance with their depth on the basis of a lump sum for each.

HARBOR DREDGING

1. GENERAL PROVISIONS

2. CHARACTER OF MATERIAL TO BE DREDGED

Generally the contractor is required to assume the responsibility of informing himself as to the nature of the material to be dredged, but such information as is available should be written into the specifications with a careful definition of the various classes of material to be removed.

3. DISPOSAL OF EXCAVATED MATERIAL

The specifications should require the contractor to comply with all laws and regulations and to furnish all permits required in connection with the handling of his equipment and the disposal of dredged material. Usually the drawings will indicate the location of suitable spoil areas, but the specifications should contain instructions requiring spoil to be evenly distributed so as to prevent the formation of shoals or obstructions to navigation. When the spoil is to be used to fill low areas, instructions should be given relative to the construction of retaining dikes, regulation of the dredge discharge to prevent the formation of mounds and depressions, and the drainage of the completed fill.

4. BLASTING

Give instructions on the conduct of blasting for rock excavation including the types of blasting caps, the maximum amount of explosives to be used in each charge, and precautions to prevent damage to vessels and adjacent structures.

5. SIDE SLOPES

Dredging plans usually indicate the side slopes of the excavation, but frequently these must be considered approximate. Therefore, provision should sometimes be made in the specifications for the final determination of side slopes during the progress of the work. Final determination of side slopes may be made from soil tests or by observation of natural slopes.

6. OVERDEPTH

Because of inaccuracies in dredging processes, it is customary to allow the contractor some latitude with reference to grades in the removal of dredged material. Therefore, the specifications should state that in the event of accidental overdredging beyond the theoretical lines the contractor will be paid for excess material removed within prescribed limits. For earth, sand, or silt dredging, a maximum overdepth of 2 feet will usually be considered in computing dredged quantities. For rock dredging a maximum allowable overdepth of 3 feet is common. Overdepth allowances apply to side slopes as well as to channel bottoms.

7. BASIS OF MEASUREMENT AND PAYMENT

Occasionally harbor dredging is accomplished under lump-sum contracts, but the majority of such work is done on a unit-price basis. This requires a description of the methods of measurement for the various classes of excavation. If measurement of quantities in their undisturbed condition is intended, provision must be made for soundings before beginning the work and for final soundings after its completion in order to compute the dredged quantities. If barge or other methods of bulk measurement are to be permitted, bulking factors must be specified for each class of excavation. The conditions, if any, under which the contractor will be paid stand-by rental for idle plant should be carefully defined when interruptions of the work are expected to permit the movement of ships.

TIMBER DOCK

1. GENERAL PROVISIONS

2. DEMOLITION

Describe existing structures and old piles to be removed, and furnish information on the disposal or re-use of materials cleared from the site.

3. PILES AND PILE DRIVING

State the type of piles to be used, quality of materials, methods of driving, required bearing capacity, and methods by which the bearing capacity is to be determined. Give detailed instructions on the removal of damaged and misdriven piles and tolerances which will be permitted with regard to location and alignment of piles. Methods for splicing and lagging piles should also be included. Provision should be made for test piles when required.

4. LUMBER

Specify the grade and quality of lumber to be used for the various elements of the structural framing, such as pile caps, stringers, bracing, and decking. All framing should be smooth and accurate, and all bearings should have full contact over the entire bearing surfaces.

5. TIMBER PRESERVATIVE

Specify the type and quality of treatment to be used for the protection of all timber work and the methods for its application. Give details of special treatment required at tenons and heads of bolts, faces of caps and stringers, notches, and holes. Also give the precautions to be followed in handling the treated timber.

6. FASTENINGS

Specify the quality of materials and workmanship for bolts, washers, nuts, straps, pile bands, clamps, spikes, and the like. When details of fastenings or connections are not shown on the drawings, they may be scheduled in this section.

7. FITTINGS

Under this heading give the specifications for cast steel bits, cleats, and bollards. They should be set in cement mortar to secure uniform bearing, and the recesses at bolt nuts should be filled with lead and caulked. Hollow bits should be filled with concrete. Design details for fittings will usually be shown on the drawings, but the specifications should cover the quality of materials and workmanship.

8. TRACK WORK

Under this heading the track rails, bed plates, splicing bars and bolts, cross-overs, and frogs and switches, together with their accessories and control equipment, should be covered in detail. Rails on curves should be bent accurately to the required radius, and the methods of installation should be described.

9. PIPING

Under this heading specify the requirements for piping for the various services to be installed on the dock, such as salt water, fresh water, compressed air, fuel oil, and gas. Specifications should cover the quality of materials and workmanship for the various kinds of pipe to be used, including wrapping, painting and insulation, joint material and the method of making joints, valves, hydrants, outlet sleeves, supports, and manhole frames and covers.

10. ELECTRICAL WORK

Specifications in this section contain the instructions for the installation of electrical conduits and ducts, telephone pull boxes and receptacles, wires and cables, lighting, and convenience outlets.

11. EQUIPMENT AND MACHINERY

This section covers such items of equipment as locomotive cranes, gantry cranes, derricks, and other accessories. These specifications will usually be based on performance and acceptance tests, and warranties should be specified.

12. BASIS OF MEASUREMENT AND PAYMENT

Both lump-sum and unit-price types of contracts are used for dock construction, and methods of measurement and payment must be specified for the unit-price form. For piling there will usually be a unit price per linear foot of pile with a different price for various increments of total length. It is advisable to provide for additions to and deductions from the total number of piles indicated on the drawings in the event that changes become necessary during the progress of the pile driving. Bracing, pile caps, stringers, decking, and other timber work will usually be paid for on the basis of a unit price per thousand feet board measure in place, with a separate unit price for each classification of timber work. Metal fittings, track work, and piping are usually paid for on a pound basis. A lump sum is frequently set up for the electrical work complete. Likewise, each principal item of equipment is usually paid for on a lump-sum basis.

REFERENCES

Ackerman, S. B., *Insurance*, 3rd Ed., The Ronald Press Co., 1948.

Allen, C. Frank, *Business Law for Engineers*, 3rd Ed., McGraw-Hill Book Co., 1929.

American Association of State Highway Officials, *Highway Materials: Part I, Specifications; Part II, Methods of Sampling and Testing.*

American Institute of Architects, *The Handbook of Architectural Practice*, 6th Ed., 1951.

American Society for Testing Materials, *Standards* and Annual Supplements thereto.

American Society of Civil Engineers, *Manual No. 8, Engineering and Contracting Procedure for Foundations*, 1934.

American Society of Civil Engineers, *Manual No. 29, Manual of Professional Practice*, 1952.

Biesterfeld, Chester H., *Patent Law*, 2nd Ed., John Wiley & Sons, 1949.

Birdseye, C. F., Abbott, B. V., and Abbott, A., *General Business and Legal Forms*, 3rd Ed., Baker, Voorhis & Co., 1924.

Blake, C. H., *Law of Architecture and Building*, 2nd Ed., Comstock Publishing Co., 1925.

Blake, C. H., *The Architect's Law Manual*, Reinhold Publishing Corp., 1928.

Campbell, D. A., *Workmen's Compensation*, Parker, Stone, & Baird Co., 1935.

Canfield, D. T., and Bowman, J. H., *Business, Legal, and Ethical Phases of Engineering*, McGraw-Hill Book Co., 1948.

Clark, William L., *Handbook of Law of Contracts*, 4th Ed., West Publishing Co., 1931.

Goldsmith, G., *Architects' Specifications*, John Wiley & Sons, 1935.

Corbin, Arthur Linton, *Corbin on Contracts*, West Publishing Co., 1952.

Crawford, Ivan C., *Legal Phases of Engineering*, The Macmillan Co., 1950.

Graske, Theodore W., *The Law of Government Defense Contracts*, Baker, Voorhis & Co., 1941.

Harper, Fowler V., *Treatise on the Law of Torts*, The Bobbs-Merrill Co., 1933.

Hayward, Norris L., *The Contractor's Legal Problems*, McGraw-Hill Book Co., 1946.

Holland, L. B., and Parker, H., *Ready-Written Specifications*, John Wiley & Sons, 1926.

Kerr, Thomas S., *Business Law*, 2nd Ed., John Wiley & Sons, 1939.

Kirby, R. S., *Elements of Specification Writing*, 4th Ed., John Wiley & Sons, 1935.

Laidlow, R. E., and Young, C. R., *Engineering Law*, University of Toronto Press, 1941.

McCullough, C. B., and McCullough, J. R., *The Engineer at Law,* The Iowa State College Press, 1946.

Mead, Daniel W., *Contracts, Specifications, and Engineering Relations,* 2nd Ed., McGraw-Hill Book Co., 1933.

Moreell, Admiral B., *Notes on the Use of the Cost-Plus-Fixed-Fee Form in Government Construction Contracts* (Booklet), Bureau of Yards and Docks, U.S. Navy Department, 1943.

Nichols, C. A., *Annotated Forms (Legal & Business),* Callaghan & Co., 1925.

Pomeroy, John Norton, *Equity Jurisprudence as Administered in U.S.A.,* Bancroft-Whitney Co., Lawyers Co-operative Publishing Co., 1941.

Rotwein, A., and Rotwein, N., *Labor Law,* Harmon Publications, 1939.

Sadler, Walter C., *Legal Aspects of Engineering,* John Wiley & Sons, 1940.

Sadler, Walter C., *The Specifications and Law on Engineering Works,* John Wiley & Sons, 1948.

Schobinger, George, and Lackey, Alexander, *Business Methods in the Building Field,* McGraw-Hill Book Co., 1940.

Seelye, Elwyn E., *Data Book for Civil Engineers: Vol. II, Specifications and Costs,* 2nd Ed., John Wiley & Sons, 1951.

Shealey, Robert Preston, *Law of Government Contracts* (Federal Contracts), 3rd Ed., Federal Publishing Co., 1938.

Simpson, L. P., and Dillavou, E. R., *Law for Engineers and Architects,* 3rd Ed., West Publishing Co., 1946.

Sleeper, H. R., *Architectural Specifications,* John Wiley & Sons, 1940.

Small, Ben John, *Streamlined Specifications Standards,* Reinhold Publishing Corp., 1952.

Spencer, W. H., and Gillam, C. W., *A Textbook of Law and Business,* 3rd Ed., McGraw-Hill Book Co., 1952.

Tomson, Bernard, *Architectural and Engineering Law,* Reinhold Publishing Corp., 1951.

Tucker, J. I., *Contracts in Engineering,* 4th Ed., McGraw-Hill Book Co., 1947.

Wait, J. C., *Engineering and Architectural Jurisprudence,* John Wiley & Sons, 1898.

Werbin, I. Vernon, *Legal Guide for Contractors, Architects, and Engineers,* McGraw-Hill Book Co., 1952.

Werbin, I. Vernon, *Legal Phases of Construction Contracts,* McGraw-Hill Book Co., 1946.

INDEX